T0258421

Encyclopedia of Gas Chromatography: Biochemical and Other Applications

Volume IV

Encyclopedia of Gas Chromatography: Biochemical and Other Applications

Volume IV

Edited by **Carol Evans**

New York

Published by NY Research Press,
23 West, 55th Street, Suite 816,
New York, NY 10019, USA
www.nyresearchpress.com

Encyclopedia of Gas Chromatography: Biochemical and Other Applications
Volume IV
Edited by Carol Evans

International Standard Book Number: 978-1-63238-131-6 (Hardback)

This book contains information obtained from authentic and highly regarded sources. Copyright for all individual chapters remain with the respective authors as indicated. A wide variety of references are listed. Permission and sources are indicated; for detailed attributions, please refer to the permissions page. Reasonable efforts have been made to publish reliable data and information, but the authors, editors and publisher cannot assume any responsibility for the validity of all materials or the consequences of their use.

The publisher's policy is to use permanent paper from mills that operate a sustainable forestry policy. Furthermore, the publisher ensures that the text paper and cover boards used have met acceptable environmental accreditation standards.

Trademark Notice: Registered trademark of products or corporate names are used only for explanation and identification without intent to infringe.

Printed in the United States of America.

Contents

Preface

Encyclopedic information regarding the biochemical applications as well as other applications of gas chromatography are elucidated in this book. The analytical technique of gas chromatography incorporates the study of different volatile molecules in associated research fields of chemistry. This technique holds numerous characteristics and advantages making it a valuable method for the identification, evaluation and anatomical assessment of organic molecules. This book presents thorough information on applications of gas chromatography to materials like essential oils, biochemicals and narcotics. The finer characteristics of various applications have been thoroughly dealt with by the contributors to enhance their simplicity and viability. This text will prove to be of great use to beginners as well as experts who may not hold sufficient knowledge about particular uses of gas chromatography. This book will find suitable applications in labs where relevant techniques of gas chromatography are practiced.

This book unites the global concepts and researches in an organized manner for a comprehensive understanding of the subject. It is a ripe text for all researchers, students, scientists or anyone else who is interested in acquiring a better knowledge of this dynamic field.

I extend my sincere thanks to the contributors for such eloquent research chapters. Finally, I thank my family for being a source of support and help.

Editor

Part 1

Biochemical Applications

Gas Chromatography as a Tool in Quorum Sensing Studies

Oscar Osorno[1,*], Leonardo Castellanos[1],
Freddy A. Ramos[1] and Catalina Arévalo-Ferro[2]
[1]Departamento de Química,
[2]Departamento de Biología, Universidad Nacional de Colombia,
Colombia

Dedicated to Professor Carmenza Duque

1. Introduction

Since cell differentiation was studied cell-to-cell signaling and cell regulated cycles were considered exclusive of eukaryotic organisms, prokaryotic organisms were regarded as isolated cells without cooperative behaviors. It took us more than 30 years, after Nealson et al., (1970) explained bioluminescence as an auto-induced regulated mechanism of bacterial groups, to assume in microbiology research that bacteria can synchronize group behaviors. In consequence, it was possible to explain that inter-cell signaling regulates some bacterial phenotypes and this phenomenon was called Quorum Sensing (QS). QS is one of the most revolutionary mechanisms discovered in the past 15 years. It involves the cell control of bacterial population by communication using chemical signals and a complex network of genetic circuits with a positive feedback regulation (for review see: Waters et al., 2005). Sensing these chemical signals bacteria can respond as groups and detect the "quorum" of a population in order to regulate different phenotypes.

In 1970 Neaslson et al., described light production in the marine organism *Vibrio fischeri* in response to secreted signal molecules, depending on the cell density. After that, 10 years later, Eberhard et al., (1981) identified those signal molecules as N-acyl Homoserine Lactones (AHLs) synthesized by the enzyme LuxI and distributed into the media by Simple Diffusion (Eberhard et al., 1981). Auto-induction occurs when AHLs accumulates in the media (in a high cell density) and binds the target in the protein LuxR, a transcriptional activator for the expression of many genes including *luxI*, generating positive feedback loop (Figure 1A). Fuqua et al., in 1994, explaining the regulation of gene expression by a cell density manner, and called this phenomenon "Quorum Sensing" (QS) (Fuqua et al., 1994; Stauff et al., 2011). Before, it was considered a peculiarity of few Gram-negative bacteria since AHLs were not found in Gram-positive ones, nowadays it is known that Gram-positive bacteria are using oligopeptides to communicate with each other. In this case, peptide transporters secrete these autoinducing peptides (AIPs) because the cell membrane is not permeable to them, and the sensing systems do not occur inside the cell as in Gram negatives. Instead, two component transduction systems are used for the detection of the

---

* Corresponding Author

AIPs and the activation of the QS system (Figure 1B) (for review see: de Kievit & Iglewski, 2000; Miller & Bassler, 2001; Thoendel et al., 2011).

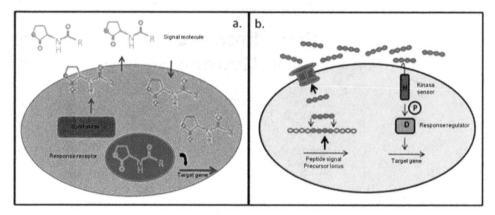

Fig. 1. General Quorum Sensing Systems: **a**. Gram-negative bacteria, **b**. Gram-positive bacteria (from Pamela Zorro, 2011).

In most Gram-negative bacteria QS circuits work as it has been described for *Vibrio fischeri*. Light production is coordinated by the *lux* regulon that comprises: (1) the *lux*ICDABE operon consisting of the *lux*I gene, encoding the 3-oxo-C6-HSL synthase, followed by the *lux*CDABE, the structural genes necessary for bioluminescence, and (2) the *lux*R gene encoding the transcriptional activator protein LuxR. At low cell densities, *lux*I is transcribed at a basal level leading to a low concentration of 3-oxo-hexanoil-homoserine lactone (3-Oxo-C_6-HSL). AHLs (Figure 2) are freely diffusible and accumulate in the medium with increasing cell density. When 3-Oxo-C_6-HSL reaches a critical threshold concentration, it binds to the LuxR-receptor protein. This complex in turn binds to a 20 bp palindromic DNA promoter element, the lux-box, and activates the transcription of *lux*ICDABE resulting in the bioluminescent phenotype and in a positive autoregulation of *lux*I (a positive feedback loop) (Engebrecht & Silverman, 1984; Kaplan & Greenberg, 1985). This was the first QS circuit described and the name of the cupple LuxI/LuxR remains as a symbol for every LuxI type protein or LuxR type protein in Gram-negative bacteria.

For Gram-positive bacteria, the circuits in general work as in the *Streptococcus pneumoniae* (ComD/ComE) Competence System. The signal molecule is the Competence Stimulating Peptide (CPS) (Figure 2), it is a processed peptide of 17 aminoacids secreted by the type 1 secretion system Com AB, an ABC-transporter. At high cell density, CPS accumulates and is detected by ComD, a sensor kinase protein. CPS induces autophosphorylation of ComD and the phosphorylation cascade until the phosphoryl group is transferred to the transcriptional activator ComE. Once ComE is activated by phosphorylation, it induces the transcription of the *com*X gene, responsible of the competence (ability to acquire exogenous DNA molecules) in *S. pneumoniae* (Miller & Bassler, 2001).

Currently different QS signaling molecules are recognized in Gram negative or Gram positive bacteria as well as the ones used for inter-species communication, as the autoinducer 2 (AI-2) to mention one (Bassler et al., 1997). However, AHLs, AI-2 and CPS are not the only nature of sensing molecules, there are diesters, butyrolactones, quinolines, branched and unsaturated fatty acids, diketopiperazines (Figure 2.), among others detected by the same mechanisms or by a hybrid of both as de QS circuit of *V. harveyi*. The hybrid QS circuits works with molecules, similar to AHLs or synthesized using similar pathways, freely diffusible through the bacterial membrane but those molecules are sensed by a kinase protein just like in Gram positive bacteria, using a cascade of phosphate to induce the expression of regulated phenotypes. The AI-2 from *Vibrio harveyi* was the first molecule described for the hybrid QS circuits, it is a furanosyl borate diester (Figure 2) synthesized by the enzyme LuxS. Interesting, this enzyme has been detected in a wide diversity of bacteria including Gram positive and Gram negative cells, in that sense AI-2 has a role of universal detection that may give the rule of crosstalk between organisms (Bassler *et al.*, 1997; Miller & Bassler, 2001; Warren et al., 2011).

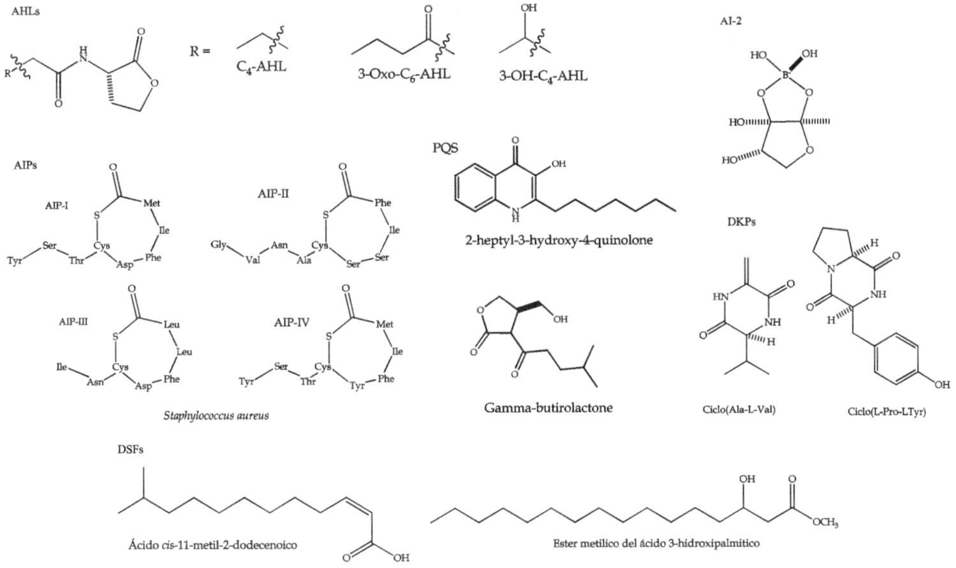

Fig. 2. Chemical structures of some Quorum Sensing signaling molecules. General structure of N-acyl-homoserine lactones (AHL), autoinducer 2 (AI-2), autoinducing peptides (AIP), *Pseudomonas* quinolone signal (PQS), Difusible signal factors (DSFs), and Diketopiperazines (DKPs).

In many bacteria Quorum Sensing regulates a variety of phenotypes including bioluminescence, transfer of tumor-inducing plasmids (Ti plasmids), antibiotic production, swarming motility and specially biofilm maturation and production of virulence factors. These last have been the most studied by pharmaceutical or chemical researches for their human health implications. It has been demonstrated that many bacteria do not express virulence factors until the population density is high enough to overwhelm host defenses and to establish the infection (Greenberg, 2003). However, further than virulence, QS allows bacteria to coordinate group behaviors therefore it has a unique ecological role in ecosystems development, bacteria can communicate sensing these molecules between the same species but also between different species. The species specificity of signaling molecules allows bacteria to be responsive only to their own cell density even when other species are present. Nevertheless, cross-species communication mechanisms have been described. In some cells it occurs because LuxR-type proteins have a range of activity with similar AHLs molecules, but the classical cross-species talk is done using signal molecules as the AI-2 (Bassler et al., 1997; Riedel et al., 2001).

This interplay has a pivotal role in development and differentiation of bacterial biofilms, therefore in the conformation of bacterial communities on biotic or abiotic surfaces. However, the interaction between organisms colonizing a surface goes beyond inter-bacteria communication, there have been reports about communication between kingdoms and it has been shown how AHLs have a role in immunomodulation, and apoptosis, as well as in development and metamorphosis of some eukaryotic organisms (Hughes & Sperandio, 2008; Joint et al., 2007). Biofilm of bacterial populations involved in the microbiome of an organism are structured by sensing diffusion gradients of QS signaling molecules, among other factors. These chemical gradients allow the community to shape up the physiological role of each population by differential regulation of gene expression in order to coordinate metabolic pathways and phenotypes (McLean et al., 2005; Valle et al., 2004). Every signaling molecule present in a community is responsible of a behavior that should be in agreement with host signaling and requirements. The influence of bacterial communication in macro-organisms surfaces has been describe for nitrogen fixation, nutrients disposition and defense that includes the production of antimicrobials and Quorum Sensing Inhibitor molecules (Knowlton & Rohwer, 2003; Shnit-Orland & Kushmaro, 2009).

Clinical implication of QS has taken this phenomenon into the pharmaceutical industry with the aim of developing new molecules able to control bacterial pathogenesis since virulence factors of bacteria are mostly QS regulated. Usage of QS Inhibitors is interesting in order to avoid virulence without release bacterial resistance to antibiotics, therefore QS became a pharmaceutical target as an important subject for anti-virulence therapies development (Köhler et al., 2010). Due to the role on human health of bacterial virulent phenotypes regulated by QS systems, there have been plenty of reports for the Quorum Sensing inhibition (QSI) activity of molecules isolated from different organisms. This will be discussed in further section, but it is important to mention that this QSI activity is a target for pharmaceutical industries nowadays since QS regulates development of biofilm formation and expression of virulence factors, both involved in chronic infection progress and medical treatments failures, in general. On the other hand, compounds with QSI activity found application in the antifoulant technologies, because macro-organisms are able for sensing bacterial QS signaling molecules to settle and metamorphose, selecting like that the proper biofilm conditions for its development. Following these concepts it is possible to assume that bacterial biofilm and the signaling molecules present in a surface, may influence the colonization of surfaces and the development of phenomena such as biofouling (Joint et al., 2007; Marshall et al.,2006; Tait et al., 2009).

Nevertheless, for further remarks, it is important to keep in mind that QS regulated phenotypes are a two-edged sword. Even if clinical role of virulence factors or biofilm development on surfaces are adverse for human civilization; researchers may not forget the functional role of QS in fitness and stability of different ecosystems. Normal associated bacteria in an organism have different functions including nutrients availability, defense, essential molecules for metabolic pathways in the host and selection of other microorganisms present in the community, among others. In general, all those interactions give the homeostasis required for an organism and changes in the composition or structure of natural communities could lead to diseases development or any other undesirable conditions in ecosystems. Furthermore, some health problems are now better undertanded like changes in the bacteria community.

The first reports about structural characterization was from the N-acyl homoserine, being consequently the most studied molecules up to now although its detection is made routinely using biosensors (Steindler & Venturi, 2007). These techniques are limited to biological variations and mutants transformation; moreover, it is almost impossible to have a biosensor for every existent signaling molecule and there are examples where these molecules used for cross-communication cannot be detected by the habitual biosensor cross streaking technique and require more specific techniques to be detected; this field is still under development. Likely, one of the reasons, is the urgent necessity of standardize more reproducible and robust methodologies for detection and quantification as the chromatographic methods developed by gas chromatography GC and high performance liquid chromatography HPLC coupled to many of the available range of compatible detectors that includes mass detectors. In the next section this issue is reviewed.

2. Analysis of signaling molecules involved in cell communication by GC

In this section we will focus on the analytical uses (detection, identification and quantification) by GC for signaling molecules like acyl homoserine lactones (AHLs), diketopiperazines (DKPs), diffusible signal factors (DSFs), autoinducer type 2 and type 3 molecules (AI-2 and AI-3), pseudomonas quinolone signals (PQS), and gamma-butirolactones. The GC methods have been used in a wide range of studies that usually search for signaling molecules in strains of different origin (i.e. phatogenics, marine, soil, food, among others). However, is remarkable how the study of crop-pathogenic bacterial strains is one of the fields that lead to the discovery of many of the signaling molecule families here mentioned. In general terms, there is an abscense of information about GC analysis of peptide signaling molecules of the Gramm positive bacteria, due to its incompatibility with gas chromatography systems by the low volatility of the molecules.

2.1 Extraction, structural and stereochemical analysis by GC of signaling molecules

The signaling molecules are usually found at low concentrations; due to that, several culture media and extraction techniques are used. In order to obtain these signaling molecules form bacterial cultures, the most used growth media is the Luria-Bertani (LB) broth, together with other complex non defined culture media such as nutrient broth (NB), sea water agar (SWA), marine agar and in general other minimal media. The choice of the culture media depends on some bacterial growth requirements. The volume of culture used in these studies ranges from 5 to 10 mL in the bottom line, up to 20 L scale cultures, depending on the extraction method and if the aims of the conducted research is analysis-detection in mixture or isolation, respectively.

As the direct injection of the samples may affect the separation or may clog the chromatographic columns due to the low volatile components of culture media, homogenates, or environmental samples, a pre-treatment that involves extraction and purification steps of the sample containing signaling molecules is highly recommended.

The bacterial culture is centrifuged an the supernatant is extracted with common solvents such as ethyl acetate, chloroform and dichloromethane to obtain the signaling molecules such as AHLs, DKP and PQS (Pesci et al., 1999; Pomini et al., 2005; Wang et al., 2010). In order to prevent degradation of AHLs the extractive solvent contains acetic acid, formic acid or TFA acid in concentration from 0.05 to 1.0 % (v/v) (Götz et al., 2007; Huang et al., 2007). For DSF the most employed solvent is hexane, a common solvent for free fatty acid and their methyl esters extraction and analysis; however, as the DSFs are consider weak acids, its extraction can be favored carring out the extraction from an acidfied (pH 4.0) supernatant with the refered organic solvents (Deng et al., 2011). The extraction of high polar AI-2 family signal molecules was achieved by a mixture of dichloromethane/methanol (1:1) (Sperandio et al., 2003). One of the main disadvantages of solvent extraction is the low selectivity of the solvents, even more if the extraction is conducted over the complex matrix of the culture media. Other disadvantages are the low yields of organic layers extractions, the concentration process that includes heating and low pressure rotatory evaporation that can lead to recovery looses for thermal degradation, artifacts formation, and evaporation. Due to that, concentration under a current of nitrogen at room temperature is commonly used and fractionation steps of the extract are required before its chromatographic analysis. The use of neutral interchange resins, such as XAD-16 and HP-20 and HP-21 for the extraction step is also reported (Shaw et al., 1997) due to its higher selectivity compared to that of the organic solvents, low solvent consume and possibility of recycling. Usually resins are added at the begin of the cultivation step, and further purification of the eluted extracts is often required, because of the high amount of culture broth absorption into the resins. The use of solvents and/or resin extraction is observed for both, small and large scale fermentations.

Solid phase extraction (SPE) is one of the most used techniques for extraction of signaling molecules, due to several advantages as higher selectivity than the above mentioned methods, high removal of culture broth, salts and other impurities, pre-concentration of the target molecules, the possibility of gradient elution that produce semi-purified fractions in a single procedure and the requirement of small samples (usually from 5 to 10 mL). All these advantages reduce detection and quantification limits at the chromatographic analysis step enhancing sensitivity and robustness of the used methods (Li et al., 2006; Shupp et al., 2005). From the wide range of options for the choice of SPE supports, reverse phase SPE cartridges are preferred over the normal phase chromatographic supports, because of the wide range of polarity of the signaling molecules to be analyzed and the strong interactions of the high polarity analyzed molecules with polar stationary phases. For AHLs, SPE cartridges with ion-exchange or H-bridge activity supports have been also tested, but there are examples where cartridges loaded with strong or mild ion-exchange abilities (basic and acid supports) retained high amounts of AHLs (Li et al., 2006), leading to looses of the usually low-concentrated molecules. The preferred SPE cartridges for AHLs were the end-capped ODS cartridges and the PSDVB cartridges, washed with water and methanol/water (15:85, v/v) to remove broth media components without target molecules looses, and further elution with isopropanol or with mixtures of hexane 25%, toluene 50%, tetrahydrofuran 75% with isopropanol, and isopropanol 100% (Li et al., 2006).

Once the extracts have been obtained detection techniques that include mutant bacterial strains used as biosensors and chromatographic techniques such as TLC, HPLC and GC are employed in order to detect, identify and quantify the signaling molecules (Charlton et al., 2000; Li et al., 2006; Shaw et al., 1997). Biosensors are used due to their high selectivity and low detection limits. However, chromatographic techniques are preferred over the bioassay detections, because of biosensors responses depends on culture conditions such as growth and pH, and the biosensors are only able to recognize a specific and, by that, a limited number of the signaling molecules.

The sample analysis has been usually conducted by GC-MS as well as by HPLC-MS methods, or even by the combination of both techniques, taking advantage of the high sensibility of the mass spectrometry (MS) detectors, the versatility of the ionization sources (EI and ESI) that allows the ionization of molecules chemically and structurally different, and the tandem techniques of the MS detectors that provides structural information that leads to the identification and differentiation of the signaling molecules (Teplitski et al., 2003; Li et al., 2006). Characteristic fragmentation patterns for signaling molecules as AHLs, DKPs and DSFs have been reported by EI and ESI analysis (Cataldi et al., 2007; Charlton et al., 2000; Morin et al., 2003; Pomini et al., 2005). In this sense, GC-MS combines the high chromatographic resolution and the specificity and sensibility of the mass detectors, required for the analysis of molecules involved in quorum sensing. Even when HPLC-MS has risen as other very interesting and widely used technique in the analysis of bacterial signaling molecules, its use in the analysis of such molecules will not be discussed here attending the scope of this book.

Injection of the samples into GC chromatograph has been made using split (Cataldi et al., 2007, Pomini et al., 2005) or in a splitless mode for AHLs and DSFs (Charlton et al., 2000; Colnaghi et al., 2007). Degradation of some AHLs at the injection port has been proposed, due to the high temperatures of the liners, leading to low sensitivity in some GC methods that analyze the sample directly (Schupp et al., 2005). This degradation could be avoided and an increase of the sensibility for the analysis of AHLs is achieved by the use of derivatives that, with the help of on-column injections, reduce the risk of thermal degradation, increase the thermal stability and the volatility of the AHLs (Charlton et al., 2000). Usually low polarity DB-1 or DB-5 columns, and equivalent columns, are employed for these chromatographic separations. Even when detection of signaling molecules has been reported by FID detectors (Soni et al., 2008) and electron capture detectors (ECD), the most used detector is the single quadrupole MS detector, in both positive and negative detection modes. In addition, for increase the sensibility of the methods for previously identified signaling molecules, selected ion monitoring (SIM) detection modes are preferred over the full SCAN detection mode. The SIM mode allows detecting signaling molecules at femtomolar scales (Charlton et al., 2000).

As was previously mentioned, chemical derivatives help to deal whit thermal stability and the volatility required for the GC chromatographic analysis of the signaling molecules. AHLs can be directly analyzed by GC-MS (Pomini et al., 2005), but an increase in sensitivity and accurateness of the GC-MS method can be reached by the use of pentafluorobenzyloxime PTFBO derivatives (Charlton et al., 2000). Other derivatives are employed for DSFs as the formation of the fatty acid methyl esters FAMES (Wang et al., 2010). However, in order to confirm molecular weight, methylation patterns and double bond position for of these DSFs other derivatives as beta-picolinic esters or N-acylpyrrolidides or dimetyldisufide (DMDS) must be prepared for the GC-MS analysis (Huang & Lee-Wong, 2007; Thiel et al., 2009). AI-2 type signaling molecules were analyzed as an equilibrium mixture transformed to quinoxaline derivatives (Thiele, 2009b). Due to his volatility, other signaling molecules such as gamma-butirolactones, DKPs and PQSs signaling molecules have been analyzed without derivatization.

AHLs possess two quiral centers, one at the carbon C-3 in the lactone ring for all AHLs and the other at the hydroxylated position at C-3´ of the side chain for the 3-OH-AHLs. The absolute configuration assignment at position 3 (C-3) in AHLs, was completed by a GC-FID method using beta-cyclodextrin as chiral stationary phase using coinjection with synthetic standards (Pomini et al., 2005; 2006; Pomini & Marsaioli, 2008).On the other hand, the absolute configuration assignment at position C-3´ for the 3-OH-AHL was completed by the analysis of the FAMEs obtained by treatment of the AHLs from the crude extract with sulphuric acid in methanol (it maintains the stereocentre at C-3´ in the acyl chain) and further chromatographyc analysis employing a beta-cyclodextrin column and coinjection with enantiomerically pure FAMEs standards (Thiel et al., 2009a). Double bond configuration for AHLs was also established by GC-MS injection and comparison with synthetic standards (Thiel et al., 2009a). This approach seems to be compatible for the stereochemical analysis of the substituted fatty acids of the DSFs. The identification of L- and D-DKPs isomers was carried out by GC-MS using a 5% pheyl column due to that DKPs corresponds more to a diasteroemeric mixtures of the cyclic dipeptides than the enantomeric mixtures (Wang et al., 2010). Even when there are not examples abot the absolute configuration analysys for other signaling molecules it seems realiable to conduch such analysis by the use of quiral columns.

The quantification of signaling molecules by GC has been proposed only for AHLs, AI-2 and DKPs. As the N-heptanoilhomoserine lactone had not be detected as signaling molecule it was consider a non natural AHL and its use as internal or external standard for AHLs quantification was accepted (Cataldi et al., 2007). However, nowdays it is known that this molecule is produced by several bacterial species (Pomini et al., 2005; Pomini & Marsaioli, 2008; Thiel et al., 2009a) and its use for quantification proposes is no longer recommended. Other quantification method was proposed for 3-oxo-AHLs, usingthe PTFBO derivatives of ^{13}C-AHLs as internal standard for quantitative analysis (Charlton et al., 2000). The main advantage of this method is that the internal standard undergoes the same losses during work up as does the analyte. Moreover, the isotopes displays the same chemical behavior differing only slightly from the analyte in terms of mass and its volatility, reactivity and distribution coefficient and chromatographic behavior are the same as those of the unlabeled compounds, with the exception of minor and negligible isotope effects. This method showed a high sensitivity (signal noise ratios of 10/1 for 1 ng), and high accurateness. Its applicability was validated in the quantification of 3-oxo-AHLs from supernatants, cell fractions and biofilms. Following the same approach for the isotope dilution quantification of AI-2 type molecules, the [5,6,7,8-^{2}H]1-(3-methylquinoxalin-2-yl)ethane-1,2-diol was used, finding again high linearity, sensitivity and accurateness showing S/N of 5 and detection limits of 0,7 ng/mL. Diketopiperazines have been quantified by the use of a L-cyclo (Phe-Pro) as external standard (Wang et al., 2010).

Extraction, detection, and quantification analysis of signaling molecules produced by bacteria, and other organisms from their culture media are of great interest and the development of novel analyitic strategies can provide information of their biological role in cell-cell communication. In the following lines advances in their structure identification and activity, obtained in studies that used GC will be presented.

2.2 AHLs

The most widely studied signaling molecules are the AHLs, the main QS modulators for Gram negative bacteria. Some of its structural features are the presence of hydroxyl or carbonil substitution at C-3 in the side-chain, the presence of double bonds and the carbon chain length, that ranges commonly from C_4 up to C_{18} and are mainly even; however, there are examples of odd AHLs as the C_7, C_{13} and C_{15} –AHLs that have been found in *Erwinia psiidi* (Pomini et al., 2005), *Rhizobium leguminosarum* (Horng et al., 2002), *Pantoea ananatis* (Pomini et al., 2006), *Edwardsiella tarda* (Morohosi et al., 2004), *Serratia marcescens* (Lithgow et al., 2000) and in marine alphaproteobacteria (Wagner-Döbler et al., 2005). The odd and even AHLs biosynthesis is suggested to be made by the same enzymes (Pomini & Marsaioli, 2008). Double bond position and geometry assignation (Thiel et al., 2009a), presence or absence of carbonyl and or hydroxyl groups at the position C-3 and its C-3 absolute configuration (Pomini et al., 2005; 2006; Pomini & Marsaioli, 2008) have been assigned by GC analysis, using both chromatographic and spectroscopic criteria. Fragmentation pattern analysis shows as characteristic fragments the ion at m/z 143 most likely due to the McLafferty rearrangement, and the ion at m/z 102 probably formed after an acylic side chain cleavage and H rearrangement (Cataldi et al., 2007; Cuadrado, 2009).

In one attempt for develop a GCMS method for AHLs analysis, the bacteria fish pathogens *Aeromonas hydrophila*, *A. salmonicida*, and the opportunistic human pathogens *Pseudomonas aeruginosa*, *P. fluorescens*, *Yersinia enterocolitica* and *Serratia liquefaciencs*, were studied (Cataldi et al., 2007). The bacteria growth medium were extracted with solvents and these extracts analized by GCMS. The series of AHL´s from 4 to 14 all even, 3-oxo-C_6-AHL and 3-oxo-C_{12}-AHL were detected in SIM mode by EI, searching for the prominent ion at m/z 143, from the crude cell free supernatant. Full scan mode did not allowed AHLs detection. The Quantification was conducted by external standard using C_7-AHL. This method allowed detecting signaling molecules that were not detected previously by the use of biosensors or other chromatographyc techniques (Cataldi et al., 2007). However, that study failed at detection of underivatized 3-oxo-AHLs which are heat labile compounds.

Brazilian strains of the crop-pathogen bacteria *Erwinia psidii*, *Pantoea ananatis* and *Pantoea* sp. were studied for its production of AHLs by an analysis protocol that involves solvent extraction, fractionation by column chromatography, identification by direct GC-MS and absolute configuration analysis by GC-FID with chiral column and synthetic standards, together with biosensor detection with *Agrobacterium tumefaciens* NTL4 (pZLR4) strain (Pomini et al., 2005; 2007; Pomini and Marsaioli, 2008). The studies allowed to detect (S)-(-)-C_6-AHL and the rare (S)-C_7-(-)-AHL with a 92% enatiomeric exces in *E. psidii*; (S)-(-)-C_6-AHL, C_7-AHL, and (S)-C_8-AHL from *P. ananatis*; and finally (S)-C_4-AHL, and traces of (S)-C_6-AHL in *Pantoea* sp. Analysis of the activity of *R*- and *S*-AHLs showed that enantiomers were both equally active against *Bacillus cereus*, while the racemic mixture was less active than the pure enantiomers (Pomini & Marsaioli, 2008)

The way how AHLs can induce the settlement of the fouler polychaete *Hydroydes elegans* was evaluated using seven bacterial strains that were originally isolated from natural biofilm that effectively induced larval settlement of the *H. elegans* and synthetic C_6-AHLs, C_{12}-AHL and 3-oxo-C_8-AHL (Huang et al., 2007). AHLs and bacterial biofilms were tested for settlement activity and characterized by GC-MS. The GC-MS analysis of the biofilms solvent-extracts showed that biofilms can produce C_6-AHLs, C_{12}-AHL. In the settlement assays the tree synthetic AHLs, but specialy C_{12}-AHL, induced some initial larval settlement behaviors such as reducing swimming speed and crawling on the bottom.

In a study using C_6- and C_8- and C_{10}-AHLs that attempted for the evaluation of the uptake, degradation, and chiral discrimination of N-acyl-D/L-homoserine lactones by barley (*Hordeum vulgare*) and yam bean (*Pachyrhizus erosus*) using UPLC, FTICRMS and tritium labeled AHLs, and chiral separation by GC-MS revealed that both plants discriminated D-AHLs stereoisomers to different extents. These results indicate substantial differences in uptake and degradation of different AHLs at the tested plants (Götz et al., 2007).

Two strains of the nitrogen-fixing bacterial symbiont *Sinorhizobium meliloti* were analyzed to determine the production of AHLs (Teplitski et al., 2003). Both strains produce C_6-AHL, 3-OH-C_6-AHL, C_8-AHL, 3-oxo-C_8-AHL, 3-OH-C_8-AHL, C_{10}-AHL, 3-oxo-C_{10}-AHL, 3-OH-C_{10}-AHL, C_{14}-AHL, 3-oxo-C_{14}-AHL, C_{16}-AHL, $C_{16:1}$-AHL, 3-oxo-$C_{16:1}$-AHL, characterized by a protocol that includes solvent extraction, SPE fractionation and HPLC-MS GC-MS analysis. The differences in AHLs produced suggest that significant differences in their patterns of quorum-sensing regulation could exist.

In a study that search in more than 100 bacterial isolates from various marine habitats for AHLs production by *Pseudomonas aeruginosa* and *Vibrio fischeri* biosensensors, 39 Alphaproteobacteria isolates induced fluorescence in either one or both of the used sensor strains (Wagner-Dobler et al., 2005). AHLs were identified by GC-MS analysis and shown to have chain lengths of C_8, C_{10}, C_{13}–C_{16}, and C_{18}. One or two double bonds were often present, while a 3-oxo or 3-OH group occurred only rarely in the side chain. Most strains produced several different AHLs. $C_{18:1}$-HSL and $C_{18:2}$-HSL were produced by *Dinoroseobacter shibae*. 7(Z)-$C_{14:1}$-AHL, which has previously been detected in *Rhodobacter sphaeroides*, was produced by *Roseovarius tolerans* and *Jannaschia helgolandensis*. The same research group identified for the first time in an *Aeromonas culicicola* strain AHLs with acyl chains carrying a methyl branch, by means of a GC-MS method that allowed distinguish these compounds from unbranched isomers. In the same publication, a strain of the marine bacterium *Jannaschia helgolandensis*a shows to produce a doubly unsaturated AHL, identified as (2E,9Z)-C16-AHL (Thiel et al., 2009b). The position and configuration of the double bonds was proven by MS spectrometric analysis and by synthesis. Absolute configuration of the detected AHLs was determined by mild cleavage with sulphuric acid and chiral chromatographic analysis.

A preconcentration methodology was developed for the analysis of AHLs using single drop microextraction or liquid-liquid microphase extraction in toluene for their analysis using GC-MS (Kumar-Malik et al., 2009). The performance of the method was determined and discussed for the chiral separation of these autoinducers using a β-cyclodextrin chiral column. A remarkable fact of this study is the demonstration, that *Burkholderia cepacia* LA3 produced D-C_{10}-AHL besides the L-C_{10}-AHL and the L-C_8-AHL enantiomers.

GC has proved to be helpful in the characterization of AHLs from the extreme acidophile *Acidithiobacillus ferrooxidans* produce, that produced C_{12}-, C_{14}-, and C_{16}-AHLs detected by CG-MS (Rivas et al., 2007). In that work, different assays with mutant strains make evident that the AHLs production was controlled by two different quorum sensing systems (Rivas et al., 2007). Other study showed that the ammonia oxidizing bacteria *Nitrosomonas europea* was a producer of C_6-, C_8- and C_{10}-AHLs characterized by independent methods including biosensors and GC-MS (Burton et al., 2005). Additionally, other study used SPE combined with GC-MS in order to detect AHLs from environmental samples such as the dried tissue of the marine sponge *Stylinos* sp. (Schupp et al., 2005). Even when the C_6-AHL and 3-oxo-C_6-AHL used in this study could be detected by GC-MS, the applied method was less sensitive than the biosensor *Agrobacterium tumefaciens* A136. This result compared to the other here mentioned shows the importance of conditions adjustment for through the method development process.

2.3 AI-2

AI-2 is an important interspecies signaling molecule, produced by both Gram-positive and Gram-negative bacteria: It is well established that AI-2 mediates intra- and interspecies communication among bacteria. The cleavage of S-ribosylhomocysteine yields homocysteine and (S)-4,5-dihydroxy-2,3-pentanedione ((S)-DPD), the metabolic precursor of AI-2. This (S)-DPD exist as an equilibrium mixture of several compounds that can be formed by cyclation, hydration and borate formation reacctions. Different bacterial species recognize different signals within this AI-2 pool. (Reading and Sperandio, 2006; Sperandio et al., 2003). AI-3 has been identified as a signaling molecule related to epinephrine and norepinephrine (Sperandio et al., 2004). Its molecular weight was stablished as 212 u, but its structure has nos been yet published.

These compounds are usually analyzed qualitatively by means of bioluminescent biosensors. Recently, a chromatographic methodology for AI-2 type molecules detection and quantification was developed, using a quinoxaline derivative, formed directly from the cultura media by the quantitative reaction of 1,2-phenylenediamine with 1,2-dicarbonyl compounds in water bufered at pH 7.2 (Thiel et al., 2009b). The quinoxaline derivative was then reacted with N-methyl-N-(trimethylsilyl)trifluoroacetamide (MSTFA) in order to obtain a polar quinoxaline derivative accesible for GC–MS analysis by formation of its trimethylsilyl derivative. This derivative was easily recognized by its characteristic fragmentation such as the ion at m/z 73 $[(CH_3)_3Si]^+$, a significant molecular ion M^+ at m/z 348, a typical fragment ion resulting from the loss of a methyl group $[M-CH_3]^+$ at m/z 333, and the characteristic ion at m/z 245 $[M-CH_2OSi(CH_3)_3]^+$. Quantification was conducted by isotope dilusion analysis using the deuterium-labeled standard $[5,6,7,8-^2H]$-phenylendiamine reaching detection limits of 0,7 ng/mL and quantification limits of 2,1 ng/mL. The (S) configuration of the natural DPD was confirmed by the coninjection of natural and both synthetic enantiomer derivatives using beta-cyclodextrin chiral columns for this analysis.

2.4 DFS´s and other signaling molecules

The role of diffusible signal factors (DSFs) as signaling molecules was originally identified in *Xanthomonas campestris* pv *campestris*, the causal agent of black rot of cruciferous plants. However, there are evidences that these signals are widespread (Deng et al., 2011). The DSFs are generally fatty acids with chain-lenghts of C-10, C-12, C-13 and C-14, with different structural patterns such as unsaturations usually at C-2 and C-9; and chain branches (i.e. iso and anteiso branch). The biological activity of DSF-family signals depends on not only their structural features, but the bacterial species on which they act.

Analysis of DSFs has been conducted by ethyl acetate extraction, solid-phase extraction with an Oasis HLB cartridge, and a combinaton of HPLC-ESI-MS and GC-MS from the culture supernatants of *Stenotrophomonas maltophilia* (Huang & Wong, 2007). The isolated compounds that facilitate the bacteria translocation were the identified as cis-Δ^2-11-methyl-dodecenoic acid and 11-methyl-dodecanoic acid, together with other six structural related fatty acids. Their structures were established based on their FAMEs and pyrrolidide derivatives. The patogenic bacteria *Xylella fastidiosa* was found to be the producer of 12-methyl-tetradecanoic acid as DSFs, based on its endoglucanase restoration activity in a mutant strain of *Xanthomonas campestris* pv *campestris*. The active fraction was obtained by solvent extraction from its culture media with hexane and further derivatized to form the FAMEs. These derivatives were analyzed by GC-MS and the fragment at m/z 74 characteristic for McLafferty rearrengement in FAMEs and other important losses were used for the structural identification of the DSF (Colnaghi et al., 2007).

Diketopiperazines (DKPs) are aminoacid dimmers commonly found in Gram negative bacteria, and usually analyzed by GC-MS. The idenfiyed molecules were the diasteromeric couples (D- and L-) cyclo-(Ala-Val), cyclo-(Pro-Phe), cyclo-(Pro-Leu) cyclo-(Pro-Val) and cyclo-(Pro-Tyr) in *Burkholderia cepacia* (Wang et al., 2010) and *Pseudomonas putida* (Degrasi et al., 2002). These molecules have been extracted with ethyl acetate and detected by GC-MS in both SIM and SCAN modes following their M^+ ions and the characteristic ions at m/z 70, 153, 154. However, these molecules can also be culture media artifacts produced by heating as demonstrated for the cyclo-(Pro-Phe), but in minor amounts than the DKP produced by the bacteria (Wang et al., 2010). In this way, the methodologies used for DKPs analysis must involve a careful evaluation of its production by the bacterial strains.

Farnesol showed to be a signaling molecule that regulates yeast-mycelium conversion in *Candida albicans* without a depletion of its growth rate (Hornby et al., 2001). Its activity as signaling molecule was determined in six different *C. albicans* strains and different culture media and growth conditions. As this signaling molecule was extracted in ethyl acetate and analyzed in a preliminar way by TLC, it was further analyzed by GC-MS by CI and EI and its BSTFA derivative (Hornby et al., 2001). Nerolidol, a farnesol isomer showed activity but in a leser extent than farnesol.

The Pseudomonas quinolone signal has been analyzed by TLC and HPLC (Pesci et al., 1999), but there are not reports of its analysis by GC methods even when these molecules seem to be succeptible of GC analysis. The analysys by GC of signaling molecules of the gamma-butirolactone family as the A-Factor found in *Sptreptomyces* species has not been described. Other signaling molecules as 2,3-diamino-2,3-bis(hydroxymethyl)-1,4-butanediol from *Streptomyces natalensis* ATCC27448, (Recio et al., 2004) result to be non compatible with GC analysis due to its high polarity. These molecules have been analysed by HPLC isolation, and NMR structural elucidation.

In this context, gas chromatography proved to be very useful and versatile in the detection, structural analysis and quantification of signaling molecules. There are still challenges to resolve as the reduction of detection limits for signaling molecules, the detection or even the discovery of new signaling molecules, the development of quantification methods for linear AHLs that do not envolve the use of C_7AHL as internal standar and quantification methods for 3-hydroxyl AHLs, among other issues, or GC methods for PQS signaling molecules.

3. Quorum sensing inhibitors analysis

The antibiotic therapy has a lot of problems, which includes the emergence of drug-resistant bacteria both in hospital and in community-acquired infection, and the slow progress in developments of new antibiotics with novel modes of action. All this problems makes necessary the development of new strategies to control bacterial infections. In this area the quorum sensing inhibitors are emerging as a novel and potentially useful strategy. Is noteworthy to mention antifungal and cancer therapies are currently exploring the use of QS inhibitors compounds as a new strategy of control (Dembitsky et al., 2011; Chai et al., 2011). The QS inhibitors may act in four different ways in bacterial systems. First inhibiting the signaling molecules biosynthesis (i.e. AHLs); second inducing degradation of the signaling molecules (Dong et al., 2002), where the AHLs lactonolysis degradation could be followed by GC-MS or HPLC-MS analysis; third blocking specific bind sites of AHLs to LuxR type proteins; and finally, by inhibition of DNA transcription (Dobrestov et al., 2009). Inhibition of QS systems regulating the expression of virulence factors as well as biofilm formation is a highly attractive field for developing novel therapeutics, because the biofilm provides to bacteria a large resistance to antibiotics. A quorum sensing inhibitor, suppress specific genetic expression of bacteria without cause of death. Since survivable mechanisms of bacteria are not induced by QSI compounds, bacterial resistance is not developed, thus it has also been considered since years for the pharmaceutical industry as a promising strategy for the design and development of antipathogenic compounds, useful in controlling microbial chronic infections (Rasmussen et al., 2006).

The GC analysis has been used for the characterization of some compounds with QSI activity, however for the understanding of mechanism of action this technique lacks of value, and the use of biosensor and *in silico* investigation is the current option. Due to volatility restrictions for GC analysis, only the QSI inhibitors from essential oils, free volatile compounds and furanone analogues will be discussed here. The no volatile compounds with QSI properties, like some flavonoids, malyngamide (Kwan et al., 2010), *S*-ribosyl-L-homocysteine analogs (Shen et al., 2006) and macromolecules (Amara et al., 2011), are not revised. For a general revision Chan et al., 2004; Dobrestov et al, 2009; Konaklieva & Plotkin, 2006; McDougald et al., 2007; Ni et al., 2009, Rasmussen et al., 2006 and the issue number 1 of 111th volume of The Chemical Reviews are suggested.

The dependence of QS on small molecule signals has inspired organic chemists to design non-native molecules that can intercept these signals and thereby perturb bacterial group behaviors. The main investigated QS inhibitors are the AHLs analogues. Lots of analogues have been synthetized, including aromatic and sulfur derivatives (Galloway et al., 2011). The use of GC in the analysis of AHLs compounds was provided in the previous section.

3.1 Furanones and analogues

The most interesting natural compounds for its QSI activity are the bromofuranones from the red alga *Delisea pulchra* (Figure 3a), these compounds are able of inhibit QS-regulated virulence genes, including the production of antibiotics, the bacterial motility and the biofilm formation (Ren et al., 2001; Steinberg & De Nys, 2002). However, most of them have not yet qualified as chemotherapeutic agents because of its toxicity, high reactivity and instability. These facts make evident the necessity of finding new, more potent and safe compounds with QSI properties. For that reason the synthesis of furanone compounds has being explored as well as its structure-relationships (Persson et al., 2005; Martinelli et al., 2004; Wright et al., 2006). However, few examples of the use of GC for furanones analysis in the search of QSI are available in the bibliography in spite of its potential as analytical tool.

Kim et al., in 2008 synthesized new furanone derivatives (Figure 3b) with structural similarities to patulin, a *Penicillium* QSI compound, in order to develop *P. aeruginosa* QS inhibitors and biofilm controlling agents. Each of the six synthesized compounds was confirmed using techniques as NMR and GC-MS. All compounds could remarkably inhibit both *Pseudomonas* QS signaling and biofilm formation. Additionally, the authors for the profound understanding about inhibition mechanisms estimate the binding energy between QS receptor, LasR, and the synthesized compounds in silico modeling systems, which showed good agreement with the experimental results. The most important goal of this work was the structural modeling which can be used as a tool to design the QS inhibitors or some other kinds of enzyme modulators (Kim et al., 2008).

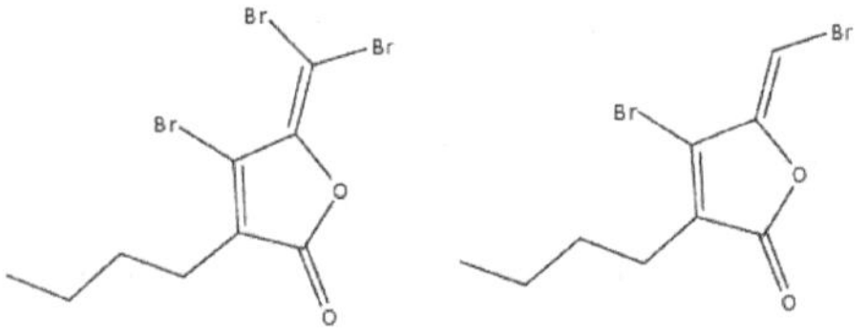

a. Some *Delisea pulchra* furanones

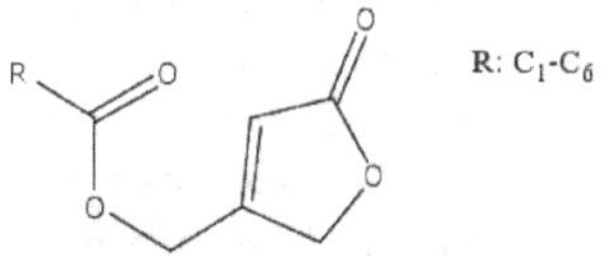

b. Furanone derivatives related to Patulin

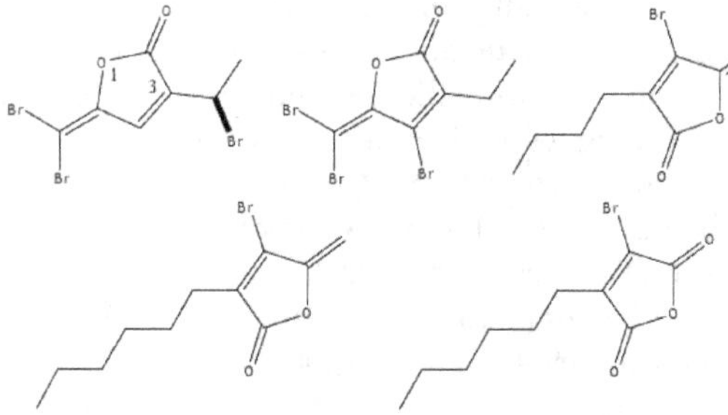

c.   Examples of the most active QSI compounds identified by Kim et al., 2008

Fig. 3. Some furanone derivatives with quorum sensing inhibition properties

Other example that used GC-MS in the characterization of synthetic compounds in the searching of new QS inhibitors is the work of Steenackers et al., 2010, who synthesized a library of 25 1'-unsubstituted and 1'-bromo or 1'-acetoxy 3-alkyl-5-methylene-2(5H)-furanones and two 3-alkylmaleic anhydrides (Figure 3c). The compounds were evaluated for the antagonistic effect against the biofilm formation by *Salmonella enterica* Typhimurium, an important pathogen for humans, and the bioluminescence of *Vibrio harveyi*, which is quorum sensing regulated. *Vibrio* species are important pathogens in the intensive rearing of marine fish and invertebrates like penaeid shrimp. Because multi-resistant *Vibrio* strains have emerged and antibiotics are no longer effective in the treatment of luminescent vibriosis, these marine industries require new alternatives for vibriosis controls, and the QSI represents an important option in this field. In this way Kim et al., found a drastic influence of 3-alkyl chain, bromination pattern and the ring structure on the biological activity of the compounds. Moreover, molecules without a 3-alkyl chain were shown to be highly toxic for both, *Salmonella* and *Vibrio*, while the 1'-unsubstituted furanones with a long 3-alkyl chain did not reduce biofilm formation (octyl chain and longer) nor bioluminescence (dodecyl chain). However, the 1'-unsubstituted furanones with ethyl, butyl or hexyl side chains inhibited biofilm formation at low concentrations, without affecting the planktonic growth at these concentrations. Similarly, the 1'-unsubstituted furanones with a butyl to decyl side chain inhibited bioluminescence without affecting the planktonic growth of the bacteria at the same concentrations.

On the other hand, the introduction of a bromine atom on the first carbon atom of the alkyl side chain drastically improved the activity of the furanones in both tested systems. The introduction of an acetoxy function in this position did in general not improve the activity. The main goal of this work was the identification of the potential of the (bromo) alkylmaleic anhydrides as a new and chemically easily accessible class of biofilm and quorum sensing inhibitors.

3.2 Microorganisms as source of QSI compounds

Many marine or terrestrial bacteria produce volatile compounds, but the specific function of these compounds keeps unknown for many cases. These volatiles could be used either in intra- or inter-specific communication and/or in the chemical defense against other organisms. Schulz et al., evaluated the antiproliferative activity of 52 volatile compounds released from bacteria characterized by GC-MS. The results showed that octanoic acid is active against hyphal fungi and yeasts, as are some *N*-phenethyl amides. Gamma-butyrolactones similar to the signaling molecules are active against fungi, yeasts, and bacteria, with a large influence of a double bond in the lactone ring. Furthermore, the expansion of the ring was found to reduce QSI activity. Pyrazines as well as ketones are largely inactive with the exception of (*Z*)-15-methylhexadec-12-en-2-one, showing broad activity. *S*-Methyl benzothioate is the only sulfur containing compound with activity against fungi and yeast, while all others are inactive. In contrast, the compounds common to many bacteria (i.e. 3-methylbutanol, 2-phenylethanol, (2*R*,3*R*)-butanediol, acetoin, geosmin, (+)-*R*-methylisoborneol, (-)-*S*-methylisoborneol and dimethyl disulfide) showed no inhibitory activity. Additionally, all compounds were investigated for their activity in AHLs mediated bacterial communication systems. The test were performed using *E. coli* MT102 (pJBA132) and *Pseudomonas putida* F117 (pKR-C12) as biosensors. The first strain shows the highest sensitivity for (3-oxo-C_6-AHL), while the second is very sensitive to C_{12}-AHL. Schulz et al., observed that 25% of the compounds were able to reduce activity of *P. putida* sensor F117, while 19% showed inhibitory activity against *E. coli* MT102 biosensor, and 13% enhanced its activity. Most of the gamma- and delta-lactones inhibited the response of the C_{12}-AHL sensor. Especially the delta-lactone of Figure 4a was highly active. In contrast, the *E. coli* MT102 biosensor showed a different behavior. In particular some gamma-lactones stimulated this sensor. This influence may be due to the structural similarity between the lactones and the AHLs. Some aliphatic ketones also proved to be active as well some related alcohols (Figure 4b). The 2-hexylpyridine (Figure 4c) showed inhibitory activity. The common compounds 3-methylbutanol, 2-phenylethanol, (+)-*R*-methylisoborneol and (-)-*S*-methylisoborneol showed some activity, reducing the effectivity of the 3-oxo-C6-AHL sensor. In summary, the observed that some compounds interfered with the quorum-sensing-systems, especially the γ-lactones while the pyrazines showed to have only low intrinsic activity. (Schulz et al., 2010).

An interesting example of the complementarily of GC-MS analysis and the use of *Chromobacterium violaceum* CV017 in the QSI bioassay was published by Dobretsov et al., in 2010. In this study the analysis of 25 marine cyanobacteria ethyl acetate/methanol (1:1) extracts collected in different locations of the world (South Florida (USA), Belize and Oman) during different seasons, showed that 19 extracts inhibited violacein pigment production of *C. violaceum*. The minimal inhibitory amount of extract varied from 1.2 to 66.4 µg per disk, these values indicated a strong QSI activity of these organic extracts. The most active extract was separated by classic chromatographic methods; finally, 0.9 mg of the pure QS inhibitory compound was isolated from 213.8 g (dry weight) of *L. majuscula* indicating a very low concentration of the active compound in the cyanobacterial extracts. The compound was identified as malyngolide (MAL) (Figure 4d) by NMR and MS analysis. All the others crude extracts were analyzed by GC-MS, using a Shimadzu GC17-A gas chromatograph coupled to a Shimadzu QP5000 mass spectrometer. A high-resolution gas chromatography column (HP-5 MS, 30 m x 0.25 mm) was used. The Injections were performed in splitless mode at 70°C. The injection port was held at 250°C with a 70°C to 290°C temperature ramp at 10°C/min. Scan mode was used to analyze ions characteristic of malyngolide, particularly the peak at m/z 239, which is characteristic for malyngolide, and is due by the loss of angular CH_2OH group. Production of MAL appears to be widespread in *Lyngbya* spp. from different locations sampled over the time, with reports of this compound from different geographic regions as diverse as Florida, Hawaii, and Guam. The QSI assays allowed to establish that active MAL concentrations ranged from 0.07 to 0.22 mM with EC_{50} = 0.11 mM, even at the highest concentration MAL doesn't effects the *C. violaceum* growth. In order to understand the mechanisms of action of MAL, Dobretsov et al. in 2010 used in addition of violacein production of CV017 the QS-dependent β-galactosidase activity of the *A. tumefaciens* NTL4 (pZLR4) (Cha et al., 1998), *Escherichia coli* JM109 strains (pSB1075; pMT505 pTIM5211; and pTIM2442) and *P. aeruginosa* PAO1. The results indicated that MAL was not recognized as an AHL signal-mimic by the reporter and does not function as an antagonist of QS in *A. tumefaciens*. Additionally, the results suggested MAL may exert its inhibitory effect on QS by reducing or partially blocking the expression of *lasR* but not by interference with the bacterial AHL-binding domain. This mode of action has not been previously characterized in any of the naturally occurring QS inhibitors and could open a new way for inhibition of bacterial QS.

On the other hand, MAL has several ecological functions in the marine environment, including feeding deterrents to opistobranch mollusks (Nagle et al., 1998) and coral reef fishes (Thacker et al., 1997) as well as an antibacterial compound (Cardellina et al., 1979; Babler et al., 1980). However, Dobretsov et al., in 2010 found that this compound was not toxic to *C. violaceum* CV017 or *E. coli* at QS inhibitory concentration, this observation allowed them to conclude that MAL activity is possibly similar to some other antibiotics, which at sublethal concentrations inhibit QS in bacteria. In order to establish if the MAL is released by *L. majuscula* into seawater, the presence of MAL in it was estimated using GC-MS, founding hat MAL could be released and accumulated at the surrounding sea water. These results suggested that MAL can block bacterial QS and help cyanobacterium to control growth of heterotrophic bacteria (Dobretsov et al., 2010).

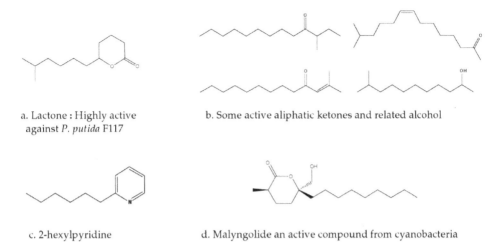

a. Lactone : Highly active   b. Some active aliphatic ketones and related alcohol
against *P. putida* F117

c. 2-hexylpyridine   d. Malyngolide an active compound from cyanobacteria

Fig. 4. Some volatiles compounds with QSI activity isolated from bacteria

3.3 Food as source of QSI

Recently, researches observed certain inhibition of the autoinducer AI-2 by food matrixes; however, the understanding of bacterial cell signaling in relation to foodborne pathogens and food spoilage organisms is limited. Widmer et al., showed that poultry meat wash (PMW) inhibited the QS of *Vibrio harveyi* BB170 strain reporter. The PMW was separated using bioguided methodology employing molecular size exclusion and reverse phase column chromatography. This procedure allowed identifying the hydrophobic fraction as the responsible for the QSI activity of the PMW. The mixture of fatty acids from the PMW were converted in their FAMEs, and extracted with hexane. This extract was further analyzed by gas chromatography, using a FID detector and passed through a fused silica capillary column, according to the AOAC oficial method Ce 1h-05. Six FAMEs were identified and quantified. After that the mayor fatty acids: linoleic acid, oleic acid, palmitic acid, and stearic acid were tested for inhibition as pure compounds, finding that all samples expressed AI-2 inhibition. Then, the fatty acids combined in concentrations equivalent to those detected by GC analysis expressed inhibition. The combined fatty acids at 100-fold natural concentrations did not demonstrate a substantial decrease in colony plate counts, despite presenting high AI-2 inhibition. The inhibition by the combined fatty acid samples was, however, significantly ($P \leq 0.05$) lower than that observed in the PMW control samples, this could be due to the absence of minor fatty acids, and other compounds, in the artificial combined fatty acids sample. These fatty acids, through modulating quorum sensing by inhibition, may offer a unique means to control for food-borne pathogens and reduce microbial spoilage (Widmer et al., 2007).

Examples of inhibitors of AI-2 are not common as the AHL quorum sensing systems inhibitors (for revision Dembitsky et al., 2011); however, some AI-2 inhibitors have been designed and synthesized. These compounds usually contain ribosyl and organic acid moieties (Alfaro et al., 2004; Shen et al., 2006). Some of these inhibitors were shown to coordinate with the metal Co^{2+} ion of LuxS using a ketone oxygen and a hydroxyl group of the inhibitor, and they may cause inhibition by preventing the hydroxamate from approaching the metal ion properly (Shen et al., 2006). The fatty acids identified by Widmer et al., 2007 did not have a similar structure to AI-2. However; the acid moiety may be binding LuxS, impeding its function. Another mode of action of these fatty acids could be the interference of transport systems employed by the bacteria for taking up AI-2 from the environment.

3.4 Essential oils as source of QSI

The essential oils have been identified as an interesting source of compounds able to disrupt the bacterial quorum sensing. The analysis of these oils is usually done by chromatographic methods; however, the isolation and identification of the compounds responsible for the QSI activity is yet an objective for the natural products researchers. The preparative GC could play an important role in this area.

Essential oils have been investigated by its QSI activity. These kinds of extracts are usually analyzed by GC-MS. This is the case of Clove, cinnamon, lavender and peppermint oils that showed activity against the biosensors *Chromobacterium violaceum* (CV12472, CVO26 and CV31532) and *Pseudomonas aeruginosa* (PAO1). Another 17 essential oils were tested without positive results; these essential oils included *Citrus* spp and *Cymbopogon* spp among others. The clove oil had the strongest activity (zones of pigment inhibition 19 mm), and its activity was found to be concentration-dependent. At sub-MICs of clove oil, 78% reductions on violacein production over control and up to 78% reduction on swarming motility in PAO1 over control were observed. The GC-MS analysis was conducted on a HP-1 Column (30 m X 25 mm X 0.25 μm). The GC–MS analysis of clove oil allowed establishing the presence of eugenol (74%), and other minor constituents identified as α-caryophyllene (4%), iso-caryophyllene (6%), caryophyllene oxide (2.4%), β-caryophyllene (5%), napthalene, 1,2,3,5,6,8a-hexahydro-4,7-dimethyl-1-(1-methyl ethyl) (7%) and 1,6-Octadiene-ol-,3,7-dimethyl acetate (1%). However, when the eugenol was assayed it did not exhibit QS inhibition activity. Additionally, neither major constituent like eugenol nor minor constituents like α-caryophyllene and α-caryophyllene share structural similarity with AHLs or known QS inhibitors like halogenated furanones (Khan et al., 2009).

A Hungarian research Group studied nine essential oils for its QSI properties, the essential oils were purchased to Phoenix Pharma Ltd. (Hungary, Budapest) in quality according to the requirements of Hungarian pharmacopoeia. As biosensors the strains *Chromobacterium violaceum* CV026, *Escherichia coli* ATTC 31298 and the partially characterized Ezf 10/17 isolated from a grapevine crown gall tumor were used. The last two strains were used as AHLs producers in order to induce pigmentation of CV026 and therefore they can be used to monitor AHLs-induced pigment production by *C. violaceum* CV026. 5-fluoro-uracil and acridine orange were selected as positive controls by its inhibition CV026 violacein production properties. The rose, geranium, lavender and rosemary oils had potent QS inhibition activity in different biological model test. Eucalyptus and citrus oils moderately reduced pigment production by *C. violaceum* CV026, whereas the chamomile, orange and juniper oils were ineffective (Szabó et al., 2010). However, in this work the authors didn't identify the responsible compounds for QSI activity from the essential oils.

Farnesol is a sesquiterpene (Figure 5a) present in many essential oils (i.e. *Pluchea dioscoridis*, *Zea mays* and *Pittosporum undulatum*), and able to inhibit the growth of some microorganisms, such as the human pathogens *Staphylococcus aureus* and *Streptococcus mutans*, and the plant pathogenic fungus *Fusarium graminearum*. Farnesol also enhances microbial susceptibility to antibiotics, indicating a putative application as an adjuvant therapeutic, and a possible role as chemical defense in the plant. This sesquiterpenoid was also identified as a quorum sensing molecule produced by the dimorphic fungus *Candida albicans*, where it prevents the fungal transition from yeast to mycelium, and disrupts biofilm formation. Some studies showed that farnesol increases the virulence of *C. albicans* in a mouse infection model and this fungus uses it in order to reduce competition with other microbes. Derengowski et al., tested the effects of farnesol on *Paracoccidioides brasiliensis* growth and morphogenesis. This fungus is the etiologic agent of paracoccidioidomycosis (PCM), a systemic human mycosis geographically confined to Latin America. The results indicated that farnesol acts as a potent antimicrobial agent against *P. brasiliensis*. The fungicide activity of farnesol on this pathogen is probably associated to cytoplasmic degeneration, in spite of the apparent cell wall integrity, as observed by transmission and scanning electron micrographs. In concentrations that did not affect fungal viability, farnesol retards the germ-tube formation of *P. brasiliensis*, suggesting that the morphogenesis of this fungal is controlled by environmental conditions (Derengowski et al., 2009).

Another well known compound present in essential oils like cinnamon and cassia is the cinnamaldehyde (Figure 5b). This compound is effective at inhibiting two types of AHLs mediated QS, and also AI-2 mediated QS. The effect of cinnamaldehyde on 3-OH-C_4-AHL and AI-2 mediated cell signalling was determined using *V. harveyi* BB886 and BB170 strains. The effect of cinnamaldehyde on LuxR-mediated transcription from the P_{luxI} promoter, which is induced by 3-oxo-C_6-AHL, was evaluated using the destabilized green fluorescent protein-based bioreporters *E. coli* ATCC 33456 pJBA89 and *E. coli* ATCC 33456 pJBA113. Niu et al., in 2006 proposed that the three carbon aliphatic side chain of cinnamaldehyde interferes with the binding of the smaller 3-hydroxy-C4- and 3-oxo-C_6-HSLs to their cognate receptors, but was not sufficiently long enough to substantially reduce the binding of 3-oxo-C_{12}-HSL to LasR. They also observed that cinnamaldehyde significantly reduced AI-2 mediated signalling. Its concentrations in common cinnamon-containing foods range from 4 to 300 ppm. Consequently, the potential influence of it on 3-OH-C4-, 3-oxo-C_6-HSL and AI-2 mediated quorum sensing could affect bacterial activity, and may be relevant to food ecology. Brackman et al., in 2008 studied some derivatives of cinnamaldehyde as quorum sensing inhibitors, the mechanism of QS inhibition was evaluated by measuring the effect on bioluminescence in several *Vibrio harveyi* mutants. The compounds were also evaluated in an *in vivo* assay measuring the reduction of *Vibrio harveyi* virulence towards *Artemia* shrimp.

Cinnamaldehyde and several derivatives were shown to interfere with AI-2 based QS by decreasing the ability of LuxR to bind to its target promoter sequence. These compounds, used in sub-inhibitory concentrations, did not only affect *in vitro* the production of multiple virulence factors and biofilm formation, but also reduced *in vivo* the mortality of *Artemia* shrimp exposed to *Vibrio harveyi* BB120. In addition, cinnamaldehyde reduced the ability to cope with stress factors like starvation and exposure to antibiotics. Since inhibitors of AI-2 based quorum sensing are rare, and considering the role of AI-2 in several processes these compounds may be useful leads towards antipathogenic drugs. These results indicated that cinnamaldehyde and cinnamaldehyde derivatives are potentially useful antipathogenic lead compounds for treatment of vibriosis, which is a major disease of marine fish and shellfish and is an important cause of economic loss in aquaculture (Brackman et al., 2008).

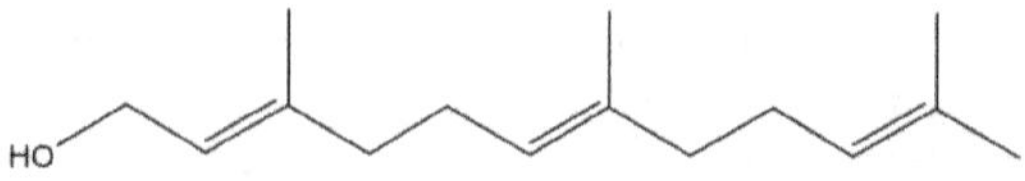

a- farnesol 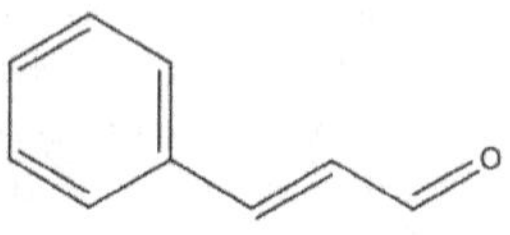   b. cinnamaldehyde

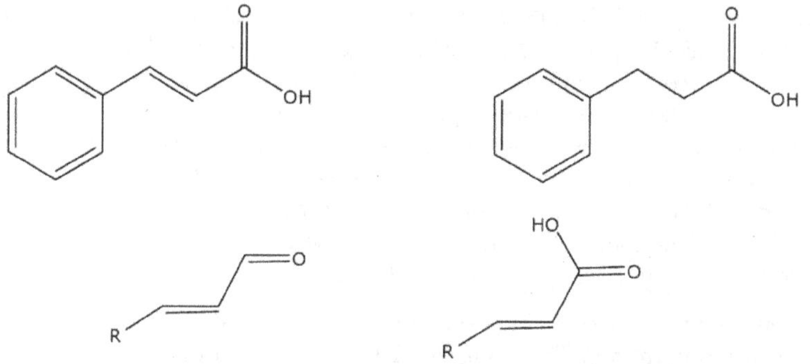

c. Cinnamic acid, 3-phenylpropanoic acid, 3-alkyl-acrylaldehydes and 3-alkyl-acrylic acids

Fig. 5. Some compounds present in essential oils with QSI activity.

In 2011 Brackman et al., studied 42 cinnamaldehyde analogs, including cinnamic acids, 3-phenylpropanoic acid derivatives, 3-alkyl-acrylaldehydes and 3-alkyl-acrylic acids derivatives (Figure 5c). These included an α,β-unsaturated acyl group capable of reacting as Michael acceptor connected to a hydrophobic moiety and a partially negative charge. Cinnamaldehyde and most analogs did not affect the growth of the different *Vibrio* strains when they were used at concentrations up to 250 mM. Several new and more active cinnamaldehyde analogs were discovered and they were shown to affect *Vibrio* spp. virulence factor production *in vitro* and *in vivo*. These compounds significantly increased the survival of the nematode *Caenorhabditis elegans* infected with *Vibrio anguillarum*, *Vibrio harveyi* and *Vibrio vulnificus*. In addition, the most active cinnamaldehyde analogs were found to reduce the *Vibrio* species starvation response, to affect biofilm formation in *V. anguillarum*, *V. vulnificus* and *V. cholerae,* to reduce pigment production in *V. anguillarum* and protease production in *V. anguillarum* and *V. cholera*. However, the mechanism of action of these compounds is not clear at all; the chemical structure of cinnamaldehyde analogs and their effect on the DNA binding ability of LuxR, led authors to purpose that cinnamaldehyde analogs can act as LuxR-ligands, thereby changing the DNA-binding ability of LuxR (Brackman et al., 2011). Compounds capable of participating in a Michael-type addition reaction were found to be active, while replacement of the aldehyde group by a carboxylic acid moiety resulted in less active compounds. Compounds lacking conjugated double bond were found to be inactive. In this way, nucleophilic amino acid side chains (e.g. thiol groups of cysteine residues) in LuxR possibly react with the electrophilic beta-position to form irreversible cinnamaldehyde-receptor conjugates.

3.5 Marine organism

The marine organisms are well known as a source of bioactive compounds, particularly cytotoxic; however, a few examples of its potential as source of quorum sensing inhibitory compounds are described in the literature, in spite of the active bromo furanones were isolated from the marine red algae *Delisea pulchra*. Nowadays, examples of systematic bioprospection of this resource are more common than five years before (Skindersoe et al., 2008; Dobrestov et al., 2010).

Peters et al., in 2003 studied the North Sea bryozoan *Flustra foliacea*. The GC-MS analysis of the dichloromethane extract of the bryozoan allowed identifying 11 compounds. GC-MS analysis were conducted using a Perkin-Elmer PE-1 column (30 m x 0.32 mm) and He (2 ml/min) as the carrier gas, the temperatura gradient program was used: increase from 90°C (at zero time) to 160°C at a rate of 6°C/min and increase from 160 to 300°C at a rate of 10°C/min. Preparative HPLC of the extract yield one diterpene and 10 bromo-tryptamine alkaloids (Figure 6). All of these compounds were tested in order to determine their activities in agar diffusion assays against bacteria derived from marine and terrestrial environments. Additionally, using the biosensors *P. putida* (pKR-C12), *P. putida* (pAS-C8), and *E. coli* (pSB403) the antagonistic effect on AHL dependent quorum-sensing systems was investigated. The most active compounds caused reductions in the signal intensities in these bioassays ranging from 50 to 20% at a concentration of 20 μg/ml. At higher concentrations, however, the compounds had additional biocidal effects.

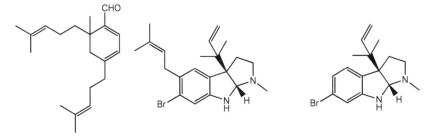

Fig. 6. Diterpene and the most active prenylated alkaloids isolated form bryozoan *Flustra foliacea*.

4. Conclusion

Studies in quorum sensing involve both, understanding of quorum sensing molecular pathways as well as signaling molecules characterization. The use of GC has shown to be one of the more versatile tools for detection, identification and quantification of many types of signaling molecules involved in quorum sensing. Additionaly, GC allowed characterizing several quorum sensing inhibitors from different organisms including bacteria, fungi, terrestrial and marine macroorganisms. The success of this chromatographic technique in characterize quorum sensing molecules (inducers and inhibitors) is supported with the development of other extraction and fractionation techniques that allowed reducing detection limits and improve accurateness, making the GC methods as sensible as the biosensors. Further improvements should be maked for detection, identification and quantification of known and unknown signaling molecules.

5. References

Amara, N.; Krom, B.P.; Kaufmann, G.F. & Meijler, M.M. (2011). Macromolecular inhibition of quorum sensing: enzymes, antibodies, and beyond. *Chemical Reviews*, Vol.111, (January 2011), No1, pp. 195–208, ISSN 0009-2665

Alfaro, J.F.; Zhang, T.; Wynn, D.P.; Karschner, E.L. & Zhou, Z.S. (2004). Synthesis of LuxS inhibitors targeting bacterial cell-cell communication. *Organic Letters*, Vol.6, No.18, (September 2004), pp. 3043–3046, ISSN 523-7052.

Babler, J.H.; Invergo, B.J. & Sarussi, S.J. (1980). Lactone formation via oxidative cyclization of ansaturated carboxylic acid application to the stereoselective synthesis of (±)-malyngolide, an antibiotic from the marine blue-green alga *Lyngbya mayuscula* Gomon. *Journal Organic Chemistry*, Vol.45, No.21, (October 1980), pp. 4241–4243., ISSN 1520-6904.

Bassler, B.L.; Greenberg, E.P. & Stevens, A.M. (1997). Cross-species induction of luminescence in the quorum-sensing bacterium Vibrio harveyi. *Journal of bacteriology*, Vol.179, No.12, (June 1979), pp. 4043-4045, ISSN 1098-5530.

Brackman, G.; Defoirdt, T.; Miyamoto, C.; Bossier, P.; Van Calenbergh, S.; Nelis, H. & Coenye, T. (2008). Cynnamaldehyde and cynnamaldehyde derivates reduce virulence in Vibrio spp. By decreasing the DNA-binding activity of the quorum sensing response regulated LuxR. *BMC Microbiology* Vol.8, (September 2008) pp. 149-149, ISSN 1471-2180.

Brackman, G.; Cos, P.; Maes, L.; Nelis, H.J. & Coenye, T. (2011). Quorum sensing inhibitors increase the susceptibility of bacterial biofilms to antibiotics in vitro and in vivo. *Antimicrobial Agents Chemotherapy*. Vol.55, No.6, (June 2011), pp. 2655-61, ISSN 1098-6596.

Burton, E.O.; Read, H.W.; Pellitteri, M.C. & Hickey, W.J. (2005). Identification of Acyl-Homoserine Lactone Signal Molecules Produced by *Nitrosomonas europaea* Strain Schmidt. *Applied and Environmental Microbiology*, Vol.71, No.8, (August 2005), pp. 4906–4909, ISSN 1098-5336.

Cardellina, J.H.; Moore, R.E.; Arnold, E.V. & Clardy, J. (1979). Structure and absolute configuration of malyngolide, an antibiotic from the marine blue-green alga *Lyngbya majuscula* Gomont. *Journal of Organic Chemistry*, Vol.44, No.23, (November 1979), pp. 4039–4042, ISSN 1520-6904.

Cataldi, T.R.I.; Bianco, G.; Palazzo L. & Quaranta, V. (2007). Occurrence of N-acyl-L-homoserine lactones in extracts of some Gram-negative bacteria evaluated by gas chromatography–mass spectrometry. *Analilical Biochemistry*, Vol.361, No.2, (February 2007), pp. 226–235, ISSN 1096-0309.

Cha, C.; Gao, P.; Chen, Y.C.; Shaw, P.D., & Farrand, S.K. (1998). Production of acyl-homoserine lactone quorumsensing signals by gram-negative plant-associated bacteria. *Molecular Plant-Microbe Interactions*, Vol.11, No.11, (November 1998), pp. 1119–1129, ISSN 0894-0282.

Chai, H.; Hazawa, M.; Shirai, N.; Igarashi, J.; Takahashi, K.; Hosokawa, Y.; Suga, H. & Kashiwakura, I. (2010) Functional properties of synthetic N-acyl-L-homoserine lactone analogs of quorum-sensing gram-negative bacteria on the growth of human oral squamous carcinoma cells. *Investigational New Drugs*, (September 2010), DOI: 10.1007/s10637-010-9544-x, ISSN 1573-0646.

Chan, W.C.; Coyle, B.J. & Williams, P. (2004). Virulence regulation and quorum sensing in staphylococcal infections: competitive AgrC antagonists as quorum sensing inhibitors. *Journal Medicinal Chemistry*, Vol.47, No.19, (September 2004) pp. 4633-4641. ISSN 1520-4804.

Charlton, T.S.; de Nys, R.; Netting, A.; Kumar, N.; Hentzer, M.; Givskov M. & Kjelleberg S. (2000). A novel and sensitive method for the quantification of N-3-oxoacyl homoserine lactones using gas chromatography-mass spectrometry: application to a model bacterial biofilm. *Environmental Microbiology*, Vol.2, No.5, (October 2000), pp. 530-541, ISSN 1462-2920.

Colnaghi, S.A.V.; Santos, da S.D.; Rodrigues, L.M. & Carrilho, E. (2007). Characterization of a putative *Xylella fastidiosa* diffusible signal factor by HRGC-EI-MS. *Journal of Mass Spectrometry*, Vol.42, No.4, (April 2007), pp. 490–496, ISSN 1096-9888.

Cuadrado, T. (2009). *Aislamiento de n-acilhomoserinlactonas de algunas bacterias procedentes del mar caribe colombiano, como evidencia de la existencia de circuitos de quorum sensing.* Universidad Nacional de Colombia –M.Sc. Thesis. Bogotá.

Degrassi, G.; Aguilar, C.; Bosco, M.; Zahariev, S.; Pongor, S. & Ventui V. (2002). Plant growth-promoting Pseudomonas putida WCS358 produces and secretes four cyclic dipeptides: cross-talk with quorum sensing bacteria sensors. *Current Microbiology*, Vol.45, No.4, (October 2002), pp. 250–254, ISSN 1432-0991.

Dembitsky, V.M.; Quntar, A.A. & Srebnik, M. (2011). Natural and synthetic small boron-containing molecules as potential inhibitors of bacterial and fungal quorum sensing. *Chemical. Reviews*, Vol. 111, No 1, (January 2011), pp. 209–237, ISSN 0009-2665

Deng, Y.; Wu, J.; Tao, F. & Zhang, L.H. (2011). Listening to a New Language: DSF-Based Quorum Sensing in Gram-Negative Bacteria. *Chemical Reviews*, Vol.111, No.1, (January 2011), pp. 160–173, ISSN 0009-2665

Derengowski, L.S.; De-Souza-Silva, C.; Braz, S.V.; Mello-De-Sousa, T.M.; Báo, S.N.; Kyaw, C.M. & Silva-Pereira, I. (2009). Antimicrobial effect of farnesol, a *Candida albicans* quorum sensing molecule, on *Paracoccidioides brasiliensis* growth and morphogenesis *Annals of Clinical Microbiology and Antimicrobials*, Vol.8, (April 2009), pp. 13-13, ISSN 1476-0711.

Dobretsov, S.; Teplitski, M.; Alagely, A.; Gunasekera, S.O. & Paul, V. (2010). Malyngolide from the cyanobacterium *Lyngbya majuscula* interferes with quorum sensing circuitry. *Environmental Microbiology Reports*, Vol.2, No.6, (June 2010), pp. 739–744, ISSN 1758-2229.

Dobretsov, S.; Teplitski, M. & Paul, V. (2009). Mini-review: quorum sensing in the marine environment and its relationship to biofouling. *Biofouling*, Vol.25, No.5, (March 2009), pp. 413-427, ISSN 1029-2454.

Dong, Y.H.; Gusti, A.R.; Zhang, Q.; Xu, J.L. & Zhang, L.H. (2002). Identification of quorum-quenching N-acyl homoserine lactonases from *Bacillus* species. *Applied Environmental Microbiology*, Vol.68, No.4, (April 2002), pp. 1754–1759, ISSN 0099-2240.

Eberhard, A.; Burlingame, A.L.; Eberhard, C.; Kenyon, G.L.; Nealson, K.H. & Oppenheimer, N.J. (1981). Structural identification of autoinducer of Photobacterium fischeri luciferase. *Biochemistry*, Vol.20, No.9, (April 2008), pp. 2444-2449, ISSN 1520-4995.

Engebrecht, J., & Silverman, M. (1984). Identification of genes and gene products necessary for bacterial bioluminescence. *Proceedings of the National Academy of Sciences of the United States of America*, Vol.81, No.13, (July 1984), 4154-4158, ISSN 1091-6490.

Fuqua, W.C.; Winans, S.C., & Greenberg, E.P. (1994). Quorum sensing in bacteria: the LuxR-LuxI family of cell density-responsive transcriptional regulators. *Journal Of Bacteriology*, Vol.176, No.2, (January 2004), pp. 269-275, ISSN 1098-5530

Galloway, W.R.J.D.; Hodgkinson, J.T.; Bowden, S.D.; Welch, M. & Spring, D.R. (2011). Quorum Sensing in Gram-Negative Bacteria: Small-Molecule Modulation of AHL and AI-2 Quorum Sensing Pathways. *Chemical Reviews*, Vol.111, (January 2011), No.1. pp 28-67, ISSN 0009-2665

Götz, C.; Fekete, A.; Gebefuegi, I.; Forczek, S.T.; Fuksová, K.; Li, X.; Englmann, M.; Gryndler, M.; Hartmann, A.; Matucha M.; Schmitt-Kopplin, P. & Schröder P. (2007). Uptake, degradation and chiral discrimination of N-acyl-D/L-homoserine lactones by barley (Hordeum vulgare) and yam bean (Pachyrhizus erosus) plants. *Analitical and Bioanalitical Chemistry*, Vol 389, (November 2007), pp. 1447–1457, ISSN 1618-2650.

Greenberg, E.P. (2003). Bacterial communication: tiny teamwork. *Nature*, Vol.424, No.6945, (July 2003), p.134, ISSN 0028-0836.

Hornby, J.M.; Jensen, E.C.; Lisec, A.D.; Tasto, J.J.; Jahnke, B.; Shoemaker, R.; Dussault, P. & Nickerson K.W. (2001). Quorum Sensing in the Dimorphic Fungus *Candida albicans* is Mediated by Farnesol. *Applied and Environmental Microbiology*, Vol.67, No.7, (July 2001), pp. 2982–2992, ISSN 1098-5336.

Horng, Y.T.; Deng, S.C.; Dykin, M.; Soo, P.C.; Wei, J.R.; Luh, K.T.; Ho, S.W.; Swift, S.; Lai, H.C. & Williams, P. (2002). The LuxR family protein SpnR functions as a negative regulator of N-acylhomoserine lactone-dependent quorum sensing in *Serratia marcescens*. *Molecular Microbiology*, Vol.45, No.6, (September 2006), pp. 1655–1671, ISSN 1365-2958.

Huang, H.L.; Dobretsov, S.; Ki, J.S.; Yang, L.H. & Qian P.Y. (2007). Presence of Acyl-Homoserine Lactone in Subtidal Biofilm and the Implication in Larval Behavioral Response in the Polychaete *Hydroides elegans*. *Microbial Ecology*, Vol.54, No.2, (August 2007), pp. 384–392, ISSN 1432-184X.

Huang, T.P. & Lee-Wong A.C. (2007). Extracellular fatty acids facilitate flagella-independent transloction by *Stenotrophomonas maltophilia*. *Research in Microbiology*, Vol.158, No.8-9, (October 2207), pp. 702-711, ISSN 1769-7123

Hughes, D. T. & Sperandio, V. (2008). Inter-kingdom signalling: communication between bacteria and their hosts. *Nature Reviews Microbiology*, Vol.6, No.2, (February 2008), pp. 111-20, ISSN 1740-1534.

Joint, I.; Tait, K. & Wheeler, G. (2007). Cross-kingdom signalling: exploitation of bacterial quorum sensing molecules by the green seaweed Ulva. *Philosophical transactions of* the *Royal Society of London. Series B, Biological sciences*, Vol.362, No.1483, (July 2007), pp. 1223-1233, ISSN 1471-2970

Kaplan, H.B., & Greenberg, E.P. (1985). Diffusion of autoinducer is involved in regulation of the Vibrio fischeri luminescence system. *Journal of Bacteriology*, Vol.163, No.3, (September 1985), pp. 1210-1214, ISSN 1098-5530.

Khan, M.S.A.; Zahin, M.; Hasan, S.; Husain, F.M. & Ahmad, I. (2009). Inhibition of quorum sensing regulated bacterial functions by plant essential oils with special reference to clove oil. *Letters in Applied Microbiology*, Vol. 49, No.3, (September 2009), pp. 354–360, ISSN 1472-765X.

de Kievit, T.R., & Iglewski, B.H. (2000). Bacterial Quorum Sensing in Pathogenic Relationships. *Infection and Immunity*, Vol.68, No.9, (September 2002), pp. 4839-4849, ISSN 1098-5522.

Kim, C.; Kim, J.; Park, H.Y.; Park, H.J.; Lee, J.H.; Kim, C.K. & Yoon, J. (2008). Furanone derivatives as quorum-sensing antagonists of Pseudomonas aeruginosa. *Applied Microbiology and Biotechnology*, Vol.80, No.1, (August 2008), pp. 37-47, ISSN 1432-0614.

Knowlton, N., & Rohwer, F. (2003). Multispecies microbial mutualisms on coral reefs: the host as a habitat. *The American naturalist*, Vol.162, No4 Suppl, (July 2003), pp S51-S62, ISSN 1537-5323.

Köhler, T.; Perron, G.G.; Buckling, A. & van Delden, C. (2010). Quorum sensing inhibition selects for virulence and cooperation in *Pseudomonas aeruginosa*. *PLoS Pathogen*, Vol.6, No.5, (May 2010), pp. 1-6. doi:10.1371, ISSN 1553-7374.

Konaklieva, M.I. & Plotkin, B.J. (2006). Chemical communication--do we have a quorum? *Mini-Reviews in Medicinal Chemistry*, Vol.6, No.7, (July 2006), pp. 817-825, ISSN 1875-5607.

Kumar-Malik, A.; Fekete, A.; Gebefuegi, I.; Rothballer, M. & Schmitt-Kopplin P. (2009). Single drop microextraction of homoserine lactones based quorum sensing signal molecules, and the separation of their enantiomers using gas chromatography mass spectrometry in the presence of biological matrices. *Microchimica Acta*, Vol.166, No.1-2, (June 2009), pp. 101–107, ISSN 1436-5073.

Kwan, J.C.; Teplitski, M.; Gunasekera, S.P.; Paul, V.J. & Luesch, H. (2010). Isolation and biological evaluation of 8-epi-malyngamide C from the Floridian marine cyanobacterium *Lyngbya majuscule*. *Journal of Natural Products*, Vol.73, No.3, (March 2010), pp. 463–466, ISSN 1520-6025.

Li, X.; Fekete, A.; Englmann, M.; Götz, M.; Rothballer, M.; Frommberger, M.; Buddrus, M.; Fekete, J.; Cai, C.; Schröder, P.; Hartmann, A.; Chena, G. & Schmitt-Kopplin, P. (2006). Development and application of a method for the analysis of *N*-acylhomoserine lactones by solid-phase extraction and ultra high pressure liquid chromatography. *Journal of Chromatography A*, Vol.1134, No. 1-2, (November 2006), pp. 186-193, ISSN 1873-3778.

Lithgow, J.K.; Wilkinson, A.; Hardmann, A.; Rodelas, B.; Wisniewski-Dyé, F.; Williams, P.; Downie J.A., 2000. The regulatory locus cinRI in *Rhizobium leguminosarum* controls a network of quorum-sensing loci. *Molecular microbiology*. Vol.37, No.1, (July 2000), pp. 81–97, ISSN 1365-2958.

Marshall, K.; Joint, I.; Callow, M.E., & Callow, J.A. (2006). Effect of marine bacterial isolates on the growth and morphology of axenic plantlets of the green alga *Ulva linza*. *Microbial ecology*, Vol.52, No.2, (August 2000). pp. 302-10, ISSN 1432-184X

Martinelli, D.; Grossmann, G.; Séquin, U.; Brandl, H. & Bachofen. R. (2004). Effects of natural and chemically synthesized furanones on quorum sensing in *Chromobacterium violaceum*. *BMC microbiology*, Vol.4, (July 2004), pp.1-10, doi:10.1186/1471-2180-4-25, ISSN 471-2180.

McDougald, D.; Rice, S. R. & Kjelleberg, S. (2007). Bacterial quorum sensing and interference by naturally occurring biomimics. *Analytical and Bioanalytical Chemistry*. Vol.387, No. 2, (January 2007), pp. 445–453, ISSN 1618-2650.

Miller, M. B., & Bassler, B.L. (2001). Quorum sensing in bacteria. *Annual Reviews in Microbiology*, Vol.55, No.1, (October, 2001) pp. 165–199, ISSN 45-3251.

Morin, D.; Grasland, B.; Vallée-Réhel, K.; Dufau, C. & Haras, D. (2003). On-line high-performance liquid chromatography–mass spectrometric detection and quantification of N-acylhomoserine lactones, quorum sensing signal molecules, in the presence of biological matrices. *Journal of Chromatography A*, Vol.1002, No1-2, (June 2003), pp. 79-92, ISSN 73-3778

Morohoshi, T.; Inaba, T.; Kato, N.; Kanai, K. & Ikeda, T. (2004). Identification of Quorum sensing signaling molecules and le LuxRI Homologs in fish pathogen *Edwarsiella tarda*. *Journal of Bioscience and Bioengineering*,Vol.98, No.4, (Julio, 2004), pp. 274–281, ISSN 1347-4421

Nagle, D.G.; Camacho, F.T. & Paul, V.J. (1998). Dietary preferences of the opisthobranch mollusc Stylocheilus longicauda for secondary metabolites produced by the tropical cyanobacterium *Lyngbya majuscula*. *Marine Biology*, Vol.132, No.2, (April 1998) pp. 267–273, ISSN 00253162.

Nealson, K.H.; Platt, T. & Hastings, J.W. (1970). Cellular control of the synthesis and activity of the bacterial luminescent system. *Journal of Bacteriology*, Vol.104, No.1, (October 1970), pp. 313-322, ISSN 0021-9193.

Ni, N.; Li, M.; Wang, J. & Wang, B. (2009). Inhibitors and antagonists of bacterial quorum sensing. *Medicinal research reviews*. Vol. 29, No.1, (January 2009), pp. 65-124. ISSN 1098-1128.

Niu, C.; Afre, S. & Gilbert, E.S. (2006). Subinhibitory concentrations of cinnamaldehyde interfere with quorum sensing *Letters in Applied Microbiology*, Vol.43, No.5, (November 2006), pp. 489-94, ISSN 1472-765X.

Persson, T.; Hansen, T.H.; Rasmussen, T.B.; Skinders, M.E.; Givskov, M. & Nielsen, J. (2005). Rational design and synthesis of new quorum-sensing inhibitors derived from acylated homoserine lactones and natural products from garlic. *Organic Biomolecular Chemistry*, Vol.3, No.2, (January 2005), pp. 253-262. ISSN 1477-0539.

Pesci, E.C.; Milbank, J.B.J.; Pearson, J.P.; Mcknight, S.; Kende, A.S.; Greenberg, E.P. & Iglewski B.H. (1999). Quinolone signaling in the cell-to-cell communication system of *Pseudomonas aeruginosa*. *Proceedings of the National Academy of Sciences of the United States of America*, Vol.96, No.20, (September 1999), pp. 11229–11234, ISSN 1091-6490

Peters, L.; König, G.M.; Wright, A.D.; Pukall, R.; Stackebrandt, E.; Eberl, L. & Riedel, K. (2003). Secondary metabolites of Flustra foliacea and their influence on bacteria. *Applied and Environmental Microbiology*, Vol.69, No.6, (June 2003), pp. 3469–3475, ISSN 1098-5336.

Pomini, A.M.; Manfio, G.P.; Araujo, W.L. & Marsaioli A.J. (2005). Acyl-homoserine Lactones from *Erwinia psidii* R. IBSBF 435T, a Guava Phytopathogen (*Psidium guajava* L.). *Journal of Agricultural and Food Chemistry*, Vol.53, No16, (August 2005), pp. 6262-6265, ISSN 1520-5118.

Pomini, A.M.; Araújo, W.L. & Marsaioli, A.J. (2006). Structural Elucidation and Biological Activity of Acyl-homoserine Lactones from the Phytopathogen *Pantoea ananatis* Serrano 1928. *Journal of Chemical Ecology*, Vol.32, No.8, (August 2006), pp. 1769–1778, ISSN 1573-1561

Pomini, A.M.; Paccola-Meirelles, L.D. & Masaroli, A.J. (2007). Acyl-homoserine lactones produced by *Pantoea* sp. isolated from the "maize white spot" foliar disease. *Journal of Agricultural and Food Chemistry*, Vol.55, No.4, (February 2007), pp. 1200-1204, ISSN 1520-5118.

Pomini, A.M. & Marsaioli A.J. (2008). Absolute Configuration and Antimicrobial Activity of Acylhomoserine Lactones. *Journal of Natural Products*, Vol.71, No.6, (June 2008), pp. 1032–1036, ISSN 1520-6025.

Rasmussen, T.B. & Givskov, M. (2006). Quorum-sensing inhibitors as anti-pathogenic drugs. *International Journal of Medical Microbiology*, Vol.296, No.6, (April 2006), pp. 149-161. ISSN 1618-0607.

Riedel, K.; Hentzer, M.; Geisenberger, O.; Huber, B.; Steidle, A.; Wu, H.; Høiby, N.; Givskov, M.; Molin, S.; Eberl, L. (2001). N-acylhomoserine-lactone-mediated communication between *Pseudomonas aeruginosa* and *Burkholderia cepacia* in mixed biofilms. *Microbiology*, Vol.147, No.12, (December 2001) pp. 3249-3262, ISSN 1350-0872

Reading, N.C. & Sperandio V. (2006). Quorumsensing: the many languages of bacteria. *FEMS Microbiology Letters*, Vol.254, No.1, (January 2006), pp 1–11, ISSN 1574-6968.

Recio, E.; Colinas, A.; Rumbero, A.; Aparicio, J.F. & Martín J.F. (2004). PI Factor, a Novel Type Quorum-sensing Inducer Elicits Pimaricin Production in *Streptomyces natalensis*. *Journal of Biologycal Chemistry*, Vol.279, No.40, (October 2004), pp. 41586–41593, ISSN 1083-351X

Ren, D.C.; Sims, J.J. & Wood, T.K. (2001). Inhibition of biofilm formation and swarming of *Escherichia coli* by (5Z)- 4-bromo-5(bromomethylene)-3-butyl-2(5H)-furanone. *Environmental Microbiology*, Vol.3, No.11, (November 2001), pp. 731–736, ISSN 1462-2920.

Rivas, M.; Seeger, M.; Jedlicki, E. & Holmes D.S. (2007). Second Acyl Homoserine Lactone Production System in the Extreme Acidophile *Acidithiobacillus ferrooxidans*. *Applied Environmental Microbiology*, Vol.73, No.10, (May 2007), pp. 3225-3231, ISSN 1098-5336.

Shaw, P.D.; Ping, G.; Daly, S.L.; Cha, C.; Cronan, J.E.; Rinehart, K.L. & Farrand, S.K. (1997). Detecting and characterizing N-acyl-homoserine lactone signal molecules by thin-layer chromatography. *Proceedings of the National Academy of Sciences of the United States of America*, Vol.94, No.12, (June 1997), pp. 6036–6041, ISSN 1091-6490.

Schulz, S.; Dickschat, J.S.; Kunze, B.; Wagner-Dobler, I.; Diestel, R. & Sasse, F. (2010). Biological activity of volatiles from marine and terrestrial bacteria. *Marine Drugs*, Vol.8, No.12, (December 2012), pp. 2976–2987, ISSN 1660-3397

Schupp, P.J.; Charlton, T.S.; Taylor, M.W.; Kjelleberg, S. & Steinberg, P.D. (2005). Use of solid-phase extraction to enable enhanced detection of acyl homoserine lactones (AHLs) in environmental samples. *Analitical and Bioanalitical Chemistry*, Vol.383, No1, (September 2005), pp. 132–137, ISSN 1618-2650

Shen, G.; Rajan, R.; Zhu, J.; Bell, C.E. & Pei, D. (2006). Design and synthesis of substrate and intermediate analogue inhibitors of S-ribosylhomocysteinase. *Journal of Medicinal Chemistry* Vol.49, No.10, (May 2010), pp. 3003–3011, ISSN 1520-4804.

Shnit-Orland, M. & Kushmaro, A. (2009). Coral mucus-associated bacteria: a possible first line of defense. *FEMS Micriobiology Ecology*, Vol.67, No.3, (March 2009), pp. 371-380, ISSN 1574-6941.

Skindersoe, M.E.; Ettinger-Epstein, P.; Rasmussen, T.B.; Bjarnsholt, T.; de Nys, R. & Givskov, M. (2008). Quorum sensing antagonism from marine organisms. *Marine Biotechnology* Vol.10, No.1, (January 2008), pp. 56-63, ISSN 1436-2236.

Soni, K.A.; Jesudhasan, P.; Cepeda, M.; Widmer, K.; Jayaprakasha, G.K.; Patil, B.S.; Hume, M.E. & Pillai, S.D. (2008). Identification of ground beef-derived fatty acid inhibitors of autoinducer-2-based cell signaling. *Journal of Food Protection*, Vol.71, No.1, (January 2008), pp.134-138, ISSN 1944-9097.

Sperandio, V.; Torres, A.G.; Jarvis, B.; Nataro, J.P. & Kaper, J.B. (2003). Bacteria–host communication: The language of hormones. *Proceedings of the National Academy of Sciences of the United States of America*, Vol.100, No.15, (July 2003), pp. 8951–8956, ISSN 1091-6490.

Stauff, D. L., & Bassler, Bonnie L. (2011). Quorum Sensing in Chromobacterium violaceum: DNA Recognition and Gene Regulation by the CviR Receptor. *Journal of Bacteriology*, Vol.193, No.15, (August 2011), pp. 3871-3878, ISSN 1098-5530.

Steenackers, H.P.; Levin, J.; Janssens, J.C.; De Weerdt, A.; Balzarini, J.; Vanderleyden, J.; De Vos, D.E. & De Keersmaecker, S.C. (2010) Structure-activity relationship of brominated 3-alkyl-5-methylene-2(5H)-furanones and alkylmaleic anhydrides as inhibitors of Salmonella biofilm formation and quorum sensing regulated bioluminescence in *Vibrio harveyi*. *Bioorganic and Medicinal Chemistry*, Vol.18, No14, (July 2010), pp. 5224-33, ISSN 1464-3391

Steindler, L. & Venturi, V. (2007). Detection of quorum-sensing N-acyl homoserine lactone signal molecules by bacterial biosensors. *FEMS Microbiology Letters*, Vol.266, No1, (January 2007), pp. 1-9, ISSN 1574-6968.

Steinberg, P.D. & de Nys, R. (2002). Chemical mediation of colonization of seaweed surfaces. *Journal of Phycology*, Vol.38, No.4, (August 2002), pp. 621–629, ISSN 1529-8817.

Szabó, M.A.; Varga, G.Z.; Hohmann, J.; Schelz, Z.; Szegedi, E.; Amaral, L. & Molnár, J. (2010). Inhibition of quorum-sensing signals by essential oils. *Phytotherapy Research*, Vol.24, No.5, (May 2010), pp. 782-786, ISSN 1099-1573.

Tait, K.; Williamson, H.; Atkinson, S.; Williams, P.; Cámara, M. & Joint, I. (2009). Turnover of quorum sensing signal molecules modulates cross-kingdom signalling. *Environmental Microbiology*, Vol.11, No.7, (July 2009), pp. 1792-1802, ISSN 1462-2920.

Teplitski, M.; Eberhard, A.; Gronquist, M.R.; Gao, M.; Robinson, J.B. & Bauer W.D. (2003). Chemical identification of N-acyl homoserine lactone quorum-sensing signals produced by Sinorhizobium meliloti strains in defined médium. *Archives of Microbiology*, Vol.180, No.6, (December 2003), pp. 494–497, ISSN 1432-072X.

Thacker, R.W.; Nagle, D.G. & Paul, V.J. (1997). Effects of repeated exposures to marine cyanobacterial secondary metabolites on feeding by juvenile rabbitfish and parrotfish. *Marine Ecology Progress Series* Vol.147, No.1, (February 1997), pp. 21–29, ISSN 01718630

Thiel, V.; Kunze, B.; Verma, P.; Wagner-Döbler I. & Schulz S. (2009a). New Structural Variants of Homoserine Lactones in Bacteria. *ChemBioChem*, Vol.10, No.11, (July 2009) pp. 1861–1868, ISSN 1439-7633

Thiel, V.; Vilchez, R.; Sztajer, H.; Wagner-Döbler I. & Schulz S. (2009b). Identification, Quantification, and Determination of the Absolute Configuration of the Bacterial Quorum-Sensing Signal Autoinducer-2 by Gas Chromatography–Mass Spectrometry. *ChemBioChem*, Vol.10, No.3, (February 2009), pp. 479–485, ISSN 1439-7633.

Thoendel, M.; Kavanaugh, J.S.; Flack, C.E.. & Horswill, A.R. (2011). Peptide Signaling in the Staphylococci. *Chemical Reviews*, Vol.111, No.1, (January 2011), pp.117-151, ISSN 0009-2665

Wagner-Döbler, I.; Thiel, V.; Eberl, L.; Allgaier, M.; Bodor, A.; Meyer, S.; Ebner, S.; Hennig, A.; Pukall, R. & Schulz, S. (2005). Discovery of Complex Mixtures of Novel Long-Chain Quorum Sensing Signals in Free-Living and Host-Associated Marine Alphaproteobacteria. *ChemBioChem*, Vol.6, No.12, (December 2005), pp. 2195–2206, ISSN 1439-7633

Wang, J.W.; Quan, C.S.;. Qi, X.S.; Li, X. & Fan S.D. (2010). Determination of diketopiperazines of *Burkholderia cepacia* CF-66 by gas chromatography–mass spectrometry. *Analytical and Bioanalytical Chemistry*, Vol.396, No.5, (March 2010), pp. 1773–1779, ISSN 1618-2650

Waters, C.M. & Bassler, B.L. (2005). Quorum sensing: cell-to-cell communication in bacteria. *Annual Review of Cell and Developmental Biology*, Vol.21, No.1, (November, 2005) pp. 319-346, ISSN 1530-8995.

Widmer, K.W.; Soni, K.A.; Hume, M.E.; Beier, R.C.; Jesudhasan, P. & Pillai, S.D. (2007). Identification of poultry meat-derived fatty acids functioning as quorum sensing signal inhibitors to autoinducer-2 (AI-2). *Journal of Food Science*, Vol.72, No.9, (November 2007), pp. M363-M368, ISSN 1750-3841.

Wright, A.D.; de Nys, R.; Angerhofer, C.K.; Pezzuto, J.M. & Gurrath, M. (2006). Biological activities and 3D QSAR studies of a series of *Delisea pulchra* (cf. fimbriata) derived natural products. *Journal Natural Products*, Vol.69, No.8, (August 2008), pp. 1180-1187, ISSN 1520-6025.

2

Gas Chromatography in Environmental Sciences and Evaluation of Bioremediation

Vladimir P. Beškoski[1], Gordana Gojgić-Cvijović[1],
Branimir Jovančićević[1,2] and Miroslav M. Vrvić[1,2]
*[1]Department of Chemistry-Institute of Chemistry,
Technology and Metallurgy, University of Belgrade,
[2]Faculty of Chemistry, University of Belgrade,
Serbia*

1. Introduction

Crude oil and its derivatives, as the key energy-generating substances and raw materials used for production, are very widely used in all domains of work and everyday life. With the advent of oil as a fuel, there was the most intense economic growth and it can be said that the entire modern civilization is based on the utilization of oil.

However, rapid growth and development of civilization in the past two centuries with the mass use of fossil fuels has led to imbalances and distortions of natural processes. In spite of improvements in technology and equipment used for oil drilling, transport and processing by the petroleum industry, oil and oil derivatives represent a significant source of environmental contamination.

Both soil and water become contaminated by oil and oil derivatives due to accidental spills in their exploitation, transportation, processing, storing and utilization. In 2010, 3.91 billion tons of crude oil was produced (BP Statistical Review, 2011) and estimations are that annually 0.1% of produced petroleum is released into the environment (Ward et al., 2003) as a result of anthropogenic activities.

Petroleum and petroleum products are complex mixtures consisting of thousands of compounds that are usually grouped into four fractions: aliphatics, aromatics, nitrogen–oxygen–sulphur (NSO) compounds and asphaltenes. Asphaltenes are generally solvent insoluble and resistant to biodegradation. Aliphatic hydrocarbons consist of normal alkanes (n-alkanes), branched alkanes (isoalkanes) and cyclic alkanes (naphthenes). Isoalkanes, naphthenes and aromatics are much less biodegradable than *n*-alkanes (Evans & Furlog 2011; Pahari & Chauhan, 2007). The fraction of saturated hydrocarbons is the dominant fraction in most oils as compared to aromatic hydrocarbons and NSO compounds (Peters et al. 2005).

When crude oil or petroleum products are accidentally released into the environment, they are immediately subject to a wide variety of weathering process (Jordan & Payne, 1980). These weathering processes can include: evaporation, dissolution, microbial degradation, dispersion and water–oil emulsification, photooxidation, adsorption onto suspended particulate materials, and oil–mineral aggregation. Petroleum compounds are substrates for microorganisms which can use these substances as the sole source of hydrocarbons (Head et al., 2006; Van Hamme et al., 2003). The susceptibility of hydrocarbons to microbial attack is ranked in the following order n-alkanes > branched alkanes > branched alkenes > n-alkylaromates of small molecular mass > monoaromates > cyclic alkanes > polycyclic aromates > asphaltenes (Alexander, 1999; Atlas & Philp, 2005; Singh & Ward, 2004).

Bioremediation is a technology of cleaning and remedying the soil through biological methods by means of non-pathogenic microorganisms that feed on the contaminating substances. The microorganisms are used to reduce the complexity of organic molecules (biotransformation), or for degradation to complete mineralization (biodegradation). Some defined bacterial species are able to degrade, to a limited extent, all hydrocarbons present in heavy fuel oil or oil sludge. Some of the polluting components may be dissolved only by the coupled metabolic activity of multiple genera of microorganisms. A consortium (mixed culture) of microorganisms can conduct these complex processes of degradation, while at the same time, being more resistant, on average, to changes in the ecosystem than just a single microbial species (Brenner et al., 2008).

To increase the rate of biodegradation of hydrocarbons in the ecosystem and to maximize the process in bioremediation technologies, three main approaches are applied: biostimulation, in which nutrients are added to stimulate the intrinsic hydrocarbon degraders, bioventilation which ensures the required quantity of the molecular oxygen – aeration, and bioaugmentation, in which microbial strains with specific degrading abilities are added to work cooperatively with normal indigenous soil microorganisms (Alvarez & Illman, 2006). The contaminated soil can be treated by bioremediation *in situ* or *ex situ*. As a natural cleaning process, bioremediation has been proven to be efficient in the removal of crude oil and oil derivatives (Ollivier & Magot, 2005), chlorinated solvents (Bamforth & Singleton, 2005; Gavrilescu, 2005), and even some heavy metals (Seidel, 2004).

In order to optimize bioremediation, continuous monitoring is required. One indicator that is critical for bioremediation of soil polluted with crude oil, and which should be monitored, is total petroleum hydrocarbons (TPH). According to the most widely used method, ISO 16703 (ISO 16703, 2004), TPH represents the total content of hydrocarbons ranging from C_{10} through C_{40} which originate from petroleum. Data observed are used to set the end-point of bioremediation in the fulfillment of legal and regulatory criteria, and together with microbiological indicators are used for assessing the biodegradation potential of contaminated soil.

Several review papers have been published recently about new gas chromatography (GC) techniques and their application (Cortes et al., 2009; Marriott et al., 2003). This review presents a brief overview of the GC techniques, especially gas chromatography–mass spectrometry (GC–MS) that are currently applied in differentiating and fingerprinting oil hydrocarbons, identifying oil spills in assessments of environmental impacts, and in following up the efficacy of bioremediation procedures. It is focused on up-to date results observed in differentiation and transformation studies on petroleum-type pollutants in underground and surface water from the locality of Pančevo Oil Refinerty, Serbia (River Danube alluvial formations), and trends in the analysis methods of oil hydrocarbon biomarkers for monitoring the industrial-level bioremediation process in soil polluted with oil derivatives (Beškoski et al., 2010; Beškoski et al., 2011; Gojgić-Cvijović et al., 2011;

Jovančićević et al., 1997). The application area of GC in environmental sciences in the context of this work relates to monitoring the changes of oil pollutants in natural environments and during bioremediation as a controlled process of microbiological transformation and degradation.

2. Gas chromatography as analytical method of choice in environmental sciences

A wide variety of analytical methods and techniques are currently used in the examination of environmental samples, which include GC, GC–MS, high-performance liquid chromatography (HPLC), size exclusion HPLC, infrared spectroscopy (IR), supercritical fluid chromatography (SFC), thin layer chromatography (TLC), ultraviolet (UV) and fluorescence spectroscopy, isotope ratio mass spectrometry, and gravimetric methods. GC technique is the most widely used and today it is very hard to imagine an environmental laboratory without at least a gas chromatograph (Wang et al., 1999).

2.1 Gas chromatography

Chromatography is the method of separation in which several chemicals to be separated for subsequent analyses are distributed between two phases. In GC, separation is based mainly on the partitioning between a gas mobile phase and a liquid stationary phase. It is estimated that more than 60 types of GC detectors have been developed. For analysis of samples from the environment the few most commonly used are flame ionization detector (FID), thermal conductivity detector (TCD), electron capture detector (ECD), nitrogen-phosphorous detector (NPD), flame photometric detector (FPD), photo ionization detectors (PID) and mass selective detector or mass spectrometer (MS) which also allows qualitative and quantitative analysis (Driscoll, 2004). The GC-FID technique is a routine technique for the quantitative analysis of all the non-polar hydrocarbons which are extracted by applied solvents (n-hexane or acetone / n-heptane), and it has various scopes, depending on the standard (ISO 16703, 2004; Jovančićevič et al., 1997; Jovančićević et al., 2007). It is also used for rapid semiquantitative assessments of the successfulness of bioremediation treatments of polluted environments (water, soil, sediments) or the decomposition of hydrocarbon materials originating from oil under natural conditions or historical pollution (Beškoski et al. 2011; Hinchee and Kitte, 1995; Jensen et al., 2000; Jovančićević et al., 2008; Milic et al., 2009). GC methods are sensitive, selective and can be used to determine the specific target compounds. In environmental science GC methods are mainly used for: identifying organic pollutants in recent sediments, following abiotic and biotic transformations of petroleum-type pollutants, improving our understanding of migration mechanisms of organic pollutants in soil/water/air environments, distinguishing the oil pollutant from the native organic substance of recent sediments, fingerprinting and differentiation of petroleum-type pollutants, following transformation of petroleum pollutant during soil bioremediation experiments, determining other organic pollutants such us POP's (persistent organic pollutants) and analysing petroleum biomarkers.

2.2 Gas chromatography coupled with mass spectrometry

GC-MS is routinely applied to identify the individual components of petroleum hydrocarbons. These methods have high selectivity and compounds can be authenticated by analysing retention times and unique mass spectra. GC-MS can confirm the presence of the target analyte and the identification of untargeted analyte and also can be used for the separation of hydrocarbons into the groups. The main drawback of these methods is that isomeric compounds can have identical, and many different compounds can have similar mass spectra. Heavy fuel oil can contain thousands of components that cannot be separated in the gas chromatograph. Different compounds may have the same ions, which complicates the identification process (Jensen et al, 2000). Determination of TPH is possible using different methods - gravimetrically (DIN EN 14345, 2004), by infrared spectroscopy (ISO/TR 11046, 1994) or GC (ISO 16703, 2004). All these methods can be used for the quantitative analysis of analytes; however, when it comes to a mixture of compounds which is often the case with environmental samples, qualitative analysis can be realized only by GC-MS.

2.3 Novel GC techniques

The common techniques listed above, have now been in use for several decades but new techniques offer greater opportunities in this area. GC coupled to high-resolution time-of-flight mass spectrometry (GC–TOF–MS) has been applied for non-target screening of organic contaminants in environmental samples (Serrano et al., 2011). GC–TOF–MS has been successfully applied for screening, identification and elucidation of organic pollutants in environmental water and biological samples (Hernández et al., 2011) and also for confirmation of pollutants in a highly complex matrix like wastewater (Ellis et al., 2007). The strong potential of GC–TOF–MS for qualitative purposes comes from the full spectrum acquisition of accurate mass, with satisfactory sensitivity.

Comprehensive two-dimensional GC (GC×GC) coupled with MS has been widely applied in environmental analyses in the last decade (Ieda et al., 2011). The GC×GC–MS method has many practical advantages, e.g. high selectivity, high sensitivity, large separation power, group type separation and total profiling. Panić and Górecki reviewed GC×GC in environmental analyses and monitoring and they indicated that the main challenge in environmental analysis is that the analytes are usually present in trace amounts in very complex matrices (Panić & Górecki, 2006). In overcoming this problem, GC×GC–MS is a very powerful and attractive system that has been successfully applied for powerful identification of polychlorinated dibenzodioxins (PCDDs), polychlorinated dibenzofurans (PCDFs) (de Vos et al., 2011), polychlorinated biphenyls (PCBs) (Focant et al., 2004; Hoh et al., 2007), polychlorinated naphthalenes (PCNs) (Korytár et al., 2005), nonyl phenol (NP) (Eganhouse, et al., 2009), benzothiazoles, benzotriazoles, benzosulfonamides (Jover et al., 2009), Cl-/Br-PAH congeners (Ieda et al., 2011), pharmaceuticals and pesticides (Matamoros et al., 2010) in complex environmental samples.

2.4 Targets of GC and GC-MS in environmental analysis

In the fraction of saturated hydrocarbons the most predominant are n-alkanes and isoprenoid aliphatic alkanes. n-Alkanes in oil may be present in various quantities, most often as C_{10}-C_{35}, and among isoprenoids the predominant are C_{19}, pristane (2,6,10,14-tetramethyl-pentadecane) and C_{20}, phytane (2,6,10,14-tetramethylhexadecane). These molecules are used as biomarkers (Beškoski et al, 2010; Jovančićević et al., 2007; Peters et al., 2005) and their analysis is applied in organic geochemistry, environmental chemistry and studies of biodegradation (Figure 1a and 1b).

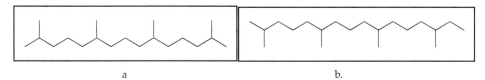

a.   b.

Fig. 1. a) Pristane, b) Phytane.

For the purpose of rigorous quantification of biodegradation and after the degradation of pristane and phytane, compounds of polycyclic hydrocarbons of the sterane-type (C_{27}-C_{29}) and terpane (tri-, tetra- and pentacyclic; C_{19}-C_{35}) may be used as a conserved internal standard (Figure 2a and 2b). They are most often found in much smaller quantities. Steranes and terpanes are very resistant to biodegradation and because of this feature they are exquisite biomarker compounds in comparison to which all other biodegradable compounds may be normalized. In addition, due to weak biodegradability, they are enriched in residual oils by the aging process, vaporization and biodegradation (Gagni & Cam, 2007; Jovančićević et al., 2008; 1997; Wang et al, 2005).

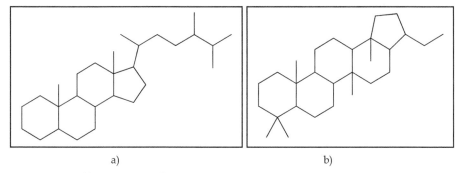

a)   b)

Fig. 2. a) Sterane, b) Triterpanes (hopanes)

The ratio between normal and isoalkanes depends on the degree of weathering since *n*-alkanes first undergo biodegradation.

Recently, compounds of sesquiterpane are increasingly being used as biomarker compounds. Most sesquiterpane compounds originate most probably from higher order plants and simple algae and bacteria. The introduction of so called new biomarker compounds resulted from the fact that in the case of lighter oil fractions, such as jet fuel and diesel, most of the high molecular biomarker compounds are removed during the process of refinement. Therefore, pentacyclic terpanes and steranes in these oil fractions are present in very small quantities, while sesquiterpane are concentrated in these fractions (Gagni & Cam, 2007; Wang et al., 2005).

For the analysis of oil and biomarkers, GC-MS is predominantly used in the selected ion monitoring mode (SIM). This method is applied in evaluating raw oil, in monitoring of biodegradation processes and in the process of bioremediation of oil that is released into the environment. Isoprenoids are identified from m/z 183, steranes from m/z 217 and triterpanes from m/z 191 selected ion chromatograms obtained from analysis in the SIM mode. Based on the abundance and distribution of biomarkers, reliable information about the origin and geological history of oil may be obtained. Therefore, the significance of these biomarkers in organic and geochemical investigations is great. Analytical data are also successfully applied in order to make a distinction between oils from various deposit layers. On the other hand, in environmental chemistry the determination of the presence of oil-type pollutants in the environment may be useful, e.g. for the distinction between native and anthropogenic organic substances (Jovančićević et al. 1997; Jovančićević et al., 2007), to follow up oil contamination, to determine the origin of long-term oil contamination and for forensic investigations of the origin of oil spills, to monitor the transformation due to atmospheric conditions (weathering), and to monitor biodegradation processes and the stage of the aging process of oil and its derivates under various conditions (Wang et al. 2005). GC and GC-MS methods are very selective and they offer a chance for a better understanding of the procedures of guided or natural biodegradation (Wang & Fingas, 2003). However, it is found that in highly degraded asphalt pavement samples, even the most refractory biomarker compounds showed some degree of biodegradation, and that the biomarkers were generally degraded in declining order of importance diasterane > C_{27} steranes > tricyclic terpanes > pentacyclic terpanes > norhopanes ~ C_{29} αββ-steranes (Wang et al., 2001).

A typical gas chromatogram of the TPH extract of soil polluted with diesel oil is presented in Figure 3. Chromatograms of petroleum hydrocarbons usually exhibit peaks of n-alkanes and a typical hump of unresolved components in different sections of the chromatogram.

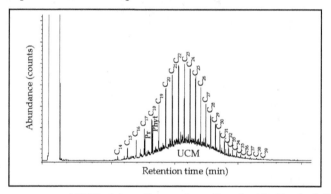

Fig. 3. Gas chromatogram of TPH of the soil recently polluted with diesel oil (Pr: Pristane; Phyt: Phytane; UCM: Unresolved Complex Mixture).

These peaks exhibit smooth distribution with one or several maxima and, possibly, alternating peak intensities of components with even and odd numbers of carbon atoms. Pristane and phytane peaks are observed in the vicinity of peaks of $C_{17}H_{36}$ and $C_{18}H_{38}$ alkanes; low intensity peaks of isoalkanes, cycloalkanes, and aromatic hydrocarbons appear between peaks of n-alkanes. Unresolved Complex Mixture (UCM), or "hump" of hydrocarbons (Rogers & Savard, 1999), is a common feature of the gas chromatograms of crude oils and certain refined products, and it is especially pronounced for weathered and biodegraded oils and oil-polluted sediment extracts. It is perhaps surprising that until recently virtually nothing was known about UCM compositions and molecular structures, even though the concentrations of these components in oils are significant (Frysinger et al., 2003; Wang & Fingas, 2003). In aged petroleum products, peaks of n-alkanes may be very weak against the hump or even absent from the chromatogram.

2.5 International GC standard methods

There are three major categories of GC-based US EPA methods. The first group is the US EPA methods for the determination of toxic organic compounds in air. The second group of GC-based methods was developed by the US EPA for water and wastewater, including 500 and 600 series methods. The two most important methods in the EPA 600 series are methods 624 and 625. Method 624 is a GC-MS method for purgeable organic compounds and method 625 measures semivolatile organics. If certain compounds are detected by method 625, they should then be confirmed with a GC method specific to that compound (methods 601, 602, 603, and 612 for 624 and methods 604, 606, 607, 609, 610, 611, and 612 for 625). The third group comprises approximately 27 GC-based methods, known as the 8000 series methods published in SW-846, and it concerns waste analysis.

3. GC for identification and determination of oil pollutant in the environment

For identification and determination of oils and petroleum products pollutants in the environment, the following structural features were used in GC and GC-MS:

- typical shape of chromatograms (fingerprints) of petroleum products;
- peaks of n-alkanes (C_6 through C_{40}) in chromatograms;
- measurement of total petroleum hydrocarbons;
- shape of the UCM;
- stable carbon isotope ratio ($\delta^{13}C$) is also included in many cases;
- close-to-unity ratio between n-alkanes with even and odd numbers of carbon atoms;
- presence of some isoalkanes, including pristane and phytane and in some cases farnesane, trimethyl-C_{13}, and norpristane isoprenoids;
- ratio between phytane and pristane and the closest $C_{17}H_{36}$ and $C_{18}H_{38}$ n-alkanes;
- presence of biomarkers (isoprenoids, steranes, triterpanes, etc.);
- predominance of methyl- and alkyl-substituted mono-, bi-, and polynuclear aromatic hydrocarbons over unsubstituted aromatic hydrocarbons;
- volatile hydrocarbons including BTEX (benzene, toluene, ethylbenzene, and three xylene isomers) and alkylated benzenes (C_3- to C_5-benzenes), volatile paraffins and isoparaffins, and naphthenes (mainly cyclopentane and cyclo-hexane compounds)
- distribution (profile) of polycyclic aromatic hydrocarbons (PAHs) and the petroleum-specific alkylated (C_1–C_4) homologues of selected PAHs;
- typical profile of sulfur-containing aromatic hydrocarbons;
- determination of NSO heterocyclic hydrocarbons for oil spill identification using ratio between different hydrocarbon groups (group composition);
- specific ratio between the concentration of PAHs and the background (Brodskii et al., 2002).

3.1 Environmental forensic - Fingerprinting of oil spills

Variability in chemical compositions results in unique chemical "fingerprints" for each oil and provides a basis for identifying the source(s) of the spilled oil. Since petroleum contains thousands of different organic compounds, successful oil fingerprinting involves appropriate sampling, analytical approaches and data interpretation strategies. Sampling is crucial, because if the sampling is not conducted properly and the sample is not representative, it will produce a result which will not accurately reflect the situation in the field. Also, in order to forensically fingerprint oil spills, it is necessary to select source-specific target analytes.

GC is in a central part of the fingerprinting as presented in the oil spill identification protocol. Figure 4 presents the modified "Protocol/decision chart for the oil spill identification methodology". The final assessments are concluded by the four operational and technical defensible identification terms: positive match, probable match, inconclusive or non-match (Wang & Fingas, 2003).

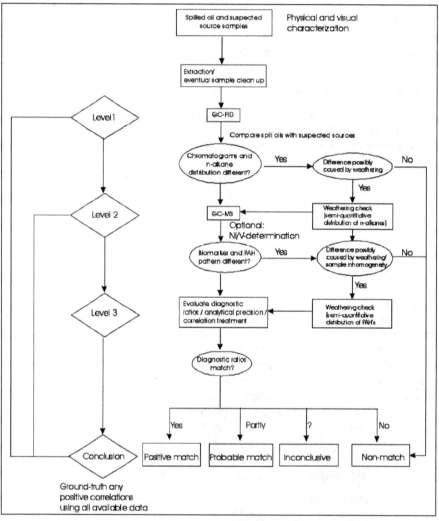

Fig. 4. Decision chart for the oil spill identification methodology (from Daling et al., 2002 as cited in Wang & Fingas, 2003).

3.2 Identification and determination of oil pollutant in the environment

GC was used for differentiation of petroleum-type pollutant and to study the fate of petroleum-type pollutants in underground and surface waters. Water and particulate matter are derived from the locality of Pančevo Oil Refinery, Serbia (River Danube alluvial formations), an area described in detail by Kaisarevic et al., (2009).

3.2.1 Determination of oil pollutant by *n*-alkane distribution and δ¹³C analysis

Identification of petroleum-type pollutants in recent sediments, soil, underground or surface waters, requires reliable and precise differentiation of native and anthropogenically released substances. The compositional differences are mainly pronounced with *n*-alkanes, one of the most abundant fractions in soluble sedimentary organic matter. An example of typical petroleum *n*-alkane distribution (in the form of a gas chromatogram of sample I1 representing a petroleum-type pollutant found in underground water from the area of Pančevo Oil Refinery), and several examples of distributions characteristic of organic substance derived from recent sediments, typical for all other samples (B2, B4, B6, B8, B10, B11, D1 and D2) originating from different localities of the River Danube alluvial formations, are shown in Figure 5 (Jovančićević et al. 1997). *n*-Alkane fraction in sample I1 was characterized by uniform distribution of odd and even homologues and a maximum at a lower member (*n*-C$_{19}$). The carbon preference index (CPI) a ratio of the sum of odd-numbered hydrocarbons to the sum of even-numbered hydrocarbons was around 1. On the other hand, the distribution of *n*-alkanes in nonpolluted sediments was characterized by domination of odd homologues (CPI considerably above 1) and a maximum at some of the higher homologue members (*n*-C$_{29}$ or *n*-C$_{31}$). The soluble sedimentary organic substance, called bitumen, was generally found in very small amounts. Its higher content in sediments may therefore indicate the presence of anthropogenic contaminants. The conclusion is that *n*-alkanes may successfully be used for differentiating the petroleum-type pollutants, as anthropogenic organic substances, from native organic matter in recent sedimentary formations.

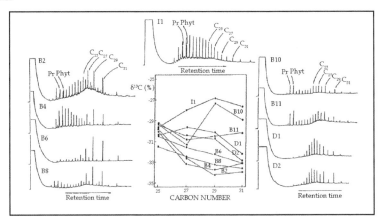

Fig. 5. Distributions of *n*-alkanes and δ¹³C$_{PDB}$ values of individual *n*-alkanes of recent sediments' bitumen fractions and one oil type pollutant from Pančevo Oil Refinery locality; Pr: Pristane; Phyt: Phytane; (from Jovančićević et al., 1997).

An alternative method to solve the problem may be δ^{13}C analysis. Being a mature organic substance in the geosphere, crude oil contains the greatest amount of the heavier carbon isotope, ^{13}C. Consequently, the ratio of ^{12}C and ^{13}C carbon isotopes δ^{13}C$_{PDB}$ in petroleum is less negative compared to organic matter in recent sediments. For example, the sample I1 presented in Figure 5 (Jovančićević et al., 1997) showed less negative δ^{13}C$_{PDB}$ with C$_{25}$, C$_{27}$, C$_{29}$ and C$_{31}$ n-alkanes compared to all other samples. As mentioned above, this particular sample was characterized by a typical petroleum n-alkane distribution, in contrast to all other samples demonstrating distributions typical for recent sediments. Hence, it is suggested that environmental native and anthropogenic organic substances may be differentiated based on comparison of their carbon isotopes ratios – in other words, δ^{13}C$_{PDB}$ may be used as a tool for revealing petroleum-type pollutants in the environment. For this purpose gas chromatographic-isotope ratio mass spectrometric (GC-IRMS) analysis of single compounds is required, or the analysis of its corresponding fractions.

In the case when differentiation of native and anthropogenic origin of an organic substance based on distribution of n-alkanes or carbon isotope analysis becomes questionable, analysis of polycyclic alkanes of sterane and triterpane types may be helpful (Jovančićević et al., 1998; Jovancicevic et al., 2007), since the distribution of these polycyclic alkanes is typical for crude oils. The use of n-alkanes and polycyclic alkanes of the sterane and triterpane types can be defined as an "organic-geochemical approach" in the identification of oil type pollutants in the environment.

3.2.2 Monitoring transformation processes of petroleum type pollutants

Petroleum-type pollutants in recent sediments, soil, underground or surface waters are exposed to microbial degradation (Jovančićević et al., 2003). The degradation intensity differs depending on a great number of biological, chemical and physicochemical parameters. The process of biodegradation of petroleum type pollutants in underground waters from Danube alluvial sediments (the locality of Pančevo Oil Refinery) was followed through a period from November 1997 to February 2000 by GC analyses of isolated alkane fractions. The corresponding gas chromatograms are shown in Figure 6.

In underground waters, a petroleum-type pollutant is exposed to microbiological degradation which is manifested through relatively fast degradation of n-alkanes. In the period from November 1997, when the first sample was taken, to February 2000, when the fifth sample was taken, important changes in the chemical composition were obvious. The relative contribution of n-alkanes as compared to pristane and phytane in sample A indicated changes defined as "initial petroleum biodegradation". The abundance of C$_{17}$ and C$_{18}$ n-alkanes was somewhat smaller than the abundance of pristane (C$_{19}$) and phytane (C$_{20}$). In the period from November 1997 to May 1998 the amount of n-alkanes relative to isoprenoids was reduced, a phenomenon typical for biodegradation intensity in geochemical literature defined as "minimal biodegradation". In September 1998 (sample C), the amount of n-alkanes was still smaller. Finally, during the next year, n-alkanes were almost completely degraded (sample D) while pristane and phytane remained nonbiodegraded. Comparison of Pr/n-C$_{17}$ and Phyt/n-C$_{18}$ ratios observed in samples A–C (winter 1997–autumn 1998) confirmed that biodegradation was considerably more intensive during the summer period than during the winter or spring periods.

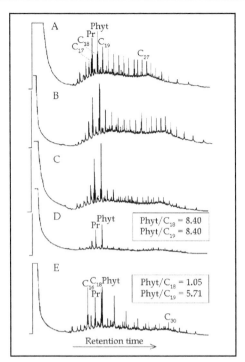

Fig. 6. Gas chromatograms of alkanes isolated from oil polluted alluvial ground waters (Pančevo Oil Refinery locality). Samples were taken in November 1997 (A), May 1998 (B), September 1998 (C), September 1999 (D) and in February 2000 (E); Pr: Pristane; Phyt: Phytane; (from Jovančićević et al., 2001).

In a relatively short period of time, from September 1999 to February 2000, the alkane fraction of the petroleum-type pollutant suffered an unexpected change. Namely, while pristane and phytane were found in the same amounts characterized by approximately the same ratios, in this fraction of the pollutant, new even carbon-number C_{16}–C_{30} n-alkanes were observed. Such a significant change raised the question of possible new contamination during this period of time, e.g., by a pollutant characterized by dominating even carbon-number n-alkanes. However, identical distributions of both steranes and triterpanes in sample D and E were excellent evidence that no new pollution occurred (Jovančićević *et al.*, 2001). Detailed analyses of extracts obtained from samples D and E were conducted. Sample E, containing even n-alkane homologues, contained, also in the alcoholic fraction, a homologous series of even carbon-number alcohols in the C_{14}–C_{20} range and a relatively significant amount of cholesterol (Figure 7). On the other hand, sample D, which did not contain any significant amounts neither of odd nor even n-alkane homologues, also did not contain any alcohols or higher fatty acids.

Even carbon-number alcohols and fatty acids observed in sample E were taken as a proof of the presence of particular microorganisms, i.e., of unicellular algae of *Pyrrophyta* type known as "fire algae". These types of microorganisms are able to synthesize even n-alkane homologues in a suitable base such as petroleum-type pollutants.

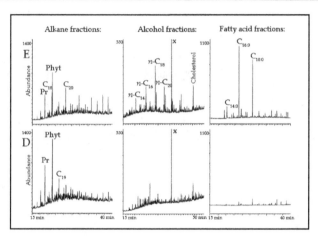

Fig. 7. Total ion chromatogram of GC-MS analyses of fractions of alkanes, alcohols and fatty acids (their methyl-esters), isolated from extracts E and D; Pr: Pristane; Phyt: Phytane; (from Jovančićević et al., 2003).

4. GC in evaluation of bioremediation

In evaluating bioremediation, GC was applied in the estimation of the bioremediation potential of microorganisms for crude oil biodegradation, and also in the monitoring of *ex situ* bioremediation of a soil contaminated by mazut (heavy residual fuel oil). In that study, GC was used for analysis of TPH and *n*-alkane composition, analysis of transformation of isoprenoids, steranes and terpanes and for determination of bioremediation effects on tricyclic aromatic hydrocarbons. In general, GC was used for analysis of TPH and biomarkers to evaluate the successfulness of applied bioremediation treatments.

4.1 Study of the bioremediation potential of microorganisms for crude oil biodegradation

The bioremediation potential of zymogenous microorganisms isolated from soil was investigated under controlled laboratory conditions using a mixture of paraffinic types of oils as a substrate (Šolević et al., 2011). The ability and efficiency of these microorganisms in crude oil bioremediation was assessed by comparing the composition of samples which were exposed to the microorganisms with a control sample which was prepared and treated in the same way, but containing no microorganisms. Biodegradation was stopped by sterilization at 120 °C for 25 min, after which samples were taken on 15, 30, 45, 60, and 75 days, while the control experiment was sampled only after 75 days and hydrocarbons were analyzed by the GC–MS.

The total ion current (TIC) chromatogram of the hydrocarbon fraction from the control and treated samples are shown in Figure 8. The dominant compounds in the hydrocarbon fraction of the control sample were *n*-alkanes and the isoprenoids pristane and phytane. These preliminary analyses, based on the TIC chromatograms showed a gradual decrease in the amount of *n*-alkanes and isoprenoids during 45 days. After 60 days of the experiment, *n*-alkanes and isoprenoids could not be observed in the TIC chromatograms indicating their possible complete degradation. At the end of the study, the TIC chromatograms were dominated by sterane biomarkers as the most abundant compounds in the fraction of hydrocarbons. The dominant compounds in the hydrocarbon fraction of the control sample were n-alkanes and the isoprenoids pristane and phytane.

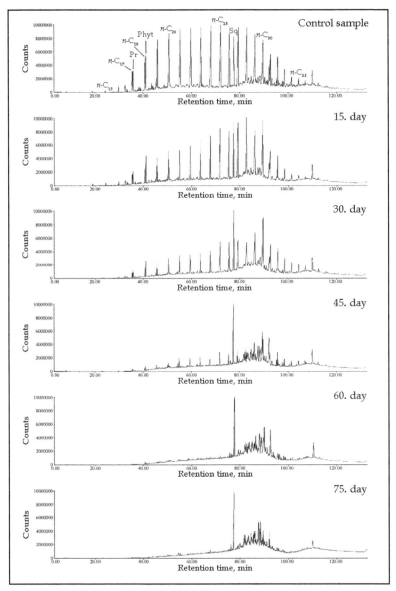

Fig. 8. Total ion chromatograms of the hydrocarbon fractions isolated from the extracts of the control sample and from the samples during the biodegradation experiment after 15, 30, 45, 60, and 75 days; Pr: Pristane; Phyt: Phytane; Sq: squalane; (from Šolevic et al., 2011).

According to GC-MS analysis by the end of the study, after 75 days of exposure to the microorganisms, the *n*-alkanes and isoprenoids had been completely degraded. The surprisingly low bioremediation potential of these microorganisms was proven in the case of polycyclic alkanes of the sterane and triterpane type and dimethyl phenanthrenes since after 75 days of the study, no significant alteration in the concentration of these compounds was observed. In the class of aromatic hydrocarbons, it was confirmed that the microorganisms had a considerably high bioremediation potential in the biodegradation of phenanthrene and methyl phenanthrenes. This fact can be surprising at first glance, but there is a reasonable explanation. Although the biodegradation of crude oil is often explained as a quasi-stepwise process in which various components are removed in a well-recognized sequence, it is well known that several compound classes are actually degraded simultaneously but at different rates, reflecting differences in the rate of their catabolism under varying conditions (Peters et al., 2005). Numerous studies have shown that aromatic compounds not only can be degraded prior or concomitantly with sterane and terpane biomarkers in reservoir oils, (Huang et al., 2004), but also can be degraded in the environmental conditions during oil spills (Wang & Fingas, 2003).

4.2 *Ex situ* bioremediation of a soil contaminated by mazut (heavy residual fuel oil)

Previously, it was reported that the oil pollutant mixture in the soil treated by *ex situ* bioremediation behaves in a complex way: different degradation rates and time evolutions are observed for fractions of the hydrocarbon mixture characterized by different molecular weights and structures (Beškoski et al., 2010; Jovančićević et al., 2008a, 2008b). Also, a stable microbial community had been formed after initial fluctuations, and the microorganisms which decompose hydrocarbons were the dominant microbial population at the end of the *ex situ* bioremediation process, with a share of more than 80% (range 10^7 colony forming units g^{-1}) (Milic et al., 2009).

While there is significant information in the literature about the microbiological degradation of defined individual hydrocarbons (Singh and Ward, 2004) there is significantly less data about the biodegradability of some commercial petroleum products, including mazut and heavy residual fuel oil (Iturbe et al., 2004). Mazut is a low quality, heavy (chain length 12–70 C atoms) residual fuel oil (ASTM D396-09a, 2009.; ISO 8217, 2005), blended or broken down with the end product being diesel in Western Europe and US and used as a source of heating fuel in Eastern Europe.

A field scale study was designed and conducted in order to evaluate the possibility of using bioremediation for treating a soil contaminated with heavy residual fuel oil such as mazut and mazut waste material (Beškoski et al., 2011). The mazut-polluted soil was excavated contaminated soil from an energy power plant which, due to a break-down, had been polluted with mazut and sediment from a mazut reservoir for a year. The bioremediation biopile contained approximately 600 m³ soil polluted with mazut, mixed with un-graded river sand and softwood sawdust. GC-FID and GC-MS were used for analysis of the process. The techniques of biostimulation, bioventilation and reinoculation of a microbial consortium were performed during a period of 150 days. A part of the materials (10 m³) prepared for bioremediation was set aside uninoculated, and maintained as an untreated control pile subjected to natural bioremediation process.

4.2.1 Change in TPH and *n*-alkane composition

The content of TPH in the soil was extracted as described in ISO 16703 (ISO 16703, 2004) and determined gravimetrically in accordance with DIN EN 14345 (DIN EN 14345, 2004). The contamination level of TPH at the start was found to be 5.2 g kg⁻¹ of soil. With application of bacterial consortium and nutrients, the TPH level was reduced to 2.1, 1.3 and 0.3 g kg⁻¹ of soil after 50, 100, and 150 days, respectively, meaning 60%, 75% and 94% of the TPH were biodegraded. The analysis of change in the group composition evidenced that the average rate of decrease during the biodegradation of mazut was 23.7 mg kg⁻¹ day⁻¹ for the aliphatic fraction, 5.7 mg kg⁻¹ day⁻¹ for the aromatic fraction and 3.3 mg kg⁻¹ day⁻¹ for the NSO-asphaltene fraction (Beškoski et al., 2010).

The chromatograms (Figure 9) gave qualitative and semiquantitative information on the changes in the composition of hydrocarbons in the samples. As judged from GC, the abundance of *n*-C₁₇ and *n*-C₁₈ *n*-alkanes at time zero was somewhat smaller than the abundance of pristine (C₁₉) and phytane (C₂₀). This indicates that the hydrocarbons left in the soil were already degraded to some extent during the natural biodegradation process. Around 50% of *n*-alkanes in the size range of C₂₉–C₃₅ were biodegraded during the first 50 days. *n*-Alkanes in the range of C₁₄–C₂₀ were degraded completely by 100 days, followed by

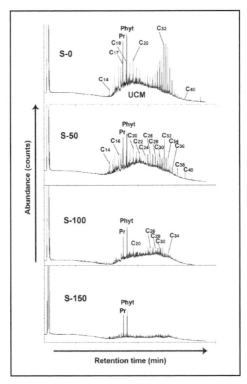

Fig. 9. Chromatograms of the TPH extracts of the treated samples during the bioremediation experiment after 0, 50, 100 and 150 days; Pr: Pristane; Phyt: Phytane; (from Beškoski et al., 2011).

complete degradation of C_{20}–C_{36} by 150 days. After 100 days the GC trace also revealed a significant reduction in the UCM.

The applied zymogenous microbial consortium biodegraded and "consumed" all components of the complex mass of hydrocarbons, although different rates of degradation were observed, as shown in Table 1.

Indicator	S-0	S-50	S-100	S-150
Dominant n-alkane	C_{32}	C_{20}	C_{20}	-
Pristane/n-C_{17}	1.47	1.42	2.16	3.8
Phytane/n-C_{18}	1.75	1.70	2.9	4.17
Biodegradation of TPH, [%]	0	60.2	74.9	94.4
Biodegradation of Pristane, [%]	0	31.8	50.9	56.8
Biodegradation of Phytane, [%]	0	32.0	42.0	55.8

Table 1. Indicators of bioremediation from gas chromatography (from Beškoski et al., 2011).

The ratios of pristane/n-C_{17} and phytane/n-C_{18} can be used to differentiate physical weathering and biodegradation (Wang et al., 1998). As the volatility of n-C_{17} and pristane and n-C_{18} and phytane are similar, weathering should be attributed to a reduced concentration of these substances over time, if their ratio remains constant. An increase in the pristane/n-C_{17} and phytane/n-C_{18} ratios over time is likely to be a consequence of bioremediation, since n-C_{17} and n-C_{18} degrade more rapidly than pristane and phytane, respectively. Biodegradation of pristane and phytane, as found in Beškoski et al., (2011), shows that these compounds are not suitable as markers for following a bioremediation process. In fact, comparison of the content of other hydrocarbons in relation to pristane and phytane could underestimate the degree of their biodegradation.

4.2.2 Transformation of isoprenoids, steranes and terpanes

Isoprenoid aliphatic alkanes, pristane and phytane, and polycyclic alkanes of sterane and triterpane types in saturated hydrocarbon fractions were analysed by GC–MS. Isoprenoids were identified from the m/z 183, steranes from m/z 217 and triterpanes from m/z 191 selected ion chromatograms obtained from analysis in the SIM mode. The most relevant peaks were identified based on organic geochemical literature data (Peters et al., 2005), or based on total mass spectra, using mass spectra databases (NIST/EPA/NIH Mass spectral library NIST2000, Wiley/NBS Registry of Mass spectral Data 7th Ed., electronic versions).

In the chromatogram m/z 183 of all the samples, the pristane and phytane peaks are clearly differentiated, being the most intensive in the sample S-0 (Figure 10). In this sample, individual peaks originating from the homologue chain > C_{20} isoprenoids are also observed, while in sample S-150 they are biodegraded.

The chromatogram of terpane in the initial sample S-0 (SIM, m/z 191, Figure 11) is dominated by peaks originating from C_{19}–C_{28} tricyclic terpanes, C_{24} tetracyclic terpane and C_{27}–C_{29} pentacyclic terpanes, with a distribution not typical for raw heavy fuel oil (data not shown), but rather which is typical for samples that had been exposed to weathering and biodegradation over extended periods of time (Peters et al., 2005). During the bioremediation process, the quantity of all tricyclic and tetracyclic terpanes decreased and in the sample S-150, the quantity of tricyclic (C_{19}, C_{24}, C_{25}, and C_{28}) and tetracyclic terpanes (C_{24}), is according to their intensity virtually no different from the peaks overlapped by the background noise. In relation to them, the quantity of pentacyclic terpanes of the hopane type also decreased at an intensity that conforms to degradation of oil and oil derivatives under natural conditions.

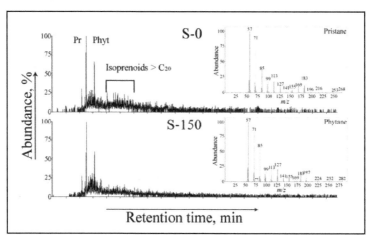

Fig. 10. Selected ion chromatograms of isoprenoids (SIM, m/z 183) of alkane fractions in samples S-0 and S-150 (full mass spectra corresponding to peaks of pristane C_{19} and phytane C_{20} for S-0 sample are also presented); Pr: Pristane; Phyt: Phytane; (from Beškoski et al., 2010).

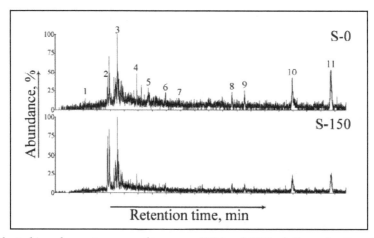

Fig. 11. Selected ion chromatograms of terpanes (SIM, m/z 191) of alkane fractions in samples S-0 and S-150. 1: C_{19}-tricyclic terpane; 2: C_{21}-tricyclic terpane; 3: C_{22}-tricyclic terpane; 4: C_{24}-tricyclic terpane; 5: C_{25}-tricyclic terpane; 6: C_{24}-tetracyclic terpane; 7: C_{28}-tricyclic terpane; 8: C_{27}-18α(H)-22,29,30-trisnorhopane (Ts); 9: C_{27}-17α(H),18α(H),21β(H)-25,28,30-trisnorhopane; 10: C_{29}-17α(H),21β(H)-hopane; 11: C_{29}-18α(H),21β(H)-30-norneohopane (from Beškoski et al., 2010).

Steranes and diasteranes are polycyclic alkanes that are degraded more quickly and easily than terpanes. Over the course of the bioremediation experiments conducted, the fate of steranes and their more stable structural isomers, diasteranes, C_{27}–C_{29}, is similar to the fate of the terpanes (Figure 12), and biodegradation, decrease and loss of resolution of the individual signals, were observed in the sample S-150.

Polycyclic alkanes, steranes and diasteranes, as well as terpanes, were biodegraded in the sample S-150 and did not differ from peaks overlapped by the background noise (Beškoski et al., 2010).

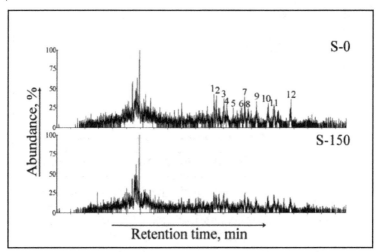

Fig. 12. Selected ion chromatograms of steranes (SIM, m/z 217) of alkane fractions in samples S-0 and S-150. 1: C_{27}-13β(H),17α(H) diasterane (20S); 2: C_{27}-13β(H),17α(H)-diasterane (20R); 3: C_{27}-13α(H),17β(H) diasterane (20S); 4: C_{27}–13α(H),17β(H) diasterane (20R); 5: C_{28}–13β(H),17α(H) diasterane (20R); 6: C_{28}-13α(H),17β(H) diasterane (20S) + C_{27}-14α(H),17α(H) sterane (20S); 7: C_{29}-13β(H),17α(H) diasterane (20S) + C_{27}-14β(H),17β(H) sterane (20R); 8: C_{27}-14β(H),17β(H) sterane (20S) + C_{28}-13α(H),17β(H) diasterane (20R); 9: C_{29}-13β(H),17α(H) diasterane (20R); 10: C_{29}-13α(H),17β(H) diasterane (20R) + C_{28}-14β(H),17β(H) sterane (20R); 11: C_{28}-14α(H),17α(H) sterane (20R); 12: C_{29}-14α(H),17α(H) sterane (20R) (from Beškoski et al., 2010).

4.2.3 The effect on tricyclic aromatic hydrocarbons

During *ex situ* bioremediation of soil contaminated by mazut the changes in the distribution of phenanthrene and its methyl isomers were investigated (Novaković et al., 2011). The results of bioremediation of soil were compared with the results of biodegradation in untreated control biopile. During the process of natural microbial degradation an expected trend was observed: the concentration of phenanthrene is reduced in relation to methyl-, dimethyl- and especially to trimethyl-phenanthrenes. However, during the process of "stimulated biodegradation" a different sequence was observed: there was a uniform increase in the relative abundance of phenanthrene compared to its methyl isomers, especially relative to trimethylphenanthrenes.

Mass selected ion chromatograms of phenanthrene, methyl-phenanthrenes, dimethyl-phenanthrenes and trimethyl-phenanthrenes obtained by GC-MS analysis of aromatic fractions isolated from extracts of samples M1 and M5, from bioremediation biopile, are shown in Figure 13a. Selected ion chromatograms of control tests (samples M1k and M5k, from untreated controle biopile) are shown in Figure 13b.

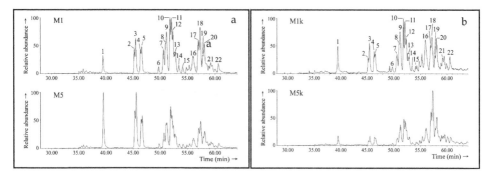

Fig. 13. Selected ion chromatograms of phenanthrene (P) (m/z 178), methyl-phenanthrenes (MP) (m/z 192), dimethyl-phenanthrenes (DMP) (m/z 206) and trimethyl-phenanthrenes (TMP) (m/z 220), obtained by GC-MS analysis (using the single ion monitoring, SIM method) of (a) aromatic fractions isolated from M1 and M5 bioremediation biopile extracts and (b) aromatic fractions isolated from untreated control biopile extracts (M1k and M5k); 1: P; 2: 3-MP; 3. 2-MP; 4. 9-MP; 5. 1-MP; 6. Et-P (Ethyl-phenanthrenes); 7. EtP + 3,6-DMP; 8. 2,6-DMP; 9. 2,7-DMP; 10. 2,7-DMP; 11. 1,6 + 2,9 + 2,5-DMP; 12. 1,7-DMP; 13. 2,3 + 1,9 + 4,9 + 4,10-DMP; 14. 1,8-DMP; 15. 1,2-DMP; 16. 2,6,10-TMP; 17. 1,3,6-TMP; 18. 1,3,7 + 1,3,9 + 2,7,10-TMP; 19. 2,3,6-TMP; 20. 1,3,8 + 2,3,7 + 2,8,10-TMP; 21. 2,3,10 + 3,8,10-TMP; 22. 1,2,3 + 1,2,8 + 1,7,10-TMP; (adapted from Novaković et al., 2011).

There is a possibility that the observed changes occurred as a result of demethylation (Huang et al., 2004) through the process of oxidative decarboxylation. In this way, an increase in the absolute concentration of phenanthrene could have occured at the expense of degradation of methyl-, dimethyl-and trimethyl-phenanthrenes. However, demethylation is a thermodynamically less favorable process, and it is more likely that the biostimulation process favored those bacterial strains in the consortium that decompose methyl-isomers (first of all trimethyl-phenanthrenes), as a consequence of better interaction of reactive methyl groups with the active centers on the surface of bacterial cells, and, thus promoted decomposition of methyl-phenanthrene derivatives (Lamberts et al., 2008).

It can be concluded that higher availability of phenanthrene and its methyl derivatives to microorganisms increases the degradability of methyl-phenanthrenes compared to phenanthrene and that the degree of degradability depends on the number of methyl groups, i.e. on the level of alkylation.

5. Conclusions

GC is an indispensable method in environmental chemistry, biotechnology, organic geochemistry and in practical aspects of environmental cleaning, protection and preservation (Evans & Furlog, 2011; Scott, 1998; Singh & Ward, 2004). With its inherent high sensitivity and high separating power GC is one of the more commonly used techniques in the analysis of environmental samples. For fundamental works in this field and for the development of biotechnologies and techniques in environmental sciences, the combination of GC with MS in all techniques is often of essential importance for the success of research and development procedures. The use of the said instrumental methods is particularly relevant in the case of oil and oil derivatives due to countless varieties of their composition and their general presence in the environment, either through their widespread usage or their occurrence as pollutants.

Our research experience in the field of fundamental and applied aspects of petroleum pollutants in the environment and their remediation by microorganisms especially by *ex situ* bioremediation methods, confirms the prominent place of GC and techniques that are based on GC among analytical methods which are commonly applied to good effect in environmental chemistry and environmental biotechnology.

Also, primarily because of the quantities of substances, the use of inert gases, and the possibility of replacing conventional organic solvents with solvents which have much smaller negative impacts on the environment, such as the use of carbon dioxide as a supercritical fluid, GC can be considered as a green analytical method. Further technical advances can be expected in the development of column fillings, probably in the area of nano-materials, and in terms of separation efficiency, as well as application of polar "green" solvents, such as water and ethanol.

6. Acknowledgments

This work was supported by the Ministry of Education and Science of the Republic of Serbia under Grant Nos. III 43004 and 176006.

7. References

Alexander, M. (1999). *Biodegradation and Bioremediation*, 2nd Ed., Academic Press, ISBN 978-0120498611, San Diego

Alvarez P.J.J. & Illman W.A. (2006) *Bioremediation and natural attenuation: process fundamentals and mathematical models*. Wiley, ISBN 978-0-471-73861-9, Hoboken

ASTM D396-09a, (2009). Standard Specification for Fuel Oils. ASTM International, West Conshohocken

Atlas, R.M. & Philp, J. (2005). *Bioremediation: Applied Microbial Solutions for Real-World Environmental Cleanup*. ASM Press, ISBN 978-1555812393, Washington, D.C.

Bamforth, S. M. & Singleton, I. (2005). Review Bioremediation of polycyclic aromatic hydrocarbons: current knowledge and future directions. *Journal of Chemical Technology & Biotechnology*, Vol.80, No.7, (July 2005), pp. 723–736, ISSN 1097-4660

Beškoski, V.P.; Gojgić-Cvijović, G.; Milić, J.; Ilić, M.; Miletić, S.; Šolević, T. & Vrvić, M.M. (2011). *Ex situ* bioremediation of a soil contaminated by mazut (heavy residual fuel oil) – A field experiment, *Chemosphere*, Vol.83, No.1, (March 2011), pp. 34-40, ISSN 0045-6535, (Figure and table reprinted with permission from Elsevier)

Beškoski, V.P.; Takić, M.; Milić, J.; Ilić, M.; Gojgić-Cvijović, G.; Jovančićević B. & Vrvić, M.M. (2010). Change of isoprenoids, steranes and terpanes during *ex situ* bioremediation of mazut on industrial level, *Journal of the Serbian Chemical Society*, Vol.75, No.11, (May 2010), pp. 1605–1616, ISSN 0352-5139 (Figures reprinted with permission from Serbian Chemical Society)

BP Statistical Review of World Energy, June 2011 In: *bp.com/statisticalreview*, 08.09.2011, Available from
http://www.bp.com/assets/bp_internet/globalbp/globalbp_uk_english/reports_ and_publications/statistical_energy_review_2011/STAGING/local_assets/pdf/sta tistical_review_of_world_energy_full_report_2011.pdf

Brenner K.; You L. & Arnold F.H. (2008). Engineering microbial consortia: a new frontier in synthetic biology, *Trends in Biotechnology*, Vol.26, No.9, (September 2008), pp. 483–489, ISSN 0167-7799

Brodskii, E.S.; Lukashenko, I.M.; Kalinkevich, G.A. & Savchuk, S.A. (2002). Identification of Petroleum Products in Environmental Samples Using Gas Chromatography and Gas Chromatography–Mass Spectrometry, *Journal of Analytical Chemistry*, Vol.57, No.6, (September 2001), pp. 486–490, ISSN 1608-3199

Cortes, H.J.; Winniford, B.; Luong, J. & Pursch, M. (2009). Comprehensive two dimensional gas chromatography review, *Journal of Separation Science*, Vol.32, No.5-6, (March 2009), pp. 883–904, ISSN 1615-9314

De Vos, J.; Dixon, R.; Vermeulen, G.; Gorst-Allman, P.; Cochran, J.; Rohwer, E. & Focant, J.F. (2011). Comprehensive two-dimensional gas chromatography time of flight mass spectrometry (GC-GC-TOFMS) for environmental forensic investigations in developing countries. *Chemosphere*, Vol.82, No.9, (February 2011), pp. 1230–1239, ISSN 0045-6535

DIN EN 14345, (2004). Characterization of Waste. Determination of Hydrocarbon Content by Gravimetry. DIN, Berlin

Driscoll, J. (2004). Gas Chromatography in Environmental Analysis, in *Environmental Instrumentation and Analysis Handbook*, Eds, Down, R.D., Lehr J.H., Wiley, ISBN 978-0471463542

Eganhouse, R.P.; Pontillo, J.; Gaines, R.B.; Frysinger, G.S.; Gabriel, F.L.P.; Kohler, H.P.E.; Giger, W. & Barber, L.B., (2009). Isomer-Specific Determination of 4-Nonylphenols Using Comprehensive Two-Dimensional Gas Chromatography/Time-of-Flight Mass Spectrometry. *Environmental Science & Technology*, Vol.43, No.24, (November 2009), pp. 9306–9313, ISSN 0013-936X

Ellis J.; Shah M.; Kubachka K.M. & Caruso J.A. (2007). Determination of organophosphorus fire retardants and plasticizers in wastewater samples using MAE-SPME with GC-ICPMS and GC-TOFMS detection. *Journal of Environmental Monitoring*, Vol.9, (September 2007), pp. 1329–1336, ISSN 1464-0333

Evans, G.M. & Furlog, J.C. (2011). *Environmental Biotechnology: Theory and Application.* 2nd Edition, Wiley, ISBN 9780470856765, Chichester

Focant, J.F.; Reiner, E.J.; MacPherson, K.; Kolic, T.; Sjödin, A.; Patterson Jr., D.G.; Reese, S.L.; Dorman, F.L. & Cochran, J. (2004). Measurement of PCDDs, PCDFs, and non-ortho-PCBs by comprehensive two-dimensional gas chromatography-isotope dilution time-of-flight mass spectrometry (GC × GC-IDTOFMS). *Talanta,* Vol.63, No.5, (August 2004), pp. 1231–1240, ISSN 0039-9140

Frysinger, G.S.; Gaines, R.B.; Xu, L. & Reddy, C.M. (2003). Resolving the unresolved Complex Mixture in Petroleum-Contaminated Sediments. *Environmental Science & Technology* , Vol.37 No.8, (March 2003), pp. 1653-1662, ISSN 0013-936X

Gagni, S. & Cam, D. (2007). Stigmastane and hopanes as conserved biomarkers for estimating oil biodegradation in former refinery plant/contaminated soil. *Chemosphere,* Vol.67, No.10, (February 2007), pp. 1975–1981, ISSN 0045-6535

Gavrilescu, M. (2005). Fate of pesticides in the environment and its bioremediation. *Engineering in Life Sciences* Vol.5, No.6, (December 2005), pp. 497-526, ISSN 1618-2863

Gojgic-Cvijovic, G.D.; Milic, J.S.; Solevic, T.M.; Beskoski, V.P.; Ilic, M.V.; Djokic, L.S.; Narancic T.M. & Vrvic M.M. (2011). Biodegradation of petroleum sludge and petroleum polluted soil by a bacterial consortium: a laboratory study, *Biodegradation* (Published online) (May 2011) DOI: 10.1007/s10532-011-9481-1, ISSN 1572-9729

Head, I.M.; Jones, D.M. & Roling, W.F.M., (2006). Marine microorganisms make a meal of oil. *Nature Reviews Microbiology,* Vol.4, No.3, (March 2006), pp. 173-182, ISSN 1740-1526

Hernández, F.; Portolés, T.; Pitarch E. & Lopez, F.J. (2011). Gas chromatography coupled to high-resolution time-of-flight mass spectrometry to analyze trace-level organic compounds in the environment, food safety and toxicology. *Trends in Analytical Chemistry,* Vol.30, No.2, (February 2011), pp. 388–400, ISSN 0165-9936

Hinchee, E.R. & Kitte, H.J. (1995). *Applied Bioremediation of Petroleum Hydrocarbons.* Battle Press, ISBN 1574770071, Columbs

Hoh, E.; Mastovska, K. & Lehotay, J.S. (2007). Optimization of separation and detection conditions for comprehensive two-dimensional gas chromatography–time-of-flight mass spectrometry analysis of polychlorinated dibenzo-p-dioxins and dibenzofurans. *Journal of Chromatography A,* Vol.1145, No.(1-2), (January 2007), 210–221, ISSN 0021-9673

Huang H.; Larter S.R.; Bowler B.F.J. & Oldenburg T.B.P. (2004). A dynamic biodegradation model suggested by petroleum compositional gradients within reservoir columns from the Liaohe basin, NE China. *Organic Geochemistry,* Vol.35, No.3, (March 2004), pp. 299-316, ISSN 0146-6380

Huang H.P.; Bowler B.F.J.; Oldenburg T.B.P. & Larter S.R. (2004). The effect of biodegradation on polycyclic aromatic hydrocarbons in reservoired oils from the Liaohe basin, NE China. *Organic Geochemistry,* Vol.35, No.(11-12), (November-December 2004), pp. 1619-1634, ISSN 0146-6380

Ieda, T.; Ochiai, N.; Miyawaki, T.; Ohura, T. & Horii, Y. (2011). Environmental analysis of chlorinated and brominated PAHs by comprehensive two-dimensional gas chromatography coupled to high-resolution time-of-flight mass spectrometry. *Journal of Chromatography A,* Vol.1218, No.21, (January 2011), pp. 3224–3232, ISSN 0021-9673

ISO 16703, (2004). Soil quality – Determination of content of hydrocarbon in the range C_{10} to C_{40} by gas chromatography. Geneve

ISO 8217, (2005). Petroleum Products – Fuels (class F) – Specifications of Marine Fuels. Geneva

ISO/TR 11046, (1994). Soil quality – Determination of mineral oil content – Method by infrared spectrometry and gas chromatographic method. Geneve

Iturbe, R.; Flores, C.; Chavez, C.; Bautista, G. & Tortes, L.G. (2004). Remediation of contaminated soil using soil washing and biopile methodologies at a field level. *Journal of Soils and Sediments,* Vol.4, No.2, (April 2004), pp. 115–122, ISSN 1439-0108

Jensen, S.T.; Arvin, E.; Svensmark, B. & Wrang, P. (2000). Quantification of Compositional Changes of Petroleum Hydrocarbons by GC/FID and GC/MS during Long-term Bioremediation Experiment, *Journal of Soil Contamination,* Vol.9, No.6, (May 2000), pp. 549–577, ISSN 1058-8337

Jordan, R.E. & Payne, J.R. (1980). *Fate and Weathering of Petroleum Spills in the Marine Environment: A Literature Review and Synopsis.* Ann Arbor Science Publishers, ISBN 978-0250403813, Michigan

Jovančićević, B.; Antić, M.; Vrvić, M.; Ilić, M.; Novaković, M.; Saheed, R.M. & Schwarzbauer, J. (2008a). Transformation of a petroleum pollutant during soil bioremediation experiments. *Journal of the Serbian Chemical Society,* Vol.73, No.5, (May 2008), pp. 577–583, ISSN 0352-5139

Jovančićević, B.; Antić, M.; Pavlović, I.; Vrvić, M.; Beškoski, V.; Kronimus, A. & Schwarzbauer, J. (2008b). Transformation of petroleum saturated hydrocarbons during soil bioremediation experiments. *Water Air & Soil Pollution,* Vol.190, No.1-4, (January 2008), pp. 299–307, ISSN 1573-2932

Jovančićević, B.; Antić, P.M.; Šolević, M.T.; Vrvić, M.M.; Kronimus, A. & Schwarzbauer, J. (2005). Investigation of interactions between surface water and petroleum type pollutant. *Environmental Science & Pollution Research,* Vol.12, No.4, (July 2005), pp. 205–212, ISSN 1614-7499

Jovančićević, B.; Polić, P. & Vitorović, D. (1998). Organic geochemical investigation of crude oils. The southeastern, part of the Pannonian Basin in Yugoslavia. *Journal of the Serbian Chemical Society,* Vol.63, No.6, (June 1998), pp. 397–418, ISSN 0352-5139

Jovančićević, B.; Polić, P.; Vitorović, D.; Scheeder, G.; Teschner, M. & Wehner, H. (2001). Biodegradation of oil-type pollutants in Danube alluvial sediments (Yugoslavia). *Fresenius Environmental Bulletin,* Vol.10, (October 2001), pp. 178–182, ISSN 1018-4619 (Figure reprinted with permission from Parlar Scientific Publications)

Jovančićević, B.; Polić, P.; Vrvić, M.; Scheeder, G.; Teschner, T. & Wehner, H. (2003). Transformations of *n*-alkanes from petroleum pollutants in alluvial ground waters. *Environmental Chemistry Letters,* Vol.1, No.1, (January 2003), pp. 73–81, ISSN 1610-3653 (Figure reprinted with permission from Springer)

Jovančićevič, B.; Tasić Lj.; Wehner, H.; Faber, E.; Šušić, N. & Polić, P. (1997). Identification of oil-type pollution in recent sediments. *Fresenius Environmental Bulletin,* Vol.6, (June 1997), pp. 667–673, ISSN 1018-4619 (Figure reprinted with permission from Parlar Scientific Publications)

Jovančićević, B.; Vrvić, M.; Schwarzbauer, J.; Wehner, H.; Scheeder, G. & Vitorović, D. (2007). Organic-geochemical differentiation of petroleum-type pollutants and study of their fate in Danube alluvial sediments and corresponding water (Pančevo Oil Refinery, Serbia). *Water Air & Soil Pollution.* Vol.183, No.1-4, (March 2007), pp. 225–238, ISSN 1573-2932

Jover, E.; Matamoros, V. & Bayona, J.M. (2009). Characterization of benzothiazoles, benzotriazoles and benzosulfonamides in aqueous matrixes by solid-phase extraction followed by comprehensive two-dimensional gas chromatography coupled to time-of-flight mass spectrometry. *Journal of Chromatography A,* Vol.1216, No.18, (May 2009), pp. 4013-4019, ISSN 0021-9673

Kaisarevic, S.; Lübcke-von Varel, U.; Orcic, D.; Streck, G.; Schulze, T.; Pogrmic, K.; Teodorovic, I.; Brack, W. & Kovacevic, R. (2009). Effect-directed analysis of contaminated sediment from the wastewater canal in Pancevo industrial area, Serbia. *Chemosphere,* Vol.77, No.7, (November 2009), pp. 907-913, ISSN 0045-6535

Korytár, P.; Leonards, P.E.G.; de Boer, J. & Brinkman, U.A.Th. (2005). Group separation of organohalogenated compounds by means of comprehensive two-dimensional gas chromatography. *Journal of Chromatography A,* Vol.1086, No.1-2, (September 2005), pp. 29-44, ISSN 0021-9673

Lamberts R.F.; Christensen J.H.; Mayer P.; Andersen O. & Johnesen A.R. (2008). Isomer-specific biodegradation of methylphenanthrenes by soil bacteria. *Environmental Science & Technology,* Vol.42, No.13, (June 2008), pp. 4790-4796, ISSN 0013-936X

Marriott, P.J.; Haglund, P. & Ong, R.C.Y. (2003). A review of environmental toxicant analysis by using multidimensional gas chromatography and comprehensive GC. *Clinica Chimica Acta,* Vol.328, No.1-2, (February 2003), pp. 1-19, ISSN 0009-8981

Matamoros, V.; Jover, E. & Bayona, J.M. (2010). Part-per-trillion determination of pharmaceuticals, pesticides, and related organic contaminants in river water by solid-phase extraction followed by comprehensive two-dimensional gas chromatography time-of-flight mass spectrometry. *Analytical chemistry,* Vol.82, No.2, (January 2010), pp. 699-706, ISSN 0003-2700

Milic, J.S.; Beskoski, V.P.; Ilic, M.V.; Ali, S.A.M.; Gojgic-Cvijovic, G. & Vrvic, M.M. (2009). Bioremediation of soil heavily contaminated with crude oil and its products: composition of the microbial consortium, *Journal of the Serbian Chemical Society,* Vol.74, No.4, (July 2008), pp. 455-460, ISSN 0352-5139

Novaković, M.; Ali Ramadan, M.M.; Šolević Knudsen, T.; Antić, M.; Beškoski, V.; Gojgić-Cvijović, G.; Vrvić M.M. & Jovančićević, B. (2011). *Ex situ* bioremediation of soil contaminated by heavy residual fuel oil (mazut) – the effect on tricyclic aromatic hydrocarbons, *Environmental Chemistry Letters,* [in print], ISSN 1610-3653

Ollivier B. & Magot M. (2005). *Petroleum Microbiology,* ASM Press, ISBN 978-1555813277, Washington

Pahari A. & Chauhan B. (2007). *Engineering Chemistry,* Infinity Science Press LLC, ISBN 978-0977858286, New Delhi

Panić, O. & Górecki, T. (2006). Comprehensive two-dimensional gas chromatography (GC×GC) in environmental analysis and monitoring. *Analytical and Bioanalytical Chemistry*, Vol.386, No.4, (July 2006), pp. 1013–102, ISSN 1618-2650

Peters, K.E.; Walters, C.C. & Moldowan, J.M. (2005). *The Biomarker Guide*. Cambridge Press, ISBN 9780521781589, Cambridge

Rogers K.M. & Savard M.M. (1999). Detection of petroleum contamination in river sediments from Quebec City region using GC-IRMS. *Organic Geochemistry*, Vol.30, No.12, (December 1999), pp. 1559-1569, ISSN 0146-6380

Scott, R.P.W. (1998). *Introduction to Analytical Gas Chromatography* (Chromatographic Science Series), Marcel Dekker, Inc., ISBN 978-0824700164, New York

Seidel, H.; Löser, C.; Zehnsdorf, A.; Hoffmann, P. & Schmerold, R. (2004). Bioremediation process for sediments contaminated by heavy metals: Feasibility study on a pilot scale. *Environmental Science & Technology*, Vol.38, No.5, (January 2004), pp.1582–1588, ISSN 0013-936X

Serrano, R.; Nácher-Mestre, J.; Portolés, T.; Amat, F. & Hernández, F. (2011). Non-target screening of organic contaminants in marine salts by gas chromatography coupled to high-resolution time-of-flight mass spectrometry, *Talanta*, Vol.85, No.2, (May 2011), pp. 877– 884, ISSN 0039-9140

Singh, A. & Ward, O.P. (2004). *Biodegradation and Bioremediation*. Springer-Verlag, ISBN 978-3642059292, Berlin

Šolević, T.; Novaković, M.; Ilić, M.; Antić, M.; Vrvić, M.M. & Jovančićević, B. (2011). Investigation of the bioremediation potential of aerobic zymogenous microorganisms in soil for crude oil biodegradation, *Journal of the Serbian Chemical Society*, Vol.76, No.3, (September 2010), pp. 425–438, ISSN 0352-5139 (Figure reprinted with permission from Serbian Chemical Society)

Van Hamme, J.D.; Singh, A. & Ward, O.P. (2003). Recent advances in petroleum microbiology. *Microbiology and Molecular Biology Reviews*, Vol.67, No.4, (December 2003), pp. 503-549, ISSN 1098-5557

Wang, Z. & Fingas, M.F. (2003). Development of oil hydrocarbon fingerprinting and identification techniques. *Marine Pollution Bulletin*, Vol.47, No.9-12, (September-December 2003), pp. 423–452, ISSN 0025-326X (Figure reprinted with permission from Elsevier)

Wang, Z.; Fingas, M.; Blenkinsopp, S.; Sergy, G.; Landriault, M.; Sigouin, L.; Foght, J.; Semple, K. & Westlake, D.W.S. (1998). Comparison of oil composition changes due to biodegradation and physical weathering in different oils. *Journal of Chromatography A*, Vol.809, No.1-2, (June 1998), pp. 89-107, ISSN 0021-9673

Wang, Z.; Yang, C.; Fingas, M.; Hollebone, B.; Peng, X.; Hansen, B.A. & Christensen, H.J. (2005). Characterization, weathering, and application of sesquiterpanes to source identification of spilled lighter petroleum products. *Environmental Science & Technology*, Vol.39, No.22, (October 2005), pp. 8700–8707, ISSN 0013-936X

Wang, Z.D.; Fingas, M. & Page, D. (1999). Oil Spill Identification. *Journal of Chromatography A*, Vol.843, No.1-2, (May 1999), pp. 369–411, ISSN 0021-9673

Wang, Z.D.; Fingas, M.; Owens, E.H.; Sigouin, L. & Brown, C.E. (2001). Long-term fate and persistence of the spilled Metula oil in a marine salt marsh environment:

degradation of petroleum biomarkers. *Journal of Chromatography A,* Vol.926, No.2, (August 2001), pp. 190–275, ISSN 0021-9673

Ward, O.; Singh, A. & Van Hamme, J. (2003). Accelerated biodegradation of petroleum hydrocarbon waste. *Journal of Industrial Microbiology & Biotechnology,* Vol.30, No.5, (April 2003), pp. 260-270, ISSN 1476-5535.

Urinary Olfactory Chemosignals in Lactating Females Show More Attractness to Male Root Voles (*Microtus oeconomus*)

Ping Sun[1,3], Xiangchao Cheng[1] and Honghao Yu[2]
[1]Animal Academy of Science and Technology,
Henan University of Science and Technology, Luoyang,
[2]Yulin University, Shannxi Yulin,
[3]State Key Laboratory of Integrated Management of Pest Insects and Rodents,
Institute of Zoology, Chinese Academy of Sciences, Beijing,
China

1. Introduction

Chemical communication among individuals of the same species serves several important functions, including sexual attraction (Achiraman and Archunan 2002, 2005, Brennan and Keverne 2004), interference with puberty, the estrus cycle, and pregnancy (Dominic 1991, Drickamer 1999). Odors also act as social behavior signals, as in territorial marking (Doty 1980, Prakash and Idris 1992), individual identification (Hurst et al 1998, Poddar-Sarkar and Brahmachary 1999), mother-young interactions (Schaal *et al.* 2003), and the initiation of aggression (Mugford and Nowell 1971), in mammals especially. The pheromones that animals secrete externally are used to communicate with conspecific receivers, which react by behavioral response or developmental process (Wyatt 2003).

A comprehensive understanding of mammalian chemical communication requires a combination of bioassay and chemical analysis to identify the pheromonal components involved (Singer *et al.* 1997, Novotny *et al.* 1999a). Gas chromatography-mass spectrometry (GC-MS) analysis is standard in semiochemical research. A systematic approach using gas chromatography to screen for pheromonal compounds from numerous scent sources has been established in mice (*Mus musculus*); volatile compounds that vary with biological characters are tagged as pheromone candidates for further verification by bioassay (Singer *et al.* 1997, Novotny *et al.* 1999a). This approach has identified a large number of pheromones, including 6-hydroxy-6-methyl-3-heptanone, a male pheromone that accelerates puberty in female mice (Novotny *et al.* 1999b); 1-iodo-2-methylundecane, an estrus-specific urinary chemosignal; and 3-ethyl-2, 7-dimethyl octane, a testosterone-dependent urinary sex pheromone in male mice (Achiraman and Archunan 2005, 2006). Pheromone components were recently discovered in the preputial glands of wild rodents, Brandt's vole (*Lasiopodomys brandtii*) (Zhang *et al.* 2007a, b).

Mammals emit chemical signals into their surroundings via urine, feces, saliva, or specialized scent glands (Dominic 1991, Novotny 2003, Hurst and Beynon 2004). Among these, urine is a major carrier of chemosignals. The odors of most females vary with their reproductive state, and all chemical constituents of urine may vary with the estrous cycle (Michael 1975, O'Connell *et al.* 1981, Rajanarayanan and Archunan 2004). The urine odor of females in estrus is usually more attractive than that of dioestrous females (Drickamer 1999, Vandenbergh 1999).

In many mammals that exhibit spontaneous estrus cycles, the natural estrous cycle is caused by hormonal changes. Thus, endocrine changes may cause excretion of different signals that communicate different reproductive states. Lactation is a period of the reproductive cycle during which hormone levels differ from those of the non-lactating state. Hence, the olfactory cues from lactating-females differ from those of non-lactating females. Several studies of rabbits have indicated that lactating rabbit females emit volatile odor cues that trigger responses in pups, especially specialized motor actions leading to sucking (Coureaud and Schaal 2000, Coureaud *et al.*2001), and the activity of these cues change with advancing lactation (Coureaud *et al.* 2006). Another interesting study indicated that compounds from lactating women might modulate the ovarian cycles of other women (Jacob *et al.* 2004) and increased sexual motivation (Spencer *et al.* 2004). Recently, a published study focused on the ability of mice to discriminate cow urinary odor from prepubertal, preovulatory, ovulatory, postovulatory, pregnancy and lactation (Rameshkumar *et al.* 2008).

The root vole (*Microtus oeconomus*), the only extant Holarctic member of the species-rich genus *Microtus*, is an interzonal small mammal that occurs in wet grasslands of both Arctic and temperate zones (Brunhoff *et al.* 2003). Our studies of social behavior and kin recognition by urine odors have suggested that social rank differences among root voles induce different behavioral patterns (Sun *et al.* 2007a). Male root voles also use a urine association mechanism to achieve discrimination of opposite-sex siblings (Sun *et al.* 2008a). Thus, urine is an important chemosignal source in root voles. Our first objective of this research is to test the hypothesis that urine from lactating and non-lactating females differs in its attractiveness to males. The evidence supported for this hypothesis would indicate that the difference in urine odors could guide preference behaviors of males. The second objective is to identify the chemosignals that cause the male behavioral responses. The individually informational coding forms were confirmed using GC-MS.

2. Materials and methods

2.1 Study animals

Wild root voles were captured from a meadow located in Menyuan County (37°29' - 37 °45' N, 101°12' – 101° 23' E), China. Laboratory colonies were established at the Northwest Institute of Plateau Biology, Chinese Academy of Sciences. The voles used in this study were housed in clear polycarbonate cages (40×28×15 cm) with wood chip beddings and cotton nesting material. A 14:10 h (light : dark) light cycle was maintained, with the light cycle commencing at 08:00. The laboratory temperature was maintained at 22±2°C, and food (BLARC, China) and water were provided *ad libitum*. The cages were cleaned and the cotton nesting material was replaced once a week. All voles used in this study were F1 – F3 generation offspring of field-captured animals and were housed with their mates until the experiments. In the behavior test, twenty 90 – 120-day-old male voles were used, while ten lactating females 10 – 15 days post partum and ten non-lactating females were used as urine donors. The females for the non-lactating condition were all in dioestrus. All of the experimental animals were used only once.

2.2 Experimental apparatus

The behavior choice maze included an odorant box (30x30x30 cm) and a neutral box (30x30x30 cm) made of organic glass (Fig. 1) and connected by a pellucid organic glass tube (20x7 cm). A switch controlled the passage of experimental voles between the odorant and neutral boxes. A culture dish containing fresh urine was placed in the center of the odorant box as an odor stimulant.

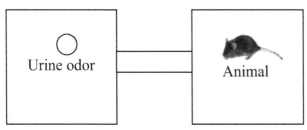

The experimental apparatus used to test behavior choice. The odorant box (30×30×30 cm, left) and neutral box (30×30×30 cm, right) were made of organic glass. The odorant box and the neutral box were connected by pellucidly organic glass tube (20×7 cm). The culture dish containing fresh urine from ether a lactating or a non-lactating vole was put in the center of the odorant box as a stimulant. A switch controlled the movement of voles between the odorant and neutral boxes.

Fig. 1. Experimental apparatus (behavior choice maze)

A vidicon (SONY Version 805E) recorded the behavior of the male voles in the choice maze. The videos were converted to digital data after the experiment.

2.3 Odorant preparation

On the day of a behavioral test, lactating and non-lactating female voles were used as urine donors. The donors were placed on the covers of the clear cages, and two layers of gauze were used to separate feces and urine. We collected the urine samples on clean, absorbent cotton, which was then placed in a culture dish as the odorant. To ensure that only fresh odors were encountered during the trials, urine more than 20 min old was discarded.

2.4 Behavioral test

In this study, behavioral tests were conducted in a room in which the temperature, lighting, and air circulation were identical to those of the breeding room. Subjects with their home cages were moved to testing room 2 h earlier before the behavioral test so they could habituate to the conditions. We tested the responses of male voles to lactating and non-lactating female urine samples. Every test lasted 10 min and was performed between 09:00 and 18:00, during the light phase. Immediately before each trial, we transferred a male subject from its home cage to the clean testing arena. The test animals were allowed to acclimatize in the apparatus for 5 min, then the odor stimulus apparatus was introduced and trial was recorded using a vidicon.

2.5 Statistical analysis of behavior

After the experiment, the videos were converted to digital data. The video files were named by a member of the lab who unfamiliar with the experiment and the videos were then rescored by the original experimenter. Behavioral variables were treated as an index of duration (sec) or frequency. Behavior variables included sniffing and self-grooming. Sniffing, a rhythmic inhalation and exhalation of air through the nose, is a behavior thought to play a critical role in shaping how odor information is represented and processed by the nervous system (Wesson et al 2008). We recorded sniffing behavior when subjects' noses within 0.5cm of the odor stimulation and displayed a rhythmic sniff. In voles, the general pattern of self-grooming consists of a cephalocaudal progression that begins with rhythmic movements of the paws around the mouth and face, over the ears, descending to the ventrum, flank, anogential area, and tail (Ferkin et al. 1996, Leonard and Ferkin 2005, Leonard et al. 2005). According to our knowledge, the root vole in our study generally groomed their anogenital area, head, and flanks (Sun et al., 2005, 2006). Thus, we recorded self-grooming when subjects rubbed, licked or scratched any of these body areas.

All statistical analyses were conducted using SPSS for Windows (version 13.0). We used the Mann-Whitney test to evaluate the hypothesis that male voles displayed different behavioral patterns in the presence of urine cues of lactating and non-lactating females. A difference was regarded as significant at $p<0.05$. All data are reported as mean $\pm$ SE.

2.6 Scent sample collection and extraction

Lactating and non-lactating female root voles were used as urine donors. On the day of GC-MS analysis, the females were removed from their home cages and placed on the covers of the clear cages. Food and water were offered. Two layers of gauze were used to separate feces and urine. Urine contaminated by feces was rejected. Urine used for analysis was collected into clean glass vials and immediately stored at -20°C.

Each sample was thawed at room temperature prior to analysis. We used dichloromethane to extract the volatile compounds from the urine samples, mixing 300 μl dichloromethane with 100 μl urine. The mixture was immediately analyzed by GC-MS.

2.7 GC-MS analysis

GC-MS was performed on a Agilent Technologies Network 7890N GC system coupled with a 5975C Mass Selective Detector with the NIST05 Mass Spectral Library (2005 version). Chemstation Software (Windows XP) was used for data acquisition and processing. The GC was equipped with a 30-m glass capillary column (0.25-mm i.d. x0.25-μm film thickness). Helium was used as the carrier gas, at a flow rate of 1.5 ml/min. For each test, 8 μl of extracted dichloromethane were injected into the GC-MS. The temperature of the injector was set at 280°C, and the electron impact ionization (EI) temperature was set at 300°C. The oven temperature was programmed as follows: 50°C initially, increasing 5°C /min to 200°C, and then 1.5°C /min to 230°C, and finally 5°C /min to 250°C, where it was maintained for 6 min. The relative amount of each component was reported as the percentage of the ion current. We identified unknown volatile compounds by probability-based matching using the computer library (NIST05 Mass Spectral Library).

2.8 GC-MS statistical analysis

As a measure of the relative abundance of a compound, the area under each peak was used for quantitative calculations. The relative amount of each component was reported as the percent of the ion current. GC peak areas that were too small to display the diagnostic MS ions of the corresponding compound were recorded as zero.

We visually examined other volatile compounds to identify those that were individual- or lactation-specific. We then quantitatively compared the basic volatile compounds by examining the distribution of the raw data of the 12 basic volatile compounds, using the Kolmogorov-Smirnov test in SPSS for Windows. As all data were normally distributed, we used an independent two-tailed t-test to analyze for stage differences in the relative abundances. All statistical analyses were conducted using SPSS for Windows (version 13.0). The critical value was set at $\alpha = 0.05$.

3. Results

3.1 Behavioral differences

The behavioral tests showed that the sniffing frequency of males was greater in response to the urine of lactating than non-lactating females urine (mean $\pm$ SE 7.3$\pm$0.79 and 5.4$\pm$0.58 Num,

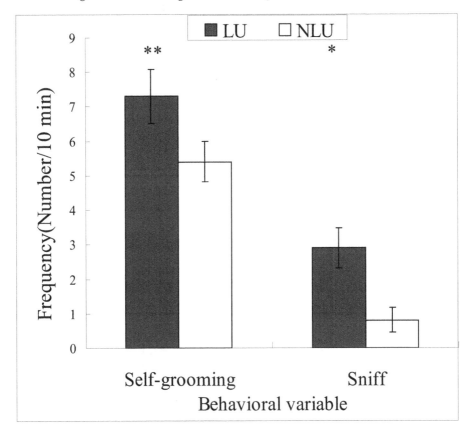

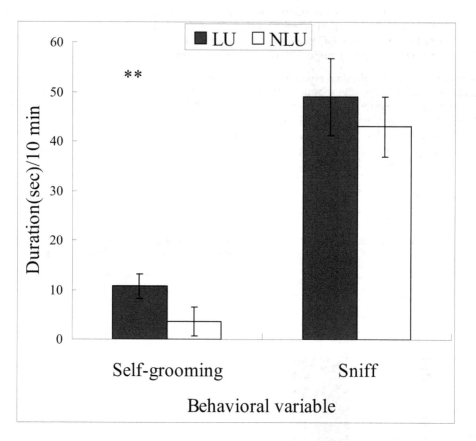

*indicates significant difference (p<0.05); ** indicates significant difference (p<0.01); LU: lactating-female urine; NLU: non-lactating-female urine.*
The responses of male root voles (Microtus oeconomus) to odors from the urine of lactating and non-lactating females (n=20). The behavior examined were frequency and duration of sniff and self-groom. The sniffing (p<0.05) and self-grooming (p<0.01) frequency of males response to lactating and non-lactating urine indicated significant difference. The self-grooming duration (p<0.01) of males response to lactating and non-lactating urine showed significant difference, but not the sniffing duration (p>0.05).

Fig. 2. Behavioral difference of male voles in response to lactating and non-lactating female urine odor

respectively; N = 10 each; p = 0.043). Males' self-grooming frequency was also greater when males were exposed to the urine odor of lactating females, compared to that of non-lactating females (2.9±0.59 and 0.8±0.36 Num, respectively; N = 10; $p < 0.01$), as was the duration of self-grooming behavior (10.7±2.54 and 3.6±2.95 sec, respectively; N = 10; $p < 0.01$; Fig. 2). Thus, the duration and frequency of self-grooming, as well as sniffing frequency, revealed that the urine of lactating females was more attractive than that of non-lactating females to male root voles.

3.2 Volatile constituents of urine

The GC-MS profiles of Table 1 and Figures 3 show the representative volatile compounds
contained in the urine of lactating and non-lactating females, with 10 – 20 detectable peaks
in each condition. In all, 34 peaks were detected in lactating and non-lactating states. The
constituents identified in the urine samples included alkanes, alkenes, alcohols, ketones,
benzo-forms, esters, acids, furans, pyrans, and other volatile compounds. Benzenes and

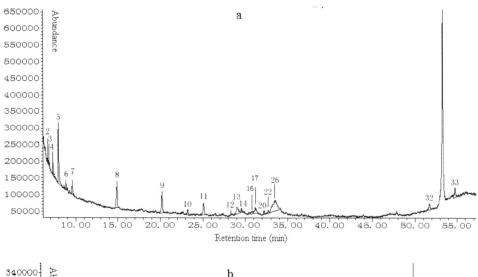

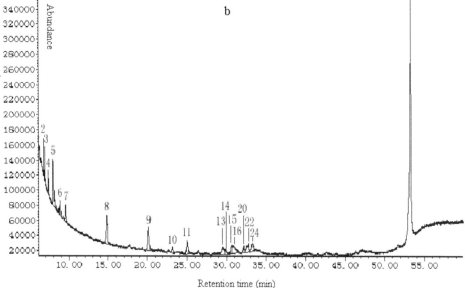

Fig. 3. Total ion chromatogram of urinary volatile compounds identified in (a) lactating and
(b) non-lactating female root voles

Peak No	Retention Time (min)	Compound	Molecular weight	Group
1	6.388	2-pentanone,4-hydroxy-4-methyl	106	Ketone
2	6.715	Ethylbenzene	106	Benzo-
3	6.869	(D)xylene	106	Benzo-
4	7.301	Styrene	104	Benzo-
5	7.943	(Z)1,3,5-hexatriene,3-methyl	94	Alkene
6	8.841	Benzene, 1-ethyl-2-methyl	120	Benzo-
7	9.585	Decane	142	Alkane
8	14.861	Dodecane	170	Alkane
9	20.179	Tetradecane	198	Alkane
10	23.181	Butylated hydroxytoluene	220	Benzo-
11	25.054	Hexadecane	226	Alkane
12	28.283	Methoxyacetic acid decyl ester	230	Ester
13	29.026	Tetradecanoic acid	228	Acid
14	29.532	Heptadecane	240	Alkane
15	30.626	(R)1-H-2benzopyran-1-1,3,4-dihydro-8hydroxy-6-methoxy-3-methyl	208	Pyran
16	30.777	1-H-2benzopyran-1-1,3,4-dihydro-8hydroxy-6-methoxy-3-methyl	208	Pyrane
17	31.249	1-Hexadecanol	242	Alcohol
18	31.344	(E)5-octadecene	252	Alkene
19	32.1	2,5-cydohexadion-1-1,2,6-bis[1,1-dimethylethyl]4-ethylidene	232	
20	32.215	3-Ethoxycarbonyl-4-hydroxyguinoline	217	
21	32.53	Etyl-4-oxo-1,4-dihydroquinoline-3-carboxylate	217	
22	32.708	Naphtho[2.3-b]furan-9(4H)-1,4a.5.6.7.8a-hexahycho-3,4a.5-frinethyl-[4aR-{4a,alpha,5,alpha,8a,beta}]	232	Furan
23	33.107	1-Naphthalenecarboxylic acid	217	Acid
24	33.267	(P)Hexaclecenoic acid	254	Acid
25	33.312	(D)octylacetophenone	232	Benzo-
26	33.531	(N)Hexadecanoic acid	256	Acid
27	36.117	Oleyl alcohol	268	Alcohol
28	36.962	(R)- [-]-ol-4-methyl-8-hexadecyn-1-ol	252	
29	37.135	(Z)7-Hexadecanoic acid, methyl ester	268	Acid
30	42.871	Squalene	410	Alkene
31	50.843	1, 19-Eicosacliene	278	Alkene
32	51.730	Eicosane	282	Alkane
33	54.73	Henicosane	296	Alkane
34	57.918	Docosane	310	Alkane

Table 1. Urinary volatile compounds identified in lactating and non-lactating female root voles

alkanes were predominant in the urine samples. Although the chromatograms showed consistent qualitative differences in each individual chemical profile, 12 basic volatile compounds were present in almost all samples, including ethylbenzene [peak number (PN) 2], (D)xylene(PN 3), styrene(PN 4), (Z)1,3,5- hexatriene,3- methyl (PN 5), 1-ethyl-2-methyl-benzene (PN 6), decane (PN 7), dodecane (PN 8), tetradecane (PN 9), butylated hydroxytoluene (PN 10), hexadecane (PN 11), heptadecane (PN 14), and 1-H-2-benzopyran-1-1,3,4-dihydro-8 hydroxy-6-methoxy-3-methyl (16).

3.3 Individual-specific constituents

Lactating and non-lactating individuals were numbered 1 – 6 and 7 – 12, respectively. In addition to the presence of 12 basic volatile compounds in urine, 20 constituents were individual-specific, including 2-pentanone, 4-hydroxy-4-methyl (PN 1), methoxyacetic acid decyl ester (PN 12), tetradecanoic acid (PN 13), (R)1-H-2benzopyran-1- 1, 3, 4- dihydro-8hydroxy-6-methoxy-3-methyl (PN 15) 1-hexadecanol (PN 17), 2, 5-cydohexadion-1- one, 2, 6-bis [1,1-dimethylethyl] 4 -ethylidene (PN 19), 3-Ethoxycarbonyl-4- hydroxyguinoline (PN 20), Etyl-4- oxo-1,4- dihydroquinoline-3- carboxylate (PN 21), Naphtho [2.3-b] furan-9 (4H) - one, 4a.5.6.7.8a - hexahycho- 3, 4a.5- frinethyl- [4aR- {4a,alpha,5,alpha,8a,beta}] (PN 22), 1-Naphththalenecarboxylic acid,3-intro- (PN 23), (P)Hexaclecenoic acid (PN 24), (D)octylacetophenone (PN 25), Oleyl Alcohol (PN 27), (R)- [-]-ol-4-methyl-8- hexadecyn-1-ol (PN 28), (Z)7-Hexadecanoic acid, methl ester (PN 29), Squalene (PN 30), 1, 19-Eicosacliene

Peak No.	Compound	Detected frequency	Individual No.
1	2-pentanone, 4-hydroxy-4-methyl	1	6
12	Methoxyacetic acid decyl ester	1	3
13	Tetradecanoic acid	4	1, 2, 3, 11
17	1-Hexadecanol	4	1, 2, 3, 10
19	2,5-cydohexadion-1-1,2,6-bis[1,1-dimethylethyl]4-ethylidene	2	5, 8
20	3-Ethoxycarbonyl-4-hydroxyguinoline	5	3, 6, 8, 10, 11
22	Naphtho[2.3-b]furan-9(4H)-1,4a.5.6.7.8a-hexahycho-3,4a.5-frinethyl-[4aR-{4a,alpha,5,alpha,8a,beta}]	4	3, 5, 8, 11
23	1-Naphthalenecarboxylic acid	1	8
24	(P)hexaclecenoic acid	5	1, 5, 8, 9, 10
25	(D)octylacetophenone	1	12
27	Oleyl Alcohol	1	1
28	(R)- [-]-ol-4-methyl-8-hexadecyn-1-ol	1	7
29	(Z)7-hexadecanoic acid, methyl ester	1	5
30	Squalene	1	4
31	1, 19-Eicosacliene	1	1
32	Eicosane	2	3, 4
33	Henicosane	2	3, 10
34	Docosane	1	3

Table 2. Individual-specific urinary volatile compounds identified in female root voles

(PN 31), Eicosane (PN 32), Henicosane (PN 33), Docosane (PN 34). These volatile compounds were presented randomly in lactating and non-lactating female urine, with each sample including one to six individual-specific volatile compounds that could identify individual females. Table 2 lists the individual urinary odor characters.

3.4 Differences in urinary volatile compounds of lactating and non-lactating females

Quantitative analyses revealed that the relative amount of the basic volatile compounds did not differ significantly between lactating and non-lactating stages. A quality comparison of the identified volatile compounds revealed that (E)5-octadecene (PN18) and (N)Hexadecanoic acid (PN 26) (Fig. 4) were specific to the lactating stage.

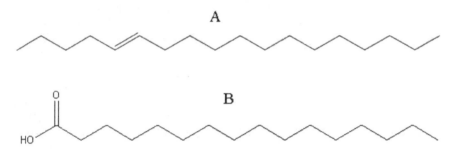

A: (E)5-octadecene, B: (N)hexadecanoic acid

Fig. 4. Molecular structural of two putative lactating pheromones.

4. Discussion

Among mammals, the olfactory information of urine is complex. The olfactory information of urine is about sex, health conditions, relatedness, individual personality, reproductive and social status (Gheusi et al. 1997, Hurst and Beynon 2004, Hurst 2004, Penn and Potts 1998). Our studies about urine odor demonstrated that the olfactory information of urine from parents is about relatedness and the retention of maternal odor of the female offspring lasted for 40 d after post weaning (Sun et al. 2007b) and of the male offspring lasted for 10 d after post weaning (Sun et al. 2008b) in root voles. During the breeding season, female voles may be pregnant, lactating, pregnant and lactating, or neither pregnant nor lactating. The female voles may be in different states of sexual receptivity, including heightened receptivity during postpartum estrus or moderate receptivity when they are not pregnant or lactating (Ferkin et at. 2004). The phenomenon of postpartum estrus presents very common in root vole. The female root voles can copulate with males within 24h after parturition (unpublished data).

The chemical constituents of mammalian urine vary with the estrous cycle (Michael 1975, Rajanarayanan and Archunan 2004, Achiraman and Archunan 2006), and the urine odor of estrous females is usually more attractive to males than that of diestrous or non-estrous females (Drickamer 1999, Dominic 1991, O'Connell et al. 1981; Vandenbergh 1999, Rajanarayanan and Archunan 2004). Since the natural estrous cycle is induced by hormonal changes, the variation in urinary volatiles may be influenced by the endocrine system. In present study, correspondingly, behavioral observations showed that the male reactions to lactating female urine differed from to non-lactating females. Therefore, our results supported the view that the variation in urinary volatiles might be influenced by the endocrine system. Male root voles engaged in more sniffing behavior and both more frequent and longer duration in self-grooming behavior when respond to lactating than to non-lactating urinary odors. Sniffing and self-grooming behaviors facilitate sexual interactions between opposite-sex conspecifics (Moffatt and Nelson 1994, Ferkin et al. 2001, Ferkin and Leonard 2005, Pierce et al. 2005). Thus, the increase of sniffing behavior in males in response to lactating female urinary odor reflects a sexual motivation. Prairie voles (Microtus ochrogaster) use self-grooming behavior to enhance sexual communication when met with opposite-sex individuals (Ferkin et al. 2001). For male root voles, self-grooming may increase the detection of their scent cues by conspecifics and may attract females (Zhao et al. 2003). Our behavioral results showed that the urine of lactating females is more attractive to males than that of non-lactating females.

We detected 34 kinds of volatile compounds in the urine of lactating and non-lactating female voles, which are fewer than the number of volatiles in the urine of male voles (Boyer et al. 1988) and mice (Schwende et al. 1986). A low number of volatile compounds have also been reported in the urine of the California mouse (Peromyscus californicus) (Jemiolo et al. 1994) and Mus musculus (Achiraman and Archunan 2006). Thirty-four kinds of volatile components those we found are distributed among basic, individual-specific, and lactation-specific volatile compounds. The basic volatile compounds included a strikingly high number of alkanes and benzo- forms. These volatile compounds were present in all individuals, suggesting that they are probably common metabolic final products in vole urine.

Our results suggest that root vole urine contains a wealth of information that codes for sex and individuality, which is consistent with previous behavioral tests (Zhao 1997, Zhao et al. 2002, 2003, Sun et al. 2007a,b). The information concerning sex and individuality, as proposed by Sun and Müller-Schwarze (1998a, b) for beavers and exemplified further by three Mustela species (Zhang et al. 2003, 2005), may be coded by analog components and/or digital forms. The two general forms of information coding-digital and analog-corresponding to information ceding by presence/absence of chemicals used for communication versus coding by varying amounts of these chemicals. In present study, we found 18 individual-specific volatile compounds that code individual information (Table 3), with information for each individual being coded by one to six volatile compounds, which are distributed uniquely and randomly. This indicates that root vole urinary odor information is in digital form.

Urinary volatile compounds that covary in quality or quantity with biological characters may be considered putative pheromones (Singer et al. 1997, Novotny et al. 1999a). Based on qualitative differences in female root vole urine, we hypothesize that (E)5-octadecene and (N)hexadecanoic acid are potential lactating pheromones. The molecular weight of these two putative pheromones is less than 300, and they have fewer than 20 carbons. Pheromones usually contain 5 – 20 carbon atoms and must be volatile to reach the receiver (Dominic 1991). The molecular weight of the urinary compound 1-iodoundecane, an estrus-specific compound in bovines, is less than 300 (Rameshkumar et al. 2000). Likewise, the preovulatory urine of female Asian elephants that is involved in attracting males contains a specific compound, (Z)7-dodecene-1-ylacetate, with 13 carbons and a molecular weight of 300 (Rasmussen *et al.* 1997). Hence, (E)5-octadecene and (N)hexadecanoic acid have the physical properties of putative urinary chemosignals.

Individual No.	Individual-Specific Compound Number	Shape simulation
1	13, 17, 24, 27, 31	■□▶◀□
2	13, 17	■□
3	12, 13, 17, 20, 32, 33, 34	▨■□♦◙∘☼
4	30, 32	●◙
5	19, 22, 24, 29	●—▶○
6	1, 20	▨♦
7	28	◊
8	19, 20, 22, 23, 24	●♦—▲▶
9	24	▶
10	17, 20, 24, 33	□♦▶∘
11	13, 20, 22	■♦—
12	25	▼

Individual-specific volatile compounds were shape-coded to express individual urinary odor characters clearly.

1 ▨ 12 ▨ 13 ■ 17 □ 19 • 20 ♦ 22 — 23 ▲ 24 ▶ 25 ▼ 27 ◀ 28 ◊ 29 ○ 30 ● 31 ◘ 32 ◙ 33 ∘ 34 ☼

Table 3. Individual information coded by individual-specific volatile compounds

(N)hexadecanoic acid is a long-chain fatty acid. Long-chain fatty acid pheromones are found in a variety of other vertebrates, including the male ferret (*Mustela furo*) (Clapperton *et al.* 1988) and leopard gecko (*Eublapharis macularius*) (Mason and Gutzke 1990). Some straight-chain fatty acids are common components of insect pheromone blends (Elsayed 2005). Such convergent uses of the same type of compound in different species indicate that these volatile compounds may possess chemical properties typical of pheromones, such as volatility, which allows them to convey airborne cues over a distance.

Hexadecanol and hexadecyl acetate have been reported to be among the major components of preputial gland secretions in both sexes of mice (Zhang *et al.* 2007a) and are pheromonal components in many insects, as well as in the bank vole (*Clethrionomys glareolus*) (Brinck and Hoffmeyer 1984; Wyatt 2003). Zhang *et al.* (2008) found that hexadecanol and hexadecyl acetate attract males in a dose-dependent manner. These investigations support the suggestion that (N)hexadecanoic acid, which is similar to hexadecanol and hexadecyl acetate, is a pheromonal compound.

In summary, we found that individual identification is coded in digital form. We confirmed that the urine of lactating females was more attractive to male root voles than that of non-lactating females and suggested that the presence of two putative pheromones was the effective component. In future studies, we plan to verify pheromonal activity via bioassay and application of synthetic chemosignals.

5. Acknowledgements

This work was supported by grants from the National Natural Science Foundation of China (Nos. 30500073 and 30870370), the China Postdoctoral Foundation (20070420525) to PS.

6. References

Achiraman S., Archunan G. 2002- Charactization of urinary volatiles in swiss male mice (*Mus musculus*): bioassay of identified compounds- J. Bio. Sci. 27: 679-686.

Achiraman S., Archunan G. 2005- 3-Ethyl 2, 7-dimethyl octane, a testosterone dependent unique urinary sex pheromone in male mouse (*Mus musculus*)- Anim. Reprod. Sci. 87: 151-161.

Achiraman S., Archunan G. 2006- 1-iodo-2methylundecane, a putative estrus-specific urinary chemo-signal of female mouse (*Mus musculus*)- Theriogenology. 66: 1913-1920.

Boyer M. L., Jemiolo B., Andreolini F., Wiesler D., NovotnyM. 1988- Urinary volatile profiles of the pine vole. *Microtus pinetorum*, and their endocrine dependency- J. Chem. Ecol. 15: 649-661.

Brennan P. A., Keverne E. B. 2004- Something in the air? New insights into mammalian pheromones- Curr. Biol. 14: 81-89.

Brinck C., Hoffmeyer I. 1984- Marking urine and preputial gland secretion of male bank voles (*Clethrionomys glareolus*): chemical analyses and behavioral test- J. Chem. Ecol. 10: 1295-1308.

Brunhoff C., Galbreath K. E., Fedorov V. B., Cook J. A., Jaarola M. 2003- Holarctic phylogeography of the root vole (*Microtus oeconomus*): implications for late Quaternary biogeography of high latitudes- Mol. Ecol. 12: 957-968.

Ciapperton B. K., Minot E. O., Crump D. R. 1988- An olfactory recognition system in the ferret *Musteal furo*. (Carnivora: Mustelidae)- Anim. Behav. 36: 541-553.

Coureaud G., Langlois D., Perrier G., Schaal B. 2006- Convergent changes in the maternal emission and pup reception of the rabbit mammary pheromone- Chemoecology. 16: 169-174.

Coureaud G., Schaal B. 2000- Attraction of newborn rabbits to abdominal odors of adult conspecifics differing in sex and physiological state- Dev. Psychobiol. 36: 271-281.

Coureaud G., Schaal B., Langlois D., Perrier G. 2001- Orientation response of newborn rabbits to odours of lactating females: relative effectiveness of surface and milk cues- Anim Behav. 61: 153-162.

Dominic C. J. 1991- Chemical communication in animals- J. Sci. Res. 41: 157-169.

Doty R. L. 1980- Scent marking in mammals. In: Denny, M. R. (Ed), Comparative psychology: research in animal behaviour. 445-460. Wiley Pub.New York.

Drickamer L. C. 1999- Sex attractants- Encycl. Reprod. 4: 444-448.

El-Sayed A. M. 2005- The pherobase: Database of insect pheromones and semiochemicals. <http://www.pherobase.com/database/compound>.

Ferkin M. H., Lee D. N., Leonard S. T. 2004-The reproductive state of female voles affects their scent marking behavior and the responses of male conspecifics to such marks- Ethology. 110: 257-272.

Ferkin M. H., Leonard S. T. 2005- Self-grooming by rodents in social and sexual contexts.- Acta Zool. Sinica. 51: 772-779.

Ferkin M. H., Leonard S. T., Heath L. A., Paz-y-Mino G. 2001- Self-grooming as a tactic used by prairie voles, *Microtus ochrogaster*, to enhance sexual communication- Ethology. 107: 939-949.

Ferkin M. H., Sorokin E. S., Johnston R. E. 1996- Self-grooming as a sexually dimorphic communicative behavior in meadow voles, *Microtus pennsylvanicus*- Anim. Behav. 51: 801-810.

Gheusi G., Goodall G., Dantzer R. 1997-Individually distinctive odours represent individual conspecifics in rats-Anim. Behav. 53: 935-934.

Hurst J. L. 2004-Scent marking and social communication. In: McGregor P. K. (Ed), Animal communication network. 220-225. Cambridge University Press. Cambridge.

Hurst J. L., Beynon R. J. 2004- Scent wars: the chemical biology of competitive signaling in mice- BioEssays. 26: 1288-1298.

Hurst J. L., Roberstson D. H. L., Tollday V., Beynon R. J. 1998- Proteins in urine scent marks of male house mice extend longevity of olfactory signals- Anim. Behav. 55: 1589-1597.

Jacob S., Spencer N. A., Bullivant S. B., Sellergren S. A., Mennella J. A., McClintock M. K. 2004- Effects of breastfeeding chemosignals on the human menstrual cycle- Hum Reprod. 19: 422-429.

Jemiolo B., Gubernick D. J., Yonder M. C., Novotny M. 1994- Chemical characterization of urinary volatile compounds of *Peromyscus californicus*, a monogamous biparental rodent- J. Chem. Ecol. 20: 2489-2499.

Lenonard S. T., Alizadeh-Naderi R., Stokes K., Ferkin M. H. 2005- The role of prolactin and testosterone in mediating seasonal differences in self-grooming behavior of male meadow voles, *Microtus pennsylvanicus*- Physiol. Behav. 85: 461-468.

Leonard S. T., Ferkin M. H. 2005- Seasonal differences in self-grooming in meadow voles, *Microtus pennsylvanicus*- Acta. Ethol. 8: 86-91.

Mason R. T., Gutzke W. H. N. 1990- Sex recognition in the leopard gecko, *Eublephairs macularius* (Sauria: Gekkonidae): Possible mediation by skin-derived semiochemicals- J. Chem. Ecol. 16: 27-36.

Michael R. P. 1975- Hormonal steroids and sexual communication in primates- J. Steroid. Biochem. 6: 161-170.

Moffatt C. A., Nelson R. J. 1994- Day length influences proceptive behavior of female prairie voles (*Microtus ochrogaster*)- Physiol Behav. 55: 1163-1165.

Mugford R. A., Nowell N. W. 1971- The preputial gland as a source of aggression promotion odour in mice- Physiol. Behav. 6: 247-249.

Novotny M. V. 2003- Pheromones, binding proteins and response in rodents- Biochem. Soc. Trans. 31: 117-122.

Novotny M. V., Jemiolo B., Wiesler D., Ma W., Harvey S., Xu F., Xie T. M., Carmack M. 1999b- A unique urinary constituent, 6-hydroxy- 6- methyl- 3-heptanone, is a pheromone that accelerates puberty in female mice- Chem. Biol. 6: 377-383.

Novotny M., Ma W., Zidek L., Daev E. 1999a- Recent biochemical insights into puberty acceleration, estrus induction, and puberty delay in the house mouse. In: R. E. Johnston, D. MÜller-Schwarze, and P. W. Sorensen (Eds). Advance in Chemical Signals in Vertebrates. 99-116. Kluwer, New York.

O´ Connell R. J., Singer A. G., Stern F. L., Jesmajian S., Agosta W. C. 1981- Cyclic variation in the concentration of sex attractant pheromone in hamster vaginal discharge- Behav. Neur. Biol. 31: 457-464.

Penn D., Potts W. K. 1998-Chemical signals and parasite-mediated sexual selection-Trends. Ecol. Evol. 13: 391-396.

Pierce J. B., Ferkin M. H., Williams T. K. 2005- Food-deprivation-induced changes in sexual behavior of meadow voles, Microtus pennsylvanicus- Anim Behav. 70: 339-348.

Poddar-Sarkar M., Brahmachary R. L. 1999- Can free fatty acids in the tiger pheromone act as an individual finger print- Curr. Sci. 76: 141-142.

Prakash I., Idris M. 1992- Scent marking behaviour. In: Prakash I., and Ghosh P. K., (Eds). Rodents in Indian Agriculture. 445-460. Scientific Publishers, India.

Rajanarayanan S., Archunan G. 2004- Occurrence of Flehmen in male buffaloes (Bubalus bubalis) with special reference to estrus- Theriogenology. 61: 861-866.

Rameshkumar K., Achiraman S., Karthikeyan K., Archunan G. 2008- Ability of mice to detect estrous odor in bovine urine: Roles of hormones and behavior in odor discrimination- Zool Sci. 29: 349-354.

Rameshkumar K., Archunan G., Jeyaraman R., Narasimhan S. 2000- Chemical characterization of bovine urine with special reference to oestrus- Vet. Res. Commun. 24: 445-454.

Rasmussen L. E. L., Lee T. D., Zhang A., Roelofs W. L., Daves G. D. 1997- Purification, identification concentration and bioactivity of (Z)-7-dodecen-1-yl acetate: sex pheromone of the female Asian elephant, Elephas maximus- Chem. Senses. 22: 417-437.

Schaal B., Coureaed G., Labgolls D., Ginies C., Semon E., Peter G. 2003- Chemical and behavioural characterization of the rabbit mammary pheromone- Natrue. 424: 68-72.

Schwende F. J., Wiesler D., Jorgenson J. W., Carmack M., Novotny M. 1986- Urinary volatile constituents of the house mouse, Mus muscullus and their endocrine dependency- J. Chem. Ecol. 12: 277-296.

Singer A. G., Beauchamp G. K., Yamazaki K. 1997- Volatile signals of the major histocompatibility complex in male mouse urine- Proc. Natl. Acad. Sci. U. S. A. 94: 2210-2214.

Spencer N. A., McClintock M. K., Sellergren S. A., Bullivant S., Jacob S., Mennella J. A. 2004- Social chemosignals from breastfeeding women increase sexual motivation- Horm Behav. 46: 362-370.

Sun L., MÜller-Schwarze D. 1998a- Anal gland secretion codes for family membership in the beaver- Behav. Ecol. Sociobiol. 44: 199-208.

Sun L., MÜller-Schwarze D. 1998b- Anal gland secretion codes for relatedness in the beaver, Castor Canadensis- Ethology. 104: 917-927.

Sun P., Zhao Y. J., Zhao X. Q., Xu S. X., Li B. M. 2005 - Kin recognition in cross-fostered colonies of root voles (Microtus oeconomus): male response to urine odor of female siblings- Zool Res. 26(5): 460 - 466.

Sun P., Zhao Y.J., Xu S.X., Zhao X.Q. 2006 - The discrimination to self and novel odours of male root voles (Microtus oeconomus) in different social status- Acta Theriol Sin. 3: 280 - 284.

Sun P., Yu H., Zhao X., Xu N., Zhao Y. 2007b-Retention of parental urine odour of post weaning female root vole- Zoological Research. 28: 141-148.

Sun P., Zhao Y. J., Zhao X. Q., Wang, D. H. 2007a- Behavioural recognition of root vole (*Microtus oeconomus Pallas*): Male of different social ranks to familiar and novel odour of conspecific males- Pol. J. Ecol. 55 (3): 571-578.

Sun P., Zhu W. Y., Zhao X. Q. 2008a-Opposite-sex sibling recognition in adult root vole, *Microtus oeconomus pallas*: phenotype matching or association- Pol. J. Ecol. 56: 701-708.

Sun P., Zhao Y. J., Zhao X. Q. 2008b-Retention to Parental Urine Odour of Post weaning Male Chaidamu Root Vole - Chinese Journal of Zoology. 43(5): 45-50.

Vandenbergh J. G. 1999- Pheromones in mammal- Encycl. Reprod. 3: 764-769.

Wesson D. W., Donahou T. N., Johnson M. O., Wachowiak M. 2008- Sniffing behavior of mice during performance in odor-guided tasks- Chem. Senses. 33: 581-596.

Wyatt T. D. 2003- Pheromones and Animal Behaviour. Cambridge University Press, Cambridge, Chapter I: pp 1.

Zhang J. X., Liu Y. J., Zhang J. H., Sun L. 2008- Dual role of preputial gland secretion and its major components in sex recognition of mice- Physiol Behav. 95(3): 388-394.

Zhang J. X., Ni J., Ren X. J., Sun L. X., Zhang Z. B., Wang Z. W. 2003- Possible coding for recognition of sexes, individuals and species in anal gland volatiles of *Mustela eversmanni* and *M. sibirica*- Chem. Senses. 28: 381-388.

Zhang J. X., Rao X. P., Sun L., Zhao C., Qin X. 2007a- Putative chemical signals about sex, individuality and genetic background in preputial gland and urine of the house mouse (*Mus musculus*)- Chem. Senses. 32: 293-303.

Zhang J. X., Rao X. P., Zhao C., Liu X., Qin X. 2007b- Possible information about gender and individuality recognition coded by insect pheromone analogs in the preputial glands in male Brandt's voles, *Lasiopodomys brandtii*- Acta Zool. Sin. 53: 616-624.

Zhang J. X., Soini H. A., Bruce K. E., Wiesler D., Woodley S. K., Baum M. J., Novotny M. V. 2005- Putative chemosignals of the ferret (*Mustela furo*) associated with individual and gender recognition- Chem. Sense. 30: 727-737.

Zhao Y. J. 1997- The strategies of social behavviours in root voles (*Microtus oeconomus*) and its fitness-PhD dissertation, Beijing, Beijing Normal University, Department of Biology, 115 pp.

Zhao Y. J., Sun R. Y., Fang J. M., Li B. M., Zhao, X. Q. 2003- Preferences of pubescent females for male dominants *vs.* subordinates in root voles- Acta Zool Sin. 49: 303-309.

Zhao Y. J., Tai F. D., Wang T. Z., Zhao X. Q., Li B. M. 2002- Effects of the familiarity on mate choice and mate recognition in *Microtus mandarimus* and *M. oeconomus*- Acta Zool Sin. 48: 167-174.

Zufall F., Leinder-Zufall T. 2007- Mammalian pheromone sensing- Curr. Opin. Neurobiol. 17: 483-489.

Comprehensive Two-Dimensional Gas Chromatography Coupled to Time-of-Flight Mass Spectrometry in Human Metabolomics

Petr Wojtowicz[1,*], Jitka Zrostlíková[2], Veronika Šťastná[1],
Eva Dostálová[1], Lenka Žídková[1], Per Bruheim[3] and Tomáš Adam[1,*]
*[1]Laboratory for Inherited Metabolic Disorders & Institute of Molecular and
Translational Medicine, Palacký University in Olomouc,
[2]LECO Application Laboratory, Prague,
[3]Department of Biotechnology, NTNU, Trondheim,
[1,2]Czech Republic
[3]Norway*

1. Introduction

Metabolomics is a discipline aiming to characterize a phenotype by means of metabolome analysis. In recent years, it has developed into an accepted and valuable tool in life sciences and its use has been growing rapidly in the study of microbial, plant, and mammalian metabolomes. It has been shown to be an effective tool in characterizing cancer cells and their response to anticancer drugs (Griffiths & Chung, 2008). To assess the effects of drugs on important pathways in clinical trials of innovative therapies, metabolomic approaches might be more cost-effective than those that measure specific molecular targets (Workman et al., 2006). The derived biomarkers applied in early clinical trials are expected to help identify appropriate patients, provide proofs of concepts, aid decision making, and ultimately reduce the high level of attrition and costs of drug development (Sarker & Workman, 2007).

The biological specimens used in human metabolomic studies are e.g. urine (Weiss et al., 2007), blood plasma (Boernsen et al., 2005), and saliva (Walsh et al., 2006). All of them can play an important role in the diagnostic processes of problem illnesses. The individual metabolome of human biofluids is defined by genetic factors but it can also be affected by diet, age, disease, etc. In this respect, the use of human cell cultures offers a good alternative, since the influence of the above-mentioned factors is minimized in a culture where a defined extracellular environment takes place (Rabinowitz et al., 2006).

As concerns the analytical techniques applied in metabololmics, chromatographic techniques coupled to mass spectrometry play an important role. For the analysis of organic acids, amino acids, and sugars gas chromatography-mass spectrometry (GC-MS) after derivatization is widely applied. Since biological materials represent a very complex matrix,

---

* Authors equally contributed to the work

classical GC-MS techniques can struggle with the high number of components present and the occurring co-elutions. In this respect, comprehensive two-dimensional gas chromatography (GC × GC) brings significant benefits and its coupling with time-of-flight mass spectrometry (GC × GC-TOF-MS) has become an emerging technique in this field (Koek et al., 2011).

Due to a polar nature of target compounds, derivatization procedure is required for GC-MS analysis. Although silylation is the most widely used approach, it has certain limitations such as formation of more products from a single analyte. Moreover, the ratios between individual silylated products can change with time. To overcome these drawbacks, other derivatization procedures such as indirect alkylation via chloroformates or acylation have been used (Husek & Simek, 2006).

This contribution is focused on the application of comprehensive two-dimensional gas chromatography coupled to time-of-flight mass spectrometry for the analysis of human biological materials (urine, plasma, and cultured skin fibroblasts) in relation to diagnosing metabolic disorders, cell metabolism quenching, and drug metabolomic impact prediction.

2. Comprehensive two-dimensional gas chromatography

2.1 Basic theory

Comprehensive two-dimensional gas chromatography (GC × GC) is a technique utilizing two columns of different selectivity connected in series by the modulation device. The modulator cuts slices from the first-dimension column effluent and re-injects them to the secondary column. Due to the difference in column polarity, each simple compound is subjected to two independent separation mechanisms. Compared to one-dimensional GC, this technique brings dramatically increased peak capacity, improved peak resolution, and up to an order of magnitude increase in compounds' detectability. In contrary to heart-cutting variety, in GC × GC all effluent from the primary column passes through the secondary column, maximizing sample resolution throughout the entire analysis (Gorecki & Harynuk, 2004; Beans & Brinkman, 2005). A theoretical and practical comparison of one-dimensional GC and GC × GC in terms of peak capacity has been published by Blumberg et al., 2008.

In GC × GC, two basic orthogonality rules should be kept: (i) independence of separation mechanisms, i.e. the two columns should possess of different selectivity; (ii) preserving of the first-dimension separation, i.e. the peaks already separated on the first column must not be mixed-up in the modulator. For this reason, the modulation must occur at frequency of at least 3-5 modulations per first dimension peak. But in practice, the full independence is not possible, so in the case of GC × GC this must be considered as the degree of GC × GC system orthogonality, or "relative" or "partial" orthogonality which can be characterized by the percent usage of the available separation space (Ryan et al., 2005; Zhu, 2009).

2.2 Technical aspects

GC × GC occurs by the subsequent re-injection of effluent from one chromatographic column into the second "orthogonal" column. As already mentioned, a minimum number of modulations per first dimension peak are required to maintain the first dimension separation which typically results in modulation periods of 1-5 s. A flash separation on the second dimension column has to be completed before the next modulation cycle starts. In this way, the separation obtained in the first dimension is preserved and additional separation on the second column is obtained. The re-injection process is called modulation and is enabled by an interface device called modulator, often referred to as the "heart" of the system. A GC × GC modulating interface can be placed at the end of the first dimension (^{1}D) or at the beginning of the second dimension (^{2}D), e.g. for thermal modulators, or between the columns (valve-based modulators). Nowadays, cryo-modulators which trap primary column effluent below ambient temperatures with the use of various cooling mechanisms are most commonly used (Edwards et al., 2011).

As a typical column set-up, a nonpolar column is used as the ^{1}D and polar one in the ^{2}D. Under these conditions, separation according to the volatility of the compounds occurs in the ^{1}D column while "polarity" separation dominates in the ^{2}D column. In principle, the column arrangement can be inverted. Both set-ups have their advantages and disadvantages and therefore the right set-up must be a result of optimization of a particular application. As concerns the column dimension, the first column is relatively long (typically 30-60 m) and normal bore (typically 0.25 mm with 0.25 µm film thickness). The second column has to be very short and narrow (typically 1-2 m of a 0.1 mm column with 0.1 µm film thickness) to perform a very fast separation. The use of such narrow-bore column results in limited sample capacity and easy ^{2}D column overloading, especially for biological samples. This limitation has been overcome by the use of wider-bore columns in the second dimension (Koek et al., 2008).

Primary GC × GC data are a series of second dimension chromatograms registered by the detector. Appropriate software reconstructs the second dimension chromatograms into three-dimensional plots or contour plots as shown in Figure 1.

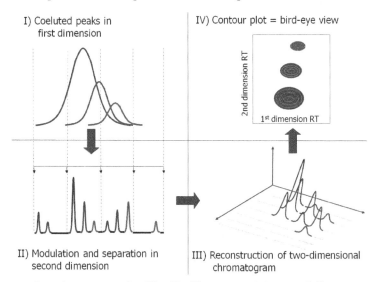

Fig. 1. Construction of a contour plot. The ^{1}D effluent containing not fully separated analytes is modulated to create a series of short ^{2}D chromatograms. They are reconstructed by the software form a three-dimensional view. For the practical purposes the contour plot view is more feasible.

As concerns the detectors applicable in GC × GC, the peak width generated by this technique must be taken into account. Typical GC × GC peaks are 0.1-0.3 s wide, thus to get a sufficient number of data points to accurately describe the shape of the peak (at least 10 points per peak) a detector must collect data at a rate of at least 100 Hz. For the coupling of GC × GC with MS detection, a high-speed TOF-MS is the technique of choice in most studies. Quadrupole MS have been also applied in GC × GC (Adahchour et al., 2006) with some compromises in the data density and mass range.

Since GC × GC generates large quantities of data, appropriate software tools become very important, not only for data acquisition but also visualization and interpretation. Depending on the data itself, various operations may be required, typically including background removal, mass spectral deconvolution, peak finding, combination of modulated peaks, peak height (area, or other characteristics) computation, identification of the found peaks by the comparison with mass spectral databases (or custom libraries), and finally export of analysis report (Reichenbach et al., 2004).

2.3 Application of GC × GC and its current trends

Since its introduction in 1991 (Liu & Philips, 1991), the GC × GC technique has gone through the years of rapid development. Today, GC × GC is widely used in many diverse areas which cover an interesting variety of applications. In general, the usage possibilities of GC × GC can be divided into three areas – fingerprints of very complex matrices, target analyses, and identification of unknown compounds. An excellent review on the GC × GC applications written by Adahchour et al. maps the usage of the technique from its introduction till 2008 in following fields – petrochemical products, environmental studies (soils and sediments, airs and aerosols, cigarette smoke), organohalogen compounds, food analysis (fats and oils, essential oils, alcoholic beverages), and also biological samples (Adahchour et al., 2008).

A significant progress can be noted in biosciences applications. In the field of human metabolomics, it covers e.g. metabolomic profiling of infant urine (Kouremenos et al., 2010; Wojtowicz et al., 2010), biomarker discovery of diabetes mellitus (Li et al, 2009), defining the "metabolome" of psychical disorders like schizophrenia (Oresic et al., 2011), sterol analysis (Mitrevski et al., 2008), analyses of tumorogenic cells (Paskanti et al., 2010), identification of anabolic agents in doping control (Mitrevski et al., 2010), or enatioselective analyses (Wadhier et al., 2011).

As regards the recent developments in GC × GC, the utilization of new stationary phases such as ionic liquids (Zapadlo et al., 2011), development of new modulators, e.g. (Panic et al., 2011), and improvement of data handling and evaluation (Wang et al., 2010; Kim et al., 2011; Koek et al., 2011) should be mentioned.

3. Derivatization via chloroformates

Analytes derivatization is employed in many analytical methods that utilize GC as a final step. The dominant reasons for derivatization in GC are either to increase analytes volatility, to improve their chromatographic behaviour by decreasing of polarity or to increase the detector sensitivity of the target analytes. The group of silylation procedures is by far the dominant derivatization methods. Silylation is almost universal technique and the silyl groups increase the total ion current which leads to increase the sensitivity using positive ion MS (Wells, 1999).

On the other hand, this derivatization procedure has several disadvantages. First, it is time-consuming and requires elevated temperature. Further, the silylation reactions result in the formation of many artifacts as well as forming of multiple peaks for the same compound or the presence of unexpected peaks in the chromatogram. Also, in the electron impact mass spectra of silylated compounds, the non-specific masses belonging to the silyl group prevail, while the molecular and other characteristic ions have low intensity, which complicates spectra interpretation. Finally, it is also important to mention the instability of the silylated compounds and its sensitivity even to the traces of moisture which makes sample preparation and storage more demanding (Little, 1999; Ong et al., 2010).

As a promising alternative derivatization technique, an indirect alkylation via chloroformates appears. The sample preparation procedure is as follows. A portion of pyridine (serving as a catalyst) and an alcohol (to form esters) is added to the sample that is present in basic-aqueous environment. The reaction itself starts with the addition of an appropriate alkyl chloroformate. The reaction is fast (seconds) and no heating is needed. An illustration of derivatization reaction is shown in Figure 2. After the reaction, the derivates are directly extracted into the water-immiscible organic solvent (chloroform, isooctane) that can be immediately (or after drying by e.g. anhydrous sodium sulfate) injected. In this way a biological material (plasma, urine, cell extract) can be derivatized without any pretreatment.

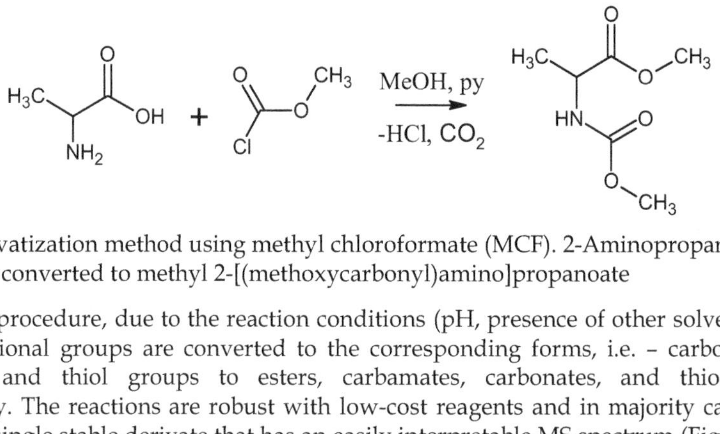

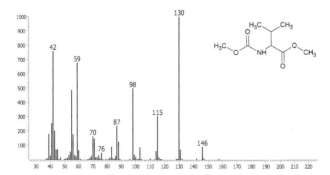

Fig. 2. Derivatization method using methyl chloroformate (MCF). 2-Aminopropanoic acid (alanine) is converted to methyl 2-[(methoxycarbonyl)amino]propanoate

Using this procedure, due to the reaction conditions (pH, presence of other solvents), many polar-functional groups are converted to the corresponding forms, i.e. – carboxy, amino, hydroxyl, and thiol groups to esters, carbamates, carbonates, and thiocarbonates, respectively. The reactions are robust with low-cost reagents and in majority cases (>95%) produce a single stable derivate that has an easily interpretable MS spectrum (Figure 3).

Fig. 3. MS spectrum of MCF derivate of valine

4. Material and methods

4.1 Chemicals and reagents

Internal standards (4-phenylbutyric acid for TMS, norvaline for MCF), ethoxyamine hydrochloride, methanol (HPLC grade), pyridine (p.a.), methyl-chloroformate (MCF, 99%, for GC), chloroform (≥99.9%, for HPLC), trypsin/EDTA (10×), N-methyl-N-trimethylsilyltrifluoroacetamide (MSTFA), containing 1% of trimethylchlorosilane, Dulbecco's Modified Eagle's Medium, and amphotericin were purchased from Sigma–Aldrich (St. Louis, USA). Fetal bovine serum was from PANBiotech (Aidenbach, Germany), sodium chloride solution (0.9%) from B Braun (Melsungen, Germany). Other chemicals for sample preparation, i.e. hydrochloric acid, sodium chloride, sodium hydroxide, anhydrous sodium sulfate, sodium bicarbonate, ethyl acetate (p.a.), and acetone (p.a.) were supplied by LACH-NER (Neratovice, Czech Republic). 5-Fluorouracil (250 mg in 5 mL) was from EBEWE Pharma (Unterach, Austria). Standard mixture of 32 amino acids and dipeptides (200 µmol/L) was from the EZfaast kit (Phenomenex, USA). All chemicals and reagents were of analytical grade or higher.

4.2 Samples and their preparation

4.2.1 Urine samples

For the analysis of organic acids, urine samples were acidic-extracted into ethyl acetate, ethoxymated and silylated by MSTFA. We analyzed 10 healthy urines and urines from patients with inherited metabolic disorder. For more detail see (Wojtowicz et al., 2010).

Analytes response was normalized to the concentration of creatinine measured by the common Jaffe rate method.

4.2.2 Fibroblasts

Human skin fibroblasts were cultured by standard protocol (Dulbecco's Modified Eagle's Medium, supplemented with 10% of fetal bovine serum and amphotericin 100 µg/mL, 37 °C, 5% CO_2, 25 cm² flasks) to confluence. The cells were harvested by quenching or trypsinization (see below).

Quenchinq procedure: The cells were quenched by spraying-out of 20 mL of 60% aqueous methanol (v/v) pre-cooled to -50 °C using plastic syringe with bent needle. The flasks with quenched cells were kept on dry ice and the cells were extracted with 1 mL of cold (-50 °C) methanol solution (80%, v/v) while scraping. The cell debris in the methanol solution was drained out with pipette and another 1 mL of cold extraction solution to wash the flask was used. Both methanol fractions were combined, sonicated (1 min), and centrifuged (1800g, 5 min) to remove the cell pellet and the supernatant was freeze-dried, silylated or alkylated via MCF. For the experiment with 5-fluorouracil (5-FU), the growing medium was supplemented with the drug (50 µmol/L) for 6, 24, and 48 hours before quenching procedure, respectively.

Trypsinization: Before trypsinization the cells were washed twice with 0.9% sodium chloride solution and then the trypsin/EDTA solution was added. After 2 min, trypsin was deactivated by adding of 5 mL of a cultivation medium. The cell suspension was centrifuged

(260g, 5 min) and the pellet was extracted twice by 1 mL of 80% methanol. Combine extracts were freeze-dried.

Freeze-dried intracellular metabolite extracts were derivatized for analysis by two-stage silylation procedure based on the method described previously (Koek et al., 2006). The dry extracts were derivatized with ethoxyamine hydrochloride (10 μL, 56 mg/mL in pyridine) and 20 μL of pyridine for 60 min at 40 °C. Subsequently, the extracts were silylated for 50 min at 40 °C with 40 μL of MSTFA and 30 μL of pyridine.

The MCF derivatization was as follows: Dry metabolite extract was dissolved in 100 μL of water and 200 μL of sodium hydroxide (0.5 mol/L). The mixture was transferred into the glass tube containing 20 μL of internal standard norvaline (0.1 mmol/L). After, 200 μL of methanol and 50 μL of pyridine as a catalyst were added and the mixture was briefly vortexed. The derivatization reaction was started by adding 20 μL of MCF and the mixture was then vortexed for 30 s. Another 20μL portion of MCF was added again followed by shaking for 30 s. To separate the MCF derivatives from the reactive mixture a 300 μL of chloroform was added and shaken 10 s followed by the addition of 300 μL of sodium bicarbonate solution (50 mmol/L) and shaking for an additional 10 s. For better layering, the tubes were centrifuged (1000g, 1 min). The upper aqueous layer was discarded and the chloroform phase was dried by adding a small portion of anhydrous sodium sulfate. The dry organic solution was transferred to a GC vial with an insert which was tightly capped and then analyzed.

4.2.3 Plasma samples

The control and patient plasma samples were from infants from routine diagnostic processes performed in the laboratory of authors. The diagnoses had been previously confirmed by biochemical, enzyme or molecular-genetic analyses in all the patients. We analyzed 10 control samples and 19 samples with amino acids defects (phenylketonuria, PKU, maple syrup urine disease, MSUD, tyrosinemia I, TYR I, homocystinuria, HCYS, carbamoyl phosphate synthetase deficiency, CPS, ornithine transcarbamylase deficiency, OTC, and non-ketotic hyperglycinemia, NKH) (Janeckova et al., 2011).

For the MCF derivatization procedure, 50 μL of plasma were pipetted into the glass tube containing 20 μL of internal standard norvaline (0.1 mmol/L). After addition of 200 μL of sodium hydroxide (0.5 mol/L), the derivatization procedure was the same as in the case of fibroblasts.

4.3 Optimized analyses conditions

A Pegasus 4D system consisting of an Agilent 7890A gas chromatograph (Agilent Technologies, Palo Alto, USA), a MPS2/CIS4/ALEX system (Gerstel, Mülheim an der Ruhr, Germany), and a Pegasus HT time-of-flight mass spectrometer (LECO Corporation, St. Joseph, USA) was used. The GC × GC system employed a dual-stage, quad-jet modulator and a secondary oven, both built-in to the Agilent GC oven. A consumable-free option of the modulator was employed. Compressed air was used for both hot and cold modulation jets. For the hot jets the air was resistively heated, while for the cold jets the air passed through a moisture filter and was cooled by immersion cooling (−80 °C).

The nonpolar/polar (BPX5, 30 m × 0.25 mm × 0.25 µm & BPX50, 2.0 m × 0.1 mm × 0.1 µm, both Supelco) column arrangement with a modulation on the first column was chosen. The columns were connected using a SilTite Mini Union (SGE, Ringwood, Australia). The oven temperature program differs due to the derivatization procedure – TMS: primary oven temperature: 40 °C (2 min), 8 °C /min to 155 °C (0.2 min), 10 °C/min to 255 °C (0.20 min), and isocratically 300 °C (5 min); secondary oven temperature: +5 °C above the primary oven temperature; modulator temperature: +50 °C above the primary oven temperature; modulation period: 3 s (hot pulse 0.6 s), solvent delay 650 s; MCF: 60 °C (1.5 min), 15 °C/min, 300 °C (5 min), secondary oven temperature: +10 °C above the primary oven temperature; modulator temperature: +30 °C above the primary oven temperature; modulation period: 4 s (hot pulse 0.8 s), solvent delay 300 s.

Other conditions were as follows:: carrier gas: helium at the corrected constant flow 1 mL/min; splitless injection (1 min, 250 °C), 0.2–1 µL due to the application; TOF-MS: electron ionization (−70 eV); ion source temperature: 250 °C; acquired mass range: m/z 35-550; acquisition rate: 100 spectra/s; detector voltage: −1500 V; transfer line temperature: 250 °C.

ChromaTOF software v. 4.24 (LECO Corporation, USA) was used for system control, data acquisition, and data processing. The NIST/EPA/NIH Mass Spectral Library (2008) was used for tentative identification of compounds, with confirmations by retention indices comparisons. System Gerstel was controlled by Maestro software v. 1.3 (Mülheim an der Ruhr, Germany).

4.4 Data processing and quantification

An automated data processing based on the so called "Reference" was performed by the ChromaTOF software. The analyte concentration was calculated from the deconvoluted total ion current (DTIC) peak area.

Multivariate statistical data analyses (Principal Component Analysis – PCA, Hierarchical Cluster Analysis – CA) were performed using Statistica 8.0 (www.statsoft.com). A heat-map was created using the Cluster v. 3.0 software (http://bonsai.hgc.jp) and visualized by the TreeView v. 1.1.5 software (http://jtreeview.sourceforge.net).

4.4.1 Deconvoluted total ion current

Reference materials of rarely occurring pathological metabolites are not always easily available and/or their cost is considerably high. Therefore, it is a common practice to use the total ion chromatogram (TIC) signal for quantification along with the internal standard use. This approach was used also in this work. However, TIC quantification can overestimated results in case of chromatographic coelutions, which cannot be completely avoided even with GC × GC.

This obstacle was overcome in our work by using deconvoluted total ion chromatogram (DTIC) for the quantification of the analytes. Deconvolution is a mathematical algorithm that is based on the absence of spectral skew and faster acquisition rates for peak apex definition in TOF-MS data. This algorithm mathematically separates mass spectra of compounds that chromatographically co-elute. In addition to producing deconvoluted

spectra, the ChromaTOF software also allows the calculation of DTIC peak area.
Deconvoluted TIC is the portion of TIC area corresponding to a particular analyte in a co-
elution. An example is shown in Figure 4.

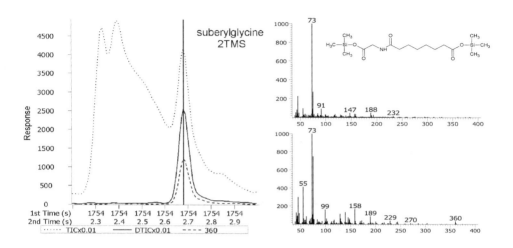

Fig. 4. Creation of DTIC for suberylglycine 2TMS (analysis of urine extract). On the right the
raw peak apex spectrum (up) and the spectrum after deconvolution (down) are shown. The
similarity between the deconvoluted spectra and the spectra from the library was 895.

4.4.2 Using the reference

When a classical GC-MS is used, the chromatogram is typically reviewed manually by
plotting characteristic masses in the segment of the expected retention time. To avoid such a
time-consuming procedure, a ChromaTOF feature called "Reference" was applied. This
procedure consisted of the following steps:

i.   The sample is processed using a general peak finding method (e.g. S/N 200).
ii.   Peaks of interest are exported to the Reference, which is a set of information containing
      the retention times and mass spectrum of each analyte, among other data. Criteria such
      as the retention time window-width in both dimensions and minimum spectral match
      are defined by the user.
iii. The Reference is applied to target search for each analyte in the unknown sample and
      for the quantification of positively identified analytes.
iv.   If some new interesting analyte is found in newly processed sample, it can be added to
      the existing Reference.

The retention time tolerances were determined based on the repeatability of retention times.
The appropriate relative standard deviations (n=10) of the retention times of selected urine
metabolites were under 0.17% and 1.74% for ^{1}D and ^{2}D, respectively (Wojtowicz et al., 2010).

Using this approach, two References have been made – one for TMS and one for MCF-based
derivatization which will be described in following sections.

5. Results and discussion

5.1 Analysis of samples derivatized by silylation

EZfaast standard solution (contains 32 amino acids and dipeptides), 10 healthy and 14 pathological urine extracts, and 15 quenched fibroblast extracts derivatized by TMS were analyzed by a GC × GC-TOF-MS method and subjected to data processing procedure. After manual inspection of these peaks and confirmation of their identity by means of retention indices, a Reference was created from these compounds. This TMS compounds Reference contains 268 analytes at this moment.

The analyzed compounds are organic acids (e.g. suberate, malonate, palmitate) and their derivates (e.g. 2-oxoisovalerate, lactate, vanillylmandelate, mevalonolactone), amino acids (e.g. glycine, valine, phenylalanine) and their derivates (e.g. 3,4-dihydroxyphenylalanine, cystine, 4-hydroxyproline), N-acetylated amino acids (e.g. N-acetyl-tyrosine, N-acetyl-lysine), amines (e.g. ethanolamine, butanediamine), pyrimidines and purines (thymin, uracil, ureate), sugars (e.g. glucose), acylglycines (hippurate, propionylglycine, hexanoylglycine), and others (e.g. succinylacetone, indoleacetate, urea, cholesterol, furan-2,5-dicarboxylate).

5.1.1 Analysis of urine acidic extract in relation to organic acidurias

A large subgroup of Inherited Metabolic Disorders called organic acidurias is characteristic by increased levels of organic acids in urine or the presence of pathological ones not appearing in healthy urine. The diagnosis of organic acidurias is commonly performed by the analysis of urine after acidic extraction. Besides organic acids, some other types of metabolites also serving as pathological markers, are extracted.

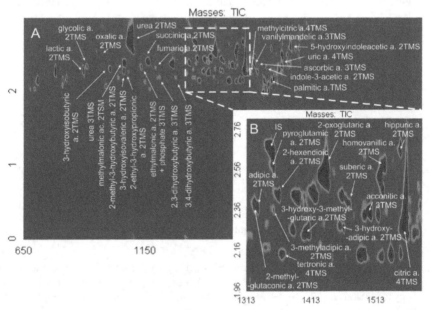

Fig. 5. GC × GC-TOF-MS contour plot from the analysis of healthy urine (A) and enlarged part (B)

Automated data processing with peak finding above S/N 200 was applied to the data, which resulted in the detection of 1353-3420 peaks in the set of studied urines. After sorting out GC column bleed peaks and the peaks belonging to the derivatization reagent, some 60% of the peaks remained. They are naturally occurring metabolites, pathological metabolites, nutrition artifacts, and drug artifacts, but most of the compounds are still not clearly identified. Our TMS Reference contains 153 identified and confirmed urine chemical species (from all 268 in the TMS Reference). In the Figure 5 is shown a 2D contour plot of healthy urine.

To evaluate the newly developed GC × GC method as well as the data processing strategy described above, we selected external quality control samples (ERNDIM, http://www.erndim.unibas.ch/) and one sample from an asymptomatic patient with medium-chain acyl-CoA dehydrogenase deficiency, who had been diagnosed by neonatal screening sixth days after birth. Since the patient was in a non-crisis state, many biochemical markers of the disease were in normal levels, so he could be missed by GC-MS. Using GC × GC-TOF-MS with the TMS Reference we found both hexanoylglycine peaks, which, although present at low concentrations, are obviously pathological markers and confirm the presence of the disease. Figure 6 illustrates the major benefit of our approach – separation of main markers of the disease from an excess of naturally occurred metabolite. For more details see (Wojtowicz et al., 2010).

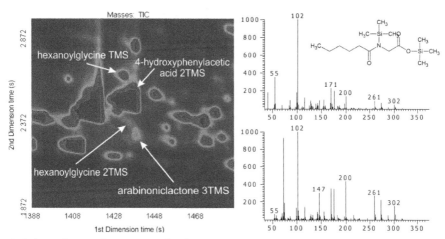

Fig. 6. Enlarged part of a contour plot from the analysis of urine from a patient with medium-chain acyl-CoA dehydrogenase deficiency. Two peaks of hexanoylglycine derivatives (1 and 2TMS) are resolved from a large peak of 4-hydroxyphenylacetic acid 2TMS. The deconvoluted spectrum (up) and the library hit belonging to the 2TMS derivate are shown (similarity 798)

5.1.2 Analysis of cultured human skin fibroblasts – comparison of trypsin treatment vs. quenching

The influence of most of external factors on the cell metabolomes is minimized in a culture where a defined extracellular environment takes place. To collect reliable metabolome data sets, culture and sampling conditions are crucial. The accurate analysis of intracellular metabolites requires a reliable sampling technique. Metabolites are generally labile species and their reactions occur on a time scale much shorter than that of large molecule synthesis or degradation, with substantial changes in small molecule concentrations possible during a time scale of seconds (Canelas et al., 2008).

The metabolomic analysis of adherent cell cultures presents a complex challenge, mostly given by limited sample sizes. The rapid quenching of the intracellular metabolism, simultaneously with the considerable removal of superabundant growing medium, are the main prerequisites. As a result of a known rapid turnover of intracellular metabolites, the sampling process should be very fast and cause minimal metabolome loss as a result of cell leakage. In this study, we compared trypsinization as the classical procedure for adherent cell harvesting (Dettmer et al., 2011) with the simple quenching procedure developed in our laboratory.

A set of studied fibroblasts has been analyzed by a GC × GC-TOF-MS method and the data was subjected to automated data processing with peak finding above S/N 200, which resulted in the detection of 431-601 peaks. Similarly as described in the section 5.1, our TMS Reference contains 72 identified and confirmed intracellular chemical species, from all 268 in the TMS Reference (Table 1). In the Figure 7 a 2D contour plot of analysis of the intracellular metabolites extracted from spray-quenched human cultured skin fibroblasts is shown.

Acetate	2,3-Dihydroxybutyrate	4-Hydroxyproline	2-Oxovalerate
Aconitate	3,4-Dihydroxybutyrate	Cholesterol	Palmitate
Adipate	Glucose	Isoleucine	Panthotenate
Alanine	Glutamine	Itaconate	Phenylalanine
2-Aminoadipate	Glutarate	Lactate	Phosphate
2-Aminobutyrate	Glycerol	Laurate	Picolinate
4-Aminobutyrate	Glycerate	Leucine	Proline
Aspargine	Glycine	Lysine	Pyruvate
Aspartate	Fumarate	Malate	Serine
Azelaic acid	Hexenedioate	Maleate	Succinate
β-Alanine	Hippurate	Mesaconate	Stearate
Benzoate	Histidine	Methionine	Threonate
Capric acid	2-Hydroxyisobutyrate	Myo-inositol	Threonine
Citraconate	3-Hydroxyisobutyrate	Oleate	Tryptophan
Citrate	2-Hydroxybutyrate	Ornithine	Tyrosine
Cystathionine	2-Hydroxyisobutyrate	2-Oxoglutarate	Uracil
Cysteine	2-Hydroxyglutarate	2-Oxo-3-methylvalerate	Urea
Diphosphate	3-Hydroxyhippurate	5-Oxoproline	Valine

Table 1. The list of 72 unique metabolites identified and confirmed in extracts of human cultured skin fibroblasts by GC × GC-TOF-MS

In order to evaluate the effect of quenching vs. trypsinization, a cellular metabolome was determined in fibroblast cultures (n=6 for each method) from a single cell line. Data from analyses were corrected to percentages of a sum. The obtained compositional data were statistically analyzed. The geometric means for the single compounds were calculated. In order to be able to deal with Gaussian distribution of single compounds for variance analysis the data were transformed by the following equation $[1/sqrt(2)*log(x/(1-x))]$, that represents a special case of so-called isometric log-ratio transformation (Egozcue et al., 2003). The results from trypsin and quenching approaches were compared as the natural

logarithms of ratios of means and variations (Figure 8) and multivariate statistics – PCA and
CA (Figure 9). It is clearly visible that the quenching technique substantially affects the
concentrations of a number of metabolites. Several metabolites (e.g. citrate, lysine) differ by
an order of magnitude. From the variations it is also evident that sample preparation by
means of conventional trypsinization provides substantially more variable data in
comparison to quenching by means of spraying.

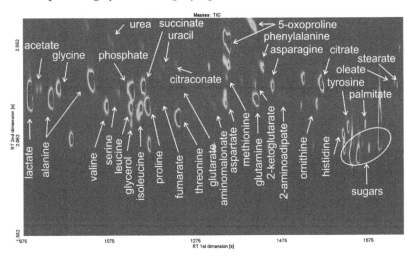

Fig. 7. GC × GC contour plot of analysis of the intracellular metabolites extracted from
spray-quenched human cultured skin fibroblasts

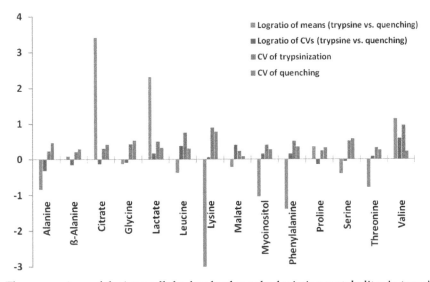

Fig. 8. The comparison of the intracellular levels of mostly deviating metabolites in trypsin
treated and quenched cells. Natural logarithms of ratios of means and coefficients of
variation (CV) are shown

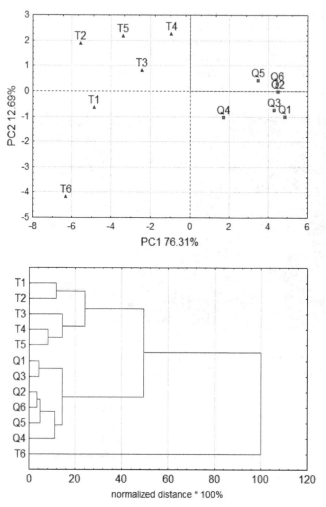

Fig. 9. PCA projection of the cases and CA dendrogram (complete-linkage, Euclidian distance) for the comparison of two cell harvesting methods (based on the listed 72 metabolites). T1-T6 trypsinization, Q1-Q6 quenching

5.2 MCF approach

EZfaast standard solution (contains 32 amino acids and dipeptides), 20 intracellular extracts from quenched fibroblasts, and 20 plasma samples derivatized by MCF were analyzed by a GC × GC-TOF-MS method and subjected to data processing procedure similarly as described in previous sections and the MCF Reference has been made.

Our MCF Reference currently contains 185 analytes – they are mostly organic acids and their derivates (the same as in the TMS approach) and compounds containing amino group (the exception is arginine because of the thermal instability of its MCF derivative that carries a free guanidine group). Sugars are another group of compounds that cannot be detected by this approach. But that can be an advantage while e.g. cell extracts usually contains an excess of sugars that can make the obtained chromatograms not easily interpretable and some important markers can be masked. Table 2 shows some MCF-Reference based data (2D retention characteristics, unique masses – the characteristic mass identified by the ChromaTOF software, and three main ions from the MS spectra).

Amino acid	1D; 2D RT [s]	RI	Unique mass	Derivate Mr	Base peak	P2/%	P3/%
Alanine	476; 2.44	1142	102	161	102	42/65	59/63
Glycine	484; 2.50	1152	71	147	88	44/62	56/55
Sarkosine	504; 2.39	1179	102	161	102	42/78	59/49
Valine	556; 2.38	1255	130	189	130	42/88	59/82
2-Aminobutyric acid	564; 2.40	1268	88	175	88	44/51	56/41
Leucine	600, 2.38	1327	88	203	88	59/46	43/43
Threonine	612; 2.58	1348	115	191	115	59/56	42/37
Proline	632; 2.64	1383	128	187	41	128/82	41/37
Asparagine	636; 2.64	1390	127	204	42	127/75	56/72
Aspargic acid	668; 2.56	1450	160	219	42	59/88	160/69
Glutamic acid	732; 2.54	1578	114	233	114	59/76	42/70
Methionine	740; 2.64	1596	61	221	61	59/48	115/40
4-Hydroxyproline	744; 2.71	1604	144	203	144	41/56	59/47
2-Aminoadipic acid	784; 2.51	1693	114	247	114	55/70	59/69
Cysteine	784; 2.69	1694	59	193	59	42/77	44/59
Phenylalanine	796; 2.68	1720	42	237	42	91/65	59/57
Glutamine	840; 3.00	1872	84	218	84	44/42	59/41
Ornithine	884; 2.71	1937	128	262	128	42/51	59/51
Lysine	920; 2.76	2033	142	218	142	59/82	44/55
Histidine	940; 3.00	2088	59	227	59	81/76	42/73
Tyrosine	980; 3.00	2203	121	252	121	59/90	42/73
Tryptophan	1052; 3.78	2424	130	276	130	77/18	51/11

Table 2. Table of 22 selected amino acids derivatized via MCF presented in our Reference. 1D and 2D RT – retention time in first and second dimension, RI – retention index (calculated on the basis of absolute RT), P2 and P3/% - second and third most abundant peak/percentage of its intensity to the base peak

5.2.1 Analysis of cultured human skin fibroblasts – effect of cultivation with 5-FU

Since its synthesis, 5-FU has become one of the most widely used anticancer drugs for a variety of common malignancies, including cancers of the colon, breast, skin, and head and neck. 5-FU has been used as a component of both first-line chemotherapy regimens and in salvage regimens. Despite extensive clinical experience with 5-FU and its effective antitumor activity, many concerns remain about the optimal use of this agent.

5-FU is a prodrug, which is subject to both anabolism and catabolism. The cytotoxic activity of 5-FU depends on its anabolism to nucleotides, which exert their effects through inhibition of thymidylate synthase activity or incorporation into RNA and/or DNA. The catabolism of 5-FU has been better understood only in recent years. The products of 5-FU catabolism have been linked to several 5-FU toxicities, including neurotoxicity (Grem, 2000; Kuhn, 2001).

This study was to show the differences in the fibroblasts metabolome after cultivation with addition of 5-FU against the non-treated controls (n=3 for each 6, 24, and 48h treatment, and without treatment, respectively), all from a single cell line. Data processing with peak

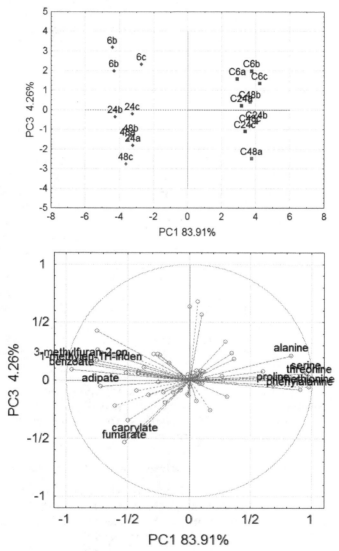

Fig. 10. PCA projection of the cases (up) and the loading plot (down) for the comparison of the effect of 5-FU on the fibroblast cultivation. Triplicates (a, b, c) for 6, 24 or 48 hours of cultivation with presence of 5-FU and controls (C) are shown

finding above S/N 200 was applied to the data, which resulted in the detection of 393-451 peaks in the set of studied fibroblasts. Our MCF Reference contains 78 identified and confirmed intracellular chemical species (from all 185 in the MCF Reference). Data from analyses were corrected to percentages of a sum, normalized to the unit standard deviation, and statistically analyzed through PCA (Figure 10). Figure 11 shows the differences for the most deviating metabolites. From the presented graphs it is clearly visible that cultivation with 5-FU influenced fibroblasts' metabolism.

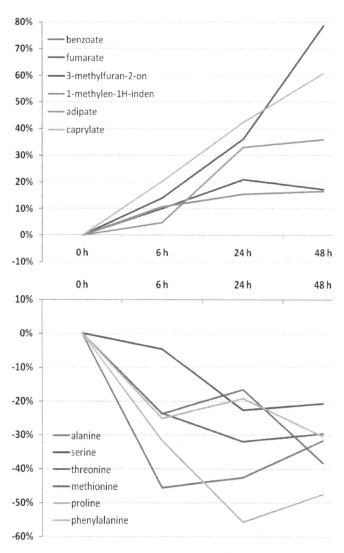

Fig. 11. Graph of the most increasing (up) and decreasing (down) intracellular metabolites influenced by 5-FU. Differences are quantified as percentages of influenced response to the non-influenced ones.

5.2.2 Analysis of human plasma in relation to metabolic disorders

In this work we focused on the diagnosis of Inherited Metabolic Disorders in plasma samples using a targeted metabolomic approach by GC × GC-TOF-MS.

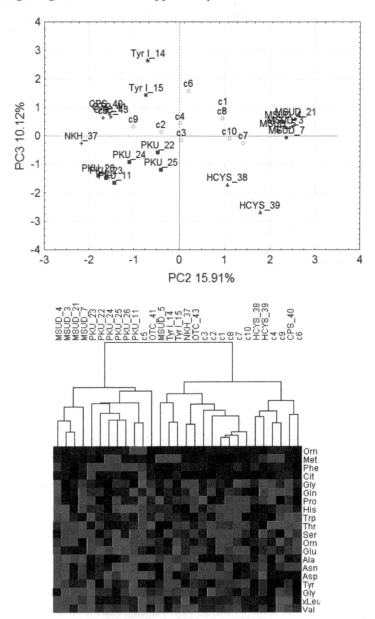

Fig. 12. PCA projection of the cases (up) and a heat map (down) visualized for selected amino acids from the analysis of human plasma

Automated data processing with peak finding above S/N 50 was applied to the data, which resulted in the detection of 408-594 peaks in the set of studied plasma samples. Our MCF Reference contains 65 identified and confirmed plasma chemical species (from all 185 in the MCF Reference).

Prior to statistical analysis the centred logratio (clr) transformation was applied. Data were evaluated using PCA and CA based on hierarchical clustering with a complete-linkage Euclidian distance method (visualized as a heat map) – Figure 12.

All the patients' samples were discriminated from the controls by appropriate metabolites in the PCA analysis. Patients with identical disease were recognized using the PCA approach and also clustered together.

6. Conclusion

This technique has been shown to be very powerful for the purpose of comprehensive sample profiling. On selected samples we demonstrated that higher separation power of GC × GC can help in removing co-elutions occurring in one dimensional approach. GC × GC is a valuable tool in metabolomic analysis of many biological matrices and enables diagnosing metabolic disorders. The great benefit is data processing that can be fully automated, what strongly simplify operator's effort and increases the sample throughput.

7. Acknowledgement

Infrastructural part of this project (Institute of Molecular and Translational Medicine) was supported from the Operational programme Research and Development for Innovations (project CZ.1.05/2.1.00/01.0030).

8. References

Adahchour, M., Beens, J. & Brinkman, U. A. T. (2008). Recent developments in the application of comprehensive two-dimensional gas chromatography. *Journal of Chromatography A*, Vol. 1186, No. 1-2, pp. 67-108, ISSN 0021-9673

Adahchour, M., Beens, J., Vreuls, R. J. J. & Brinkman, U. A. T. (2006). Recent developments in comprehensive two-dimensional gas chromatography (GC×GC). *TrAC Trends in Analytical Chemistry*, Vol. 25, No. 6, pp. 540-553, ISSN 01659936

Beens, J. & Brinkman, U. A. T. (2005). Comprehensive two-dimensional gas chromatography - a powerful and versatile technique. *The Analyst*, Vol. 130, No. 2, pp. 123-127, ISSN 1364-5528

Blumberg, L. M., David, F., Klee, M. S. & Sandra, P. (2008). Comparison of one-dimensional and comprehensive two-dimensional separations by gas chromatography. *Journal of Chromatography A*, Vol. 1188, No. 1, pp. 2-16, ISSN 0021-9673

Boernsen, K. O., Gatzek, S. & Imbert, G. (2005). Controlled protein precipitation in combination with chip-based nanospray infusion mass spectrometry. An approach for metabolomics profiling of plasma. *Analytical Chemistry*, Vol. 77, No. 22, pp. 7255-7264, ISSN 0003-2700

Canelas, A. B., Ras, C., ten Pierick, A., van Dam, J. C., Heijnen, J. J. & Van Gulik, W. M. (2008). Leakage-free rapid quenching technique for yeast metabolomics. *Metabolomics,* Vol. 4, No. 3, pp. 226-239, ISSN 1573-3882

Dettmer, K., Nurnberger, N., Kaspar, H., Gruber, M. A., Almstetter, M. F. & Oefner, P. J. (2011). Metabolite extraction from adherently growing mammalian cells for metabolomics studies: optimization of harvesting and extraction protocols. *Analytical and Bioanalytical Chemistry,* Vol. 399, No. 3, pp. 1127-1139, ISSN 1618-2650

Edwards, M., Mostafa, A. & Gorecki, T. (2011). Modulation in comprehensive two-dimensional gas chromatography: 20 years of innovation. *Analytical and Bioanalytical Chemistry,* DOI 10.1007/s00216-011-5100-6, ISSN 1618-2650

Egozcue, J.J., Pawlowsky-Glahn, V., Mateu-Figueras G. & Barceló-Vidal, C. (2003). Isometric logratio transformations for compositional data analysis. *Mathematical Geology,* Vol. 35, No. 3, pp. 279-300, ISSN 0882-8121

Gorecki, T., Harynuk, J. & Panic, O. (2004). The evolution of comprehensive two-dimensional gas chromatography (GC x GC). *Journal of Separation Science,* Vol. 27, No. 5-6, pp. 359-379, ISSN 1615-9306

Grem, J. L. (2000). 5-Fluorouracil: forty-plus and still ticking. A review of its preclinical and clinical development. *Investigation New Drugs,* Vol. 18, No. 4, pp. 299-313, ISSN 0167-6997

Griffiths, J. R. & Chung, Y. L. (2008). Using metabolomics to monitor anticancer drugs. *Oncogenes Meet Metabolism: From Deregulated Genes to a Broader Understanding of Tumour Physiology,* Vol. 4, pp. 55-78, ISSN 0947-6075

Husek, P. & Simek, P. (2006). Alkyl chloroformates in sample derivatization strategies for GC analysis. Review on a decade use of the reagents as esterifying agents. *Current Pharmaceutical Analysis,* Vol. 2, No. 1, pp. 23-43, ISSN 1573-4129

Janeckova, H., Hron, K., Wojtowicz, P., Hlidkova, E., Baresova, A., Friedecky, D., Zidkova, L., Hornik, P., Behulova, D., Prochazkova, D., Vinohradska, H., Peskova, K., Bruheim, P., Smolka, V., Stastna, S. & Adam, T. (2011). Targeted metabolomic analysis of plasma samples for the diagnosis of inherited metabolic disorders. *Journal of Chromatography A,* DOI 10.1016/j.chroma.2011.09.074, ISSN 1873-3778

Kim, S., Fang, A. Q., Wang, B., Jeong, J. & Zhang, X. (2011). An optimal peak alignment for comprehensive two-dimensional gas chromatography mass spectrometry using mixture similarity measure. *Bioinformatics,* Vol. 27, No. 12, pp. 1660-1666, ISSN 1367-4803

Koek, M. M., Muilwijk, B., van der Werf, M. J. & Hankemeier, T. (2006). Microbial metabolomics with gas chromatography/mass spectrometry. *Analytical Chemistry,* Vol. 78, No. 4, pp. 1272-1281, ISSN 0003-2700

Koek, M. M., Muilwijk, B., van Stee, L. L. & Hankemeier, T. (2008). Higher mass loadability in comprehensive two-dimensional gas chromatography-mass spectrometry for improved analytical performance in metabolomics analysis. *Journal of Chromatography A,* Vol. 1186, No. 1-2, pp. 420-429, ISSN 0021-9673

Koek, M. M., van der Kloet, F. M., Kleemann, R., Kooistra, T., Verheij, E. R. & Hankemeier, T. (2011). Semi-automated non-target processing in GC x GC-MS metabolomics analysis: applicability for biomedical studies. *Metabolomics,* Vol. 7, No. 1, pp. 1-14, ISSN 1573-3890

Kouremenos, K. A., Pitt, J. & Marriott, P. J. (2010). Metabolic profiling of infant urine using comprehensive two-dimensional gas chromatography: Application to the diagnosis of organic acidurias and biomarker discovery. *Journal of Chromatography A*, Vol. 1217, No. 1, pp. 104-111, ISSN 1873-3778

Kuhn, J. G. (2001). Fluorouracil and the new oral fluorinated pyrimidines. *The Annals of Pharmacotherapy*, Vol. 35, No. 2, pp. 217-227, ISSN 1060-0280

Li, X., Xu, Z., Lu, X., Yang, X., Yin, P., Kong, H., Yu, Y. & Xu, G. (2009). Comprehensive two-dimensional gas chromatography/time-of-flight mass spectrometry for metabonomics: Biomarker discovery for diabetes mellitus. *Analytica Chimica Acta*, Vol. 633, No. 2, pp. 257-262, ISSN 1873-4324

Liu, Z. & Phillips, J. B. (1991). Comprehensive Two-Dimensional Gas Chromatography using an On-Column Thermal Modulator Interface. *Journal of Chromatographic Science*, Vol. 29, No. 6, pp. 227-231, ISSN 0021-9665

Little, J. L. (1999). Artifacts in trimethylsilyl derivatization reactions and ways to avoid them. *Journal of Chromatography A*, Vol. 844, No.1-2, pp. 1-22. ISSN 0021-9673

Mitrevski, B. S., Brenna, J. T., Zhang, Y. & Marriott, P. J. (2008). Application of comprehensive two-dimensional gas chromatography to sterols analysis. *Joutnal of Chromatography A*, Vol. 1214, No. 1-2, pp. 134-142, ISSN 0021-9673

Mitrevski, B. S., Wilairat, P. & Marriott, P. J. (2010). Comprehensive two-dimensional gas chromatography improves separation and identification of anabolic agents in doping control. *Journal of Chromatography A*, Vol. 1217, No. 1, pp. 127-135, ISSN 1873-3778

Ong, C. N., Xu, F. G. & Zou, L. (2010). Experiment-originated variations, and multi-peak and multi-origination phenomena in derivatization-based GC-MS metabolomics. *Trends in Analytical Chemistry*, Vol. 29, No. 3, pp. 269-280, ISSN 0165-9936

Oresic, M., Tang, J., Seppanen-Laakso, T., Mattila, I., Saarni, S. E., Saarni, S. I., Lonnqvist, J., Sysi-Aho, M., Hyotylainen, T., Perala, J. & Suvisaari, J. (2011). Metabolome in schizophrenia and other psychotic disorders: a general population-based study. *Genome Medicine*, Vol. 3, No. 3, pp. 19-32, ISSN 1756-994X

Panic, O., Gorecki, T., McNeish, C., Goldstein, A. H., Williams, B. J., Worton, D. R., Hering, S. V. & Kreisberg, N. M. (2011). Development of a new consumable-free thermal modulator for comprehensive two-dimensional gas chromatography. *Journal of Chromatography A*, Vol. 1218, No. 20, pp. 3070-3079, ISSN 1873-3778

Pasikanti, K. K., Norasmara, J., Cai, S. R., Mahendran, R., Esuvaranathan, K., Ho, P. C. & Chan, E. C. Y (2010). Metabolic footprinting of tumorigenic and nontumorigenic uroepithelial cells using two-dimensional gas chromatography time-of-flight mass spectrometry. *Analytical and Bioanalytical Chemistry*, Vol. 398, No. 3, pp. 1285-1293, ISSN 1618-2642

Rabinowitz, J. D., Munger, J., Bajad, S. U., Coller, H. A. & Shenk, T. (2006). Dynamics of the cellular metabolome during human cytomegalovirus infection. *PLoS Pathogens*, Vol. 2, No. 12, pp. 1165-1175, ISSN 1553-7366

Reichenbach, S. E., Ni, M. T., Kottapalli, V. & Visvanathan, A. (2004). Information technologies for comprehensive two-dimensional gas chromatography. *Chemometrics and Intelligent Laboratory Systems*, Vol. 71, No. 2, pp. 107-120, ISSN 0169-7439

Ryan, D., Morrison, P. & Marriott, P. (2005). Orthogonality considerations in comprehensive two-dimensional gas chromatography. *Journal of Chromatography A*, Vol. 1071, No. 1-2, pp. 47-53, ISSN 0021-9673

Sarker, D. & Workman, P. (2007). Pharmacodynamic biomarkers for molecular cancer therapeutics. *Advances in Cancer Research*, Vol. 96, pp. 213-268, ISSN 0065-230X

Wadhier, M. C., Almstetter, M. F., Nurnberger, N., Gruber, M. A., Dettmer, K. & Oefner, P. J. (2011). Improved enantiomer resolution and quantification of free D-amino acids in serum and urine by comprehensive two-dimensional gas chromatography-time-of-flight mass spectrometry. *Journal of Chromatography A*, Vol. 1218, No. 28, pp. 4537-4544, ISSN 0021-9673

Walsh, M. C., Brennan, L., Malthouse, J. P. G., Roche, H. M. & Gibney, M. J. (2006). Effect of acute dietary standardization on the urinary, plasma, and salivary metabolomic profiles of healthy humans. *American Journal of Clinical Nutrition*, Vol. 84, No. 3, pp. 531-539, ISSN 0002-9165

Wang, B., Fang, A. Q., Heim, J., Bogdanov, B., Pugh, S., Libardoni, M. & Zhang, X. A. (2010). DISCO: Distance and Spectrum Correlation Optimization Alignment for Two-Dimensional Gas Chromatography Time-of-Flight Mass Spectrometry-Based Metabolomics. *Analytical Chemistry*, Vol. 82, No. 12, pp. 5069-5081, ISSN 0003-2700

Weiss, R. H., Kind, T., Tolstikov, V. & Fiehn, O. (2007). A comprehensive urinary metabolomic approach for identifying kidney cancer. *Analytical Biochemistry*, Vol. 363, No. 2, pp. 185-195, ISSN 0003-2697

Wells, R. J. (1999). Recent advances in non-silylation derivatization techniques for gas chromatography. *Journal of Chromatography A*, Vol. 843, No. 1-2, 1-18. ISSN 0021-9673

Wojtowicz, P., Zrostlikova, J., Kovalczuk, T., Schurek, J. & Adam, T. (2010). Evaluation of comprehensive two-dimensional gas chromatography coupled to time-of-flight mass spectrometry for the diagnosis of inherited metabolic disorders using an automated data processing strategy. *Journal of Chromatography A*, Vol. 51, No. 1217, pp. 8054-8061, ISSN 1873-3778

Workman, P., Aboagye, E. O., Chung, Y. L., Griffiths, J. R., Hart, R., Leach, M. O., Maxwell, R. J., McSheehy, P. M. J., Price, P. M. & Zweit, J. (2006). Minimally invasive pharmacokinetic and pharmacodynamic technologies in hypothesis-testing clinical trials of innovative therapies. *Journal of the National Cancer Institute*, Vol. 98, No. 9, pp. 580-598, ISSN 0027-8874

Zapadlo, M., Krupcik, J., Kovalczuk, T., Majek, P., Spanik, I., Armstrong, D. W. & Sandra, P. (2011). Enhanced comprehensive two-dimensional gas chromatographic resolution of polychlorinated biphenyls on a non-polar polysiloxane and an ionic liquid column series. *Journal of Chromatography A*, Vol. 1218, No. 5, pp. 746-751, ISSN 0021-9673

Zhu, S. K. (2009). Effect of column combinations on two-dimensional separation in comprehensive two-dimensional gas chromatography: Estimation of orthogonality and exploring of mechanism. *Journal of Chromatography A*, Vol. 1216, No. 15, pp. 3312-3317, ISSN 0021-9673

5

Fatty Acids Analysis of Photosynthetic Sulfur Bacteria by Gas Chromatography

María Teresa Núñez-Cardona
Universidad Autónoma Metropolitana-Xochimilco,
México

1. Introduction

The fatty acid profile of bacteria has been extensively studied for taxonomic classification purposes, as bacteria in general contain particular and rare fatty acids, compared with animal and plant tissues (Purcaro et al., 2010). Lipids and fatty acid composition are quite useful for characterization of different species and groups of photosynthetic bacteria. Fatty acids can be identified by gas chromatography using standard methods (Imhoff & Caumette, 2004) because physical and chemical factors (e.g., illumination) influence bacterial fatty acid composition, culture conditions, and the culture media used for bacterial growth. Fatty acid analysis as a chemotaxonomic tool has been applied in heterotrophic bacteria, but its use in photosynthetic sulfur bacteria it is very limited. Actually, most studies on photosynthetic purple and green sulfur bacteria fatty acid composition have been done with relatively few genera and have often given information of the major fatty acids.

Purple and green sulfur photosynthetic bacteria are included in a physiological group called anoxygenic photosynthetic (or phototrophic) bacteria. They are anaerobes and do not produce oxygen from photosynthesis. They use reduced substrates, such as sulfide, hydrogen, ferrous iron and a great number of simple organic substrates, as photosynthetic donors (Imhoff, 2008). Photosynthesis in these microorganisms occurs thanks to bacteriochlorophylls, which could be type *a*, *b*, *c*, *d*, or *e*. Purple and green photosynthetic sulfur bacteria are Gram-negative and they are not phylogenetically related: purple sulfur bacteria are γ–Proteobacteria, and green sulfur bacteria are in a separated branch.

Photosynthetic Purple sulfur bacteria comprise the families Chromatiaceae and Ectothiorhodpiraceae. They are able to grow photoautotrophically and photoheterotrophically. Bacteriochlorophylls *a* or *b* and carotenes of the spirilloxhantin, rhodopinal, spheroidene and okenone series could be present as major pigments; these are located in membranes. According to Imhoff (2008), both families could be distinguished by chemotaxonomic properties through fatty acid composition.

Chlorobiaceae is a unique family of anoxygenic green photosynthetic sulfur bacteria. They are phototrophic and obligate anaerobes, and contain bacteriochlorophyll *c*, *d*, or *e* in light–harvesting complexes located in the chlorosomes; their carotenoids could be clorobactene and isorenieratene. This family comprises the genera *Chlorobium*, *Chlorobaculum*, *Prosthecochloris* and *Chloroherpeton* (Imhoff & Thiel, 2010).

Information on fatty acid composition of purple and green photosynthetic sulfur bacteria is scarce, and this chapter shows new data on these important cellular components. Additionally, it includes general information about fatty acid nomenclature, the distribution of fatty acids in prokaryotic cells, and published information on fatty acids of photosynthetic purple and green sulfur bacteria; it also describes a standardized gas chromatography technique for fatty acid analysis of these photosynthetic bacteria using collection and wild strains, the last of which were isolated from Tampamachoco, a coastal lagoon localized in Veracruz, México.

2. Fatty acids

2.1 Fatty acids nomenclature

Fatty acids are the organic compounds most widely distributed in living organisms and are included within substances called lipids. They are rarely found free in nature; generally they are bound to a large variety of molecules, of which glycerol is the most common (Ratledge & Wilkinson, 1988). Fatty acids that are ester-linked to glycerol are typical constituents of all Bacteria (Tindall et al., 2010). Lipids represent the most complex biological molecules and are crucial for cellular functions; the chemical and physical properties of these compounds determine a variety of roles in biological processes. According with Guschina & Harwood (2008) lipids can be divided into two main groups: nonpolar lipids (acyl glycerols, sterols, free fatty acids, hydrocarbons, alcohols, wax, and steryl esters) and polar lipids (phosphoglycerides, glycosyglycerides, and sphingolipids).

Chemically, fatty acids are linear chain monocarboxylated compounds, and their general formula is R-COOH, in which group R is an unbranched linear chain containing between 8 and 26 carbon atoms (in bacteria). R groups are, therefore, characterized by the size of the chain, degree of unsaturation, the geometry of the double bond position, and by the presence of other substituents. They can be described as saturated or unsaturated (monounsaturated and polyunsaturated) depending on whether group R contains a double bond or not.

The nomenclature of these acids follows the $X:Y$ pattern, where X is the total number of carbon atoms (not the total carbon atoms of the main chain), Y is the number of double bonds [e.g., they can be di-enoic (two double bonds), tri-enoic (three double bonds) and so on]. Usually they are differentiated into monounsaturated fatty acids (one double bond) or polyunsaturated (two or more double bonds). To locate the double bonds, the wZc or t method is followed in which Z indicates the number of carbon atoms in the aliphatic tail of the molecule, and c and t represent the *cis* and *trans* geometry, respectively. Branching is indicated by the *br* prefix. When the methyl group is on the penultimate carbon atom (distal from the carboxyl group) it is denoted as an *iso* fatty acid (sometimes as w-1). Localization of hydroxyl (OH), cyclopropane (*Cy*, *Cyc* or *cyclo*) or methyl (*me*) groups, usually precede the $X:Y$ formula (Ratledge & Wikinson, 1988; Haack et al., 1994).

2.2 Fatty acids in prokaryotic cells

Among living organisms, microorganisms are probably the ones with the largest diversity regarding their fatty acid composition. They present from the simplest to the polyunsaturated, as in many filamentous cyanobacteria, mycobacteria, and some eukaryotic microorganisms.

Microbial fatty acids can be completely saturated or contain several double bonds, although the presence of one or two double bonds is the most common. The site of these double bonds indicates their biosynthesis pathway. According to Scheuerbrandt & Bloch (1962), the aerobic route for the synthesis of unsaturated acids is the same for all living organisms, but the anaerobic pathway is for bacteria. These authors suggested that adopting the oxidative pathway to synthesize unsaturated fatty acids throughout evolution allowed living organisms to abandon the anaerobic route, which is the most "primitive" and yet, perhaps more complex

The linear-chain fatty acids of bacteria are usually of the same class as those found in eukaryotic organisms, being part of membrane lipids with lengths that vary between 8 and 26 carbon atoms depending on the bacterial species. Bacterial monounsaturated fatty acids are primarily of type ω7 (unsaturation is located on carbon 7 from the methyl terminal), although type ω9, which is characteristic of higher organisms, can also be found.

In polyunsaturated fatty acids, the double bond is formed by dehydrogenation of one carbon-carbon bond, and is generally located in the center of the molecule (ω7 or ω9). Both the length of the fatty acid chain and the incorporation of double bonds (Abbas & Card, 1980) participate in membrane fluidity. Hence the synthesis of longer-chain fatty acids causes the membrane to remain constant (Melchior, 1982 cited in Jantzen & Bryan, 1985). Polyunsaturated fatty acids are common in mycobacteria containing phenolic acids of 36-38 carbons. They have also been observed in marine bacteria, such as *Shewanella gelidimarina*, *Sh. hanedai*, *Sh. putrefaciens*, *Colwellia psychrerythraea*, *Flexibacter polymorphus* (Russell & Fukunaga, 1990), and in isolates of the Antarctic, in which these chemical compounds have chains of 18 to 20 carbon atoms (Nichols et al., 1997).

Dienoic fatty acids and others with a larger number of double bonds are more frequently found in eukaryotic organisms, but they have also been detected in some heterotrophic bacteria and cyanobacteria. The latter produce linoleic acid, which has also been found by Rabinowitch et al. (1993) in *Escherichia coli*, grown in both aerobic and anaerobic conditions.

The C18:2 (octadecadienoic) polyunsaturated fatty acid is produced by *Aerobacter aerogenes* and *Pseudomonas* sp., in which it contributes 18.2% and 24%, respectively, to the total of their fatty acids, and in *Serratia marcescens*, contributing only 2.6-5.8% of the total extracted (Cho & Salton, 1966). Although the γ–linolenic fatty acid, or C18:3 (6, 9, 12), is a rare fatty acid outside the animal kingdom, it is commonly found in the oils of some seeds. Initially it was described in mycelial lipids of *Phycomyces blakesleeanus* (Lechevalier & Lechevalier, 1988), and together with α–linolenic, or C18:3 (9, 12, 15), is part of the characteristic fatty acids of some cyanobacteria. The eicosapentaenoic acid (C20:5ω3) is produced by several strains of *Shewanella* in percentages from 2 to 40% of the total fatty acids (Yazawa, 1996; Brown et al., 1997; Nichols et al., 1997).

Yano et al. (1994) found docosahexaenoic acid (C22:6ω3 or DHA) in a strain pertaining to the *Vibrio* genus and in the bacterial strain SCRC-21406 isolated from the gut of the deep-sea fish *Glossanodon semifasciatus*. Watanabe et al. (1997) reported the presence of this fatty acid in a proportion of 23%, and extremely small amounts (less than 3%) of other polyunsaturated fatty acids bound to their main phospholipids.

The polar lipids of several *Spirulina* strains contain γ–linolenic acid, particularly concentrated in their galactolipids (Durand-Chastel, 1997). Some *Synecococcus* strains also contain linoleic acid (18:2), a characteristic they share with other cyanobacteria like *Gleocapsa*, *Chloroglea*, and *Microcystis* (Kenyon, 1972).

Bacterial membranes of fatty acids consist of hydrophobic parts of glycerolipids and of lipid A. The hydrophobic parts exist in both Gram-positive and Gram-negative bacteria, whereas lipid A is only present in Gram-negative bacteria. This lipid usually contains several 3-OH fatty acid chains that can be used as a marker for detecting Gram-negative bacteria (Li et al., 2010).

Hydroxy fatty acids that carry an OH group, either in position two or three (2-OH-fatty acid or 3-OH-fatty acid), despite the fact that they are found in animal cells as D isomers, are frequently found in Gram-negative bacteria. They are generally a component of lipid A in their lipopolysaccharide, and, in some cases, they are also part of amino acids bound to lipids (Lechevalier & Moss, 1977). The presence of lipopolysaccharide in Gram-negative bacteria gives rise to the presence of hydroxy fatty acids; thus the presence of 10:0 3OH, 12:0 3OH, and/or 14:0 3OH fatty acids indicates that the organism is Gram-negative bacteria. Conversely, the absence of the lipopolysaccharide and hydroxy fatty acids indicates that the organism is Gram-positive (Kunistky et al., 2006).

Cyclopropane fatty acids are formed by the addition of a methyl group from S-adenosylmethionine in the double bond of a pre-existing fatty acid (Cronan et al., 1979; Grogan & Cronan Jr., 1997). In general, these fatty acids are part of the phospholipids of many bacterial species, and they have been related to the aging process of cells because they are produced in the stationary stage during their culture (Grogan & Cronan Jr., 1997). Because of their instability, Jantzen & Bryan (1985) recommend that, for their quantification, it is convenient to consider that their monounsaturated precursors are C16:1 for *Cyc*17:0, and C18:1 for *Cyc*19:0.

Fatty acids with alkyl branched chains are found in several bacterial groups. Methyl branches can be located in the last position (*iso*) or in the next-to-last (*anti-iso*), and, like unsaturated fatty acids their function is to regulate membrane fluidity (Jantzen & Bryan, 1985).

In Gram-negative bacteria, the most frequent saturated fatty acids are C16:0, C18:0, and C14:0; the most important unsaturated fatty acids are C16:1 *cis* or *trans* 9 and the C18:1 *cis* or *trans* 11, hydroxyl acids, which are part of their lipopolysaccharides (Janse, 1997). Gram-positive bacteria are characterized by the presence of important amounts of branched fatty acids; Coryneforms and Actynomicetes contain tubeculostearic and mycolic acids, respectively, and are exclusive to both genera (Ratledge & Wilkinson, 1988).

It is important to mention that although branched fatty acids are characteristic of Gram-positive bacteria, some Gram-negative can present them, such as *Cytophaga*, *Flavobacterium*, and *Desulfobacter*. It has been observed that they can make up almost 99% of the total fatty acid content in *Flavobacterium thermophilum*.

2.3 Fatty acid distribution in purple and green photosynthetic sulfur bacteria

2.3.1 Fatty acids in Chromatiaceae family

Anoxygenic photosynthetic bacteria as a group are unique among microorganisms because they do not exhibit the trans-3-hexadecenoic acid characteristic of the phosphatidylglycerol of photosynthetic eukaryotic organisms. The first studies on fatty acid composition of

Chromatiaceae were performed by Newton & Newton (1957) on *Chromatium vinosum* (strain D), and subsequently also analyzed by Haverkate et al. (1965) and Imhoff (1988). *Thiocapsa roseopersicina* (*Thiocapsa floridana*) was initially analyzed by Tacks & Holt (1971); they recorded the presence of C12:0, C13:0, C14:0, C16:1, C16:0 and C18:1 in intact cells and membranes. Of these fatty acids, the last three supplied 20, 28, and 45%, respectively.

Later studies, performed in *Chromatium purpuratum* 5500 (Imhoff & Thiemann 1991), *Thiocapsa roseopersicina* strains 9314 and 6311, by Tacks & Holt (1971) and Imhoff (1988), respectively; *Thiocapsa* sp. by Fredrickson (1986), *Chromatium minus* 1211, *Chromatium warmigii* 6512, and *Thiocystis gelatinosa* 2611 (Imhoff unpublished data, cited in Imhoff & Bias-Imhoff, 1995), indicated that the major fatty acids found in these strains were C16:0 (19.7-30.3%), C16:1 (25.4-36.5%), C18:1 (32.3-43.8%), small amounts of C14:0 (traces-1.3%) and C18:0 (0.8-2.0%).

Glycerophospholipid analyses in *Thiocapsa* sp. 5811, *Amoebobacter roseus* DSM235, *Chromatium gracile* DSM203, *Chromatium vinosum* (ATCC 17899), (DSM215), and in *Thiocapsa* sp. BF6400 and *Chr. gracile* BF7200, revealed the presence of at least 19 different fatty acids in the members of the Chromatiaceae family (Macarrón, 1998). Alternatively, Núñez-Cardona et al. (2008) reported data on the composition of fatty acids extracted from massive cultures of *Amoebobacter roseus* (DSM 235), *Chromatium vinosum* DSM183 (*Allochromatium vinosum*), and a wild strain isolated (Bk18) of a high mountain lake in Colombia, all members of the Chromatiaceae family. Table 1 depicts the pattern of the fatty acid characteristics of some purple sulfur bacteria pertaining to the Chromatiaceae family.

Strain	C12:0	C13:0	C14:0	C16:0	C16:1	aC17:0	C17:0	C18:0	C18:1	C20:1	Sum (%)	Ref.
Amoebobacter roseus DSM 235*			0.91	18.99	29.72			2.42	44.17	0.57	96.8	1
Amoebobacter roseus DSM 235	2.04	0.45	0.37	12.87	26.86	2.29	0.4	0.35	49.65	0.69	95.3	2
Amoebobacter sp. Bk18	1.83	0.29	0.33	12.41	27.67	2.31	0.3	0.68	47.58	0.54	93.4	2
Chromatium gracile DSM 203*			0.25	16.29	27.69			1.84	51.08	0.26	97.4	1
Chromatium gracile BF7200*			0.6	19.34	25.69			3.04	47.58	0.68	96.9	1
Chromatium minus 1211			0.3	20.5	30.4			1.2	43.8		96.2	3
Chromatium purpuratum BN 5500 (5% salts)			0.6	30.3	28.2			1.7	38.5	0.2	99.5	3
Chomatium vinosum ATCC 17899*			0.61	19.68	34.48			2.8	38.74	0.18	96.5	1
Chromatium vinosum DSM 183	1.25	0	0.79	13.62	31.77	0	0.27	0	47.14	0.43	94.8	2
Chromatium vinosum D			0.7	18.9	36.2			0.7	39.3		95.8	3
Chromatium vinosum D			+	28.7	32.6			+	38.7		100.0	3
Chromatium warmingii 6512			1.3	25.7	36.5			0.8	32.3		96.6	3
Thiocapsa roseopersicina 6311			0.4	22.1	25.4			0.9	43.8		92.6	3
Thiocapsa roseopersicina 9314			+	20	28			+	45		93.0	3
Thiocapsa sp.			tr.	19.7	27.9			2	34.6		84.1	4
Thiocapsa sp. 5811*			1.04	19.46	27.64			4.87	44.07	0.34	97.4	1
Thiocapsa sp. BF 6400*			0.8	17.42	19.8			2.22	55.12	0.9	96.3	1
Thiocystis gelatinosa 2611			0.6	22.9	30.5			1.2	39.5		94.7	3
Thiocystis gelatinosa DSM 215*			1.28	24.19	36.07			2.55	33.63	0	97.7	1

Table 1. Fatty acid composition of different strains of photosynthetic purple sulfur bacteria (Chromatiaceae family) *Linked to phospholipids, 1) Macarrón (1998), 2) Núñez-Cardona et al. (2008), 3) Data cited from Imhoff & Bias-Imhoff (1995), 4) Fredrickson et al. (1986). Ref=reference.

Recently, Sucharita et al. (2010) analyzed the fatty acid composition of the strains *Maricromatium fluminis* JA418, *Mch. indicum* JA100, *Mch. purpuratum* DSM 1591 (or

Chromatium purpuratum DSM 1591) and *Mch. gracile* DSM 203 (*Chromatium gracile* DSM 203). They concluded that C16:0, C16:1 and C18:1, are the main fatty acids for these strains and also they detailed the minor fatty acid components but only in *Mch.indicum* (Table 2).

Fatty acid	JA418	JA124	JA100	DSM203	DSM 1591
C12:0	2.4	12.5	3	9.2	1.8
C12:0 anteiso	-	-	2	-	-
C14:0 anteiso	-	-	3.1	-	-
C14:0	-	-	1.8	-	-
C14:0 2-OH	-	-	2.1	-	-
C16:0	25.7	-	2.7	15.6	20.5
C16:0 nOH*	-	-	1.7	-	-
C16:0iso	-	-	1.8	-	-
C16:1w9c	-	-	3.6	-	-
C18:0	-	-	-	4.2	-
18:1 iso OH	-	-	3.7	-	-
C18:3w6,9,12c	-	33.4	2.7	-	-
C16:1w7c	28.4	-	-	10.5	29.4
C16:1w7c alcohol	-	-	-	2.1	-
C18:1w7c	37	-	1.8	49.2	40.4
C18:1w5c	0.3	54	68.5	-	-
Total (%)	93.8	99.9	98.5	90.8	92.1

Table 2. Fatty acids (%) of *Marichromatium* genus (Sucharita et al., 2010).

According to Janse (1997), in Gram-negative bacteria, the most frequent saturated fatty acids are C16:0, C18:0, and C14:0; the most common unsaturated are C16:1 *cis* or *trans* 9 and C18:1 *cis* or *trans* 11. The hydroxyl acids, which are part of lipid A that contains several 3-OH fatty acid chains, can be used as markers for detecting Gram-negative bacteria (Li et al., 2010). This fact has been reported only in *Mch. indicum* and analyzed using the MIS system by Sucharita et al. (2008).

It is evident that information available on fatty acids of purple photosynthetic sulfur bacteria (Chromatiaceae family) has serious limitations, but it is also clear that their main cellular components are C16:0, C16:1 and C18:1, as has been reported by Imhoff (2008).

2.3.2 Fatty acids in Ectothiorhodospiraceae family

Table 3 includes information on species of *Ectothiorhodospira*, *Halorhodospira*, and *Thiorhodospira* genera, the last two have been analyzed very recently, and fatty acids with chains between 14 and 22 carbon atoms have been observed. Fatty acid C20:1 was only detected in *Ec. shaposhnikovii* and *Halorhodospira halophila* (DSM 244) and C22:0 in *Hal. halophila* BN 9626. This information reveals that major fatty acids for the Ectothiorhospiraceae family are C18:1 and C16:0 except for *Thiorhodospira sibirica* which has more C16:1 than C16:0.

Fatty acid analyses of Ectothiorhodospiraceae family species have been reported on *Ectothiorhodospira halophila* 9630 (Assellineau & Trüper, 1982) and *Ec. halophila* 9628, both grown in culture media at different salt concentrations (Imhoff &Thiemann, 1991), and *Ec. halophila* SL1 (Imhoff, unpublished data, cited in Imhoff & Bias-Imhoff, 1995), *Ec. halochloris* 9850 (Asselineau & Trüper, 1982; Imhoff, 1988), *Ec. abdelamalekii* 9840 (Asselineau & Trüper, 1982; Imhoff unpublished data, cited in Imhoff & Bias-Imhoff, 1995), and *Ec. mobilis* grown at different temperatures (Imhoff & Thiemann, 1991). Additionally, fatty acids of *Ec. shaposhnikovii* (Asselineau & Trüper, 1982; Imhoff, unpublished data cited in Imhoff & Bias-

Imhoff, 1995) and *Ec. vacuolata* DSM 2111 (Asselineau & Trüper, 1982; Imhoff, unpublished data) have been analyzed.

Strain/fatty acid	C12:0	C14:0	C16:1	C16:0	C18:1	C18:0	C19:0d8,9***	Cyc19:0	C20:1	C22:0	Sum	Reference (condition)
Ec. halophila 9628		0.8	1.2	25.4	63.8*	8.6					36	1 (10% salts)
Ec. halophila 9628		0.7	1.6	22	69.4*	5.6					29.9	1 (20% salts)
Ec. halophila 9628		0.9	1.2	24.9	63.8*	6.9					33.9	1 (30% salts)
Ec. halophila SL1			0.9	11.4	8.7	63.8		7.3			92.1	1
Ec. halophila 9630		1	1	16	33	15		31			97	1
Ec. halochloris 9850		0.4	0.4	18.9	69.7	7.5		1			97.9	2
Ec. halochloris 9850		tr	tr	19	51	7		18			95	1
Ec. abdelmaleki 9840		0.3	1.8	24.7	58.9	5.4		6.3			97.4	1
Ec. abdelmaleki 9840		0.5	3	16	57	4		10			90.5	1
Ec. mobilis 9903		0.2	5.5	14	72.3	4.8		tr			96.8	1 (5% salts, 27°C)
Ec. mobilis 9903		0.2	4.6	20.7	62.2	9.6		0.4			97.7	1 (5% salts, 36°C
Ec. mobilis 9903		0.5	3.4	23.1	59.7	10.9		0.3			97.9	1 (5% salts, 42°C)
Ec. shaposhnikovii N1		tr	8	17	68	tr		nd			93	1
Ec. shaposhnikovii N1		0.1	6.8	13.1	74.7*	2					22	1
Ec. vacuolata DSM 2111			4.8	15.3	73.5*	2.4					22.5	1
Ec. vacuolata DSM 2111		tr	6	20	60	7		nd			93	1
Ectothiorhodospira salini			8	18.0	56			12			94	3
Ec. magna			12.58	14.2	70.58	1.53		N/D			86.31	4
Ec. shaposhnikovii	1.3-1.7		4.1-4.6	24.4-26.4	57.1-59.0	5.6-6.8		N/D			92.5-98.5	4
Ec. vacuolata	N/D		3.6	21.6	65.3	5.5		N/D			96	4
Ec. shaposhnikovii**		2.5	4.6	21.6	59.4	5.6	tr*		0.5		94.2	5
Ec. mobilis**		1.2	2.7	35.7	16.5	7.6	33.8				97.5	5
Hal. halophila BN9626 **		4.6	tr	18	54.9	13.9	2.9			3.9	98.2	5
Hal. halophila DSM244 **		0.3	0.6	10.8	62.8	11.7	5.8			1.8	93.8	5
Thiorhodospira sibirica		0.3	21.3	18.1	52.9	3.7					96.3	4

Table 3. Fatty acids recorded in Ectothiorhodospiraceae family members. 1) Imhoff & Thieman (1991), 2) Imhoff (1988), 3) Ramana et al. (2010), 4) Bryanseva et al. (2010), 5) Yakimov et al. (2001). N/D=not determined, nd=not detected,tr= less 0.05%, *C18:1 included Cyc19, ** Values are percentages of the total fatty acids isolated from cellular phospholipids, ***Phospholipid fatty acid. *Ec. Ectothiorhodospira, Hal. Halorhodospira*

Information available on fatty acid profiles of *Ectothiorhodospira* genus members, shows that the major fatty acids for the analyzed species are C16:0 (11.4-35.7%) and the C18:1 (16.5-74.7%). They also contain C14:0 (traces-2.5%), C16:1 (traces-12.5%), and Cyc19:0 (tr-31%); although in some strains, the latter has not been detected or is added to its precursor, C18:1.

2.3.3 Fatty acids in Chlorobiaceae family

Regarding the fatty acid composition of green sulfur bacteria pertaining to the Chlorobiaceae family, strains of the *Chlorobium* and *Pelodyction* genera have been analyzed (Table 4). It is important to note that *Pelodyction* was reclassified, and actually is a member of the *Chlorobium* genus (Imhoff & Thiel, 2010).

In green photosynthetic sulfur bacteria, the presence of at least 12 fatty acids with 12 to 18 carbon atoms has been reported, and C14:0 (traces-27.10%), C16:1 (37.3-64.0%), C16:0 (3.52-29%), C15:0 (2.74-14.14), C18:1 (traces-15.59%), C18:0 (traces-3.30%), and Cyc17:0 (0.70-21%)

have been cited as the major fatty acids. It is possible that the information reported in table 4, which contains the fatty acid composition of green bacteria, is the only information of this type available at the present.

Strain	C12:0	C14:0	C14:1	aC15:0	C15:0	C16:1	C16:0	aC17:0	Cyc17:0	C17:0	C18:1	C18:0	Ref.
Cb. limicola f.thiosulf. 6330		13.00	2			57	17		3			tr	1
Cb. limicola f.thiosulf. 6230		21.00	2			43	10		21		1	tr	1
Cb. limicola f.thiosulf. 6230	1.1	27.10				37.3	20.3		7.3		3.1	tr	2
Cb. phaeobacteroides BF8600*		7.65		0.48	5.44	44.98	26.81	0.34		0.33	5.47	2.07	3
Cb. phaeobacteroides 2430		16.00	1			64	15		1		1	tr	1
Cb. phaeovibrioides 2631		10.00	tr			51	29		2		2	tr	1
Cb. vibrioides f.thiosulf. 1930		12.00	tr			52	23		3		2	1	1
Cb. vibrioides f.thiosulf. 8327	2.1	24.40				42.8	23		0.7		1.2	3.3	2
Cb. vibrioides f.thiosulf. 2230	2.9	23.80				43.4	22.1		2.6		3.1	tr	2
Cb.vibrioforme BF8200*		8.80	0.19		0.83	46.59	19.46			0.39	15.59	2.41	3
Cb. limicola DSM 245*		16.61	0.3		14.14	38.26	19.93	-		0.28	2.13	1.6	3
Cb. limicola BF8010*		11.10	0.5	0.48	9.38	41.81	24.25	0.25		0.22	2.82	0.87	3
Chlorobium sp. Chlo*		tr		0.59	2.74	nd	3.52		6.09	7.3	7.37	tr	4
Pelodictyon luteolum 2530		14.00	1			47	21		11		tr	tr	1

Table 4. Fatty acids of green photosynthetic sulfur bacteria. 1) Kenyon & Gray (1974), 2) Knudsen et al. (1982), 3) Macarrón (1998), 4) Fredrickson et al. (1986). *Linked to phospholipids, Cb. Chlorobium; f. thiosulf form thiosulfatophilum, tr=trace, Ref.=reference.

2.4 Fatty acids as a chemotaxonomic tool in purple and green sulfur bacteria

With the development of techniques and equipment to analyze the chemical properties of living organisms, new tools arose that markedly helped choose those properties that would enable the distinction, separation, and, finally, the classification of microorganisms (Stanier, 1968), providing data for more in-depth studies on their phylogenetic relations.

One of the first authors to propose the use of the chemical characteristics of microorganisms for taxonomic purposes was Wolochow (1959 cited in Abel et al., 1963). He suggested that microorganisms could be differentiated from "higher" organisms based on their chemical properties, as these properties are unique, providing a better theoretical basis for their classification. This concept was later amplified by Abel et al. (1963), who pointed out the importance of qualitative and quantitative analyses of several selected compounds, which would lead to a method for differentiation among species. These authors suggested that the chemical composition of living organisms could be regulated by natural or evolutionary relations and that this information could be obtained whenever culturing bacteria under defined conditions.

Based on previous studies performed by Scheuerbrandt & Bloch (1962) using liquid gas chromatography, the anaerobic and aerobic routes for the biosynthesis of fatty acids were proposed, enabling the distinction between prokaryotic and eukaryotic microorganisms to be made. It was set forth that no microorganisms can use both routes at the same time and that the presence of polyunsaturated fatty acids is an evolutionary event. This idea was later revised by Erwin & Bloch (1964), who stated that the composition of lipid molecules is determined not only genetically but also by the conditions of the environment; hence, factors such as temperature, nutritional and culture conditions (aerobic, anaerobic, etc.) can modify the lipidic pattern of living organisms.

From the start of bacterial classification it was recognized that structural characteristics by themselves were not enough for successful classification (Cohn 1872, cited in Stanier, 1968). Variations in the composition of fatty acids of bacterial cells have been used for their identification since the introduction of gas chromatography more than 50 years ago by Abel et al. (1963). Both the qualitative and quantitative composition was useful in differentiating microorganisms (Sasser, 2001). In bacteria, these cellular components have the advantage of not being present in their plasmids (Boom & Cronan Jr., 1989).

According to De Gelder (2008), chemotaxonomic techniques provide phenotypic properties, one of which is the fatty acid methyl ester (FAME) analysis that determines the fatty acid composition of the membrane and outer membrane of cells. Using this technique, fatty acids are extracted from cell hydrolysates and derivatized to volatile methyl esters which can be detected by gas chromatography. FAME allows grouping of many organisms according to the similarity of their fatty acid pattern, which is useful for identification. Quantitative data on bacterial fatty acid composition have allowed for the distinction and separation of families, genera, and even species.

In heterotrophic bacteria, the fatty acid composition has been widely studied for chemotaxonomic purposes. However, for purple and green photosynthetic sulfur bacteria, the use of this taxonomic tool is limited; most of the available data focus on major, aside from having been performed in limited species. Some of the data have been included in the description of new species, dealing mostly with the major fatty acids and with no attention given to the minor ones, despite that the latter could provide more information to separate the genera, and perhaps their species, providing more power to the use of this chemotaxonomic tool.

Currently published data on photosynthetic bacteria fatty acid composition are incomplete; data do not mention aspects like age of the bacterial culture and incubation conditions (temperature, light, etc). They neither specify the analytical techniques such as the extraction method, the total area of chromatogram integration, size of the samples, etc. nor how data were handled (normalization or standardization of data, etc.).

These deficiencies have made it quite difficult to compare information between diverse laboratories; consequently, the use of this tool becomes limited for chemotaxonomic classification and for its possible application to the phylogeny study of these ancestral photosynthetic microorganisms.

The advancement of knowledge on fatty acids in photosynthetic bacteria seems to have slowed down in the last decade, and with some exceptions, species descriptions do not include their fatty acid pattern. This is due to the fact that it is stated in the standard recommendations for the description of new species of anoxygenic photosynthetic bacteria, proposed by Imhoff & Caumette (2004), which were established in accordance with 30b of the International Code of Nomenclature of Bacteria and the support of the International Committee on Systematics of Prokaryotes, that fatty acid analysis is optional for the identification of photosynthetic bacteria.

In this respect, Tindall et al. (2010) presented a series of tests that must be used to determine the location of a strain as a member of an existing taxon, and they provide the methodologies and how those methodologies should be used and implemented; among these is the chemical characterization of cells with consideration of fatty acid composition for this objective. As a general recommendation, they point out the need to study the chemotaxonomic features of the most closed taxa for their comparison, particularly when new genera are being proposed.

With the aim of achieving reproducible profiles, Janse (1997) proposed a series of elements not only to apply a determined technique, but also to analyze the generated information. These are as follows: 1) use of cells of the same age, 2) a standard culture medium for both aerobic and anaerobic bacteria, 3) a gas chromatograph with a flame ionization detector (FID), an integrator, and a computer with a printer, 4) fatty acid solutions of known composition (standard), and 5) a statistical program for cluster analysis and analysis of its main components. Aside from the aforementioned, De Gelder (2010) points out the need to count with an extensive FAME database. With the Microbial Identification System (MIS), the techniques and methods for the analysis and identification of fatty acids, and their statistical analysis are well standardized and automated for heterotrophic bacteria. This system enables researchers to have, in a short time, accurate information on the composition and distribution of these chemical compounds in different microorganisms. By means of MIS, more than 300 fatty acids and related compounds (aldehydes, hydrocarbons, dimethyl acetyls) have been identified (Sasser, 2001).

The MIS system (MIDi Co) has been used for fatty acid analysis of non-photosynthetic bacteria and the description of species as in *Rheinheimera soli* (Ryu et al., 2008) and *Alcalilimnicola halodurans* (Yakimov et al., 2001), which are phylogenetically related to the Chromatiaceae and Ectothiorhodospiraceae families. Additionally, fatty acid profiles of purple non-sulfur bacteria have been obtained with the MIS system, an example being *Rhodovulum phaeolacus* (Lakshmi et al., 2011).

In photosynthetic purple sulfur bacteria, MIS has been applied to fatty acid analysis of different species, for example in different strains of *Marichromatium* (Sucharita et al., 2010) and *Ectothiorhodospira* (Bryantseva et al., 2010) genera. Nevertheless, one of the limitations of the MIS system is that it can only detect fatty acids with fewer than 20 carbons, and according to Tindall et al. (2010), the MIS system provides a comprehensive, but incomplete, database with some discrepancies that need to be clarified or compounds that are currently not included in the database.

Due to the variability in fatty acid analysis, the use of chromatographic techniques as taxonomic tools requires standardization of the culture conditions, the physiological age of cells, and the analysis techniques to obtain reproducible results. The need for a standard technique for fatty acid analysis of sulfur photosynthetic bacteria is real, and a technique for this purpose is included at the present chapter.

3. Fatty acid analysis of photosynthetic purple and green sulfur bacteria

3.1 Bacteria strains used for fatty acid analyses

Fatty acid analyses were done with collection strains (as reference) of phototrophic purple sulfur bacteria (Chromatiaceae family) these were: *Amoebobacter purpureus* (DSM 4197),

Chromatium purpuratum (DSM 1591), *Chromatium salexigens* (DSM 4395), *Chromatium tepidum* (DSM 3771), *Chromatium vinosum* (DSM 185), *Thiocapsa halophila* (DSM 6210), *Thiocapsa marina* (DSM 5811), and two strains of green phototrophic sulfur bacteria (Chlorobiaceae family): *Chlorobium limicola* (DSM 249) and *Chlorobium phaeobacteroides* (DSM 266). In addition, nine wild strains of purple and one green photosynthetic sulfur bacteria, isolated from the coastal lagoon Tampamachoco (Veracruz, Mexico) were included.

For the Isolation of wild strains of phototrophic purple sulfur bacteria, water samples from Tampamachoco lagoon (Veracruz, México) were collected using a 5-L van Dorn sampler. Test tubes containing Chromatiaceae and Chlorobiaceae specific culture media (Pfennig & Trüper, 1981; Nuñez-Cardona et al., 2008) were enriched with the water samples. Culture medium composition was: 1.0 g KH_2PO_4, 0.5 g NH_4Cl_2, 0.40 g $MgCl_2 \cdot 6H_2O$, 0.05 g $CaCl_2 \cdot 6H_2O$, 1.0 mL trace element solution SL12, 1.0 mL vitamin B_{12} (2.0 mg L^{-1} distilled water),

Code	Strain	Origin	Main pigments	Shape	(NaCl %)
DSM 4197	*Amb. purpureus (1)*	Schleinsee (Germany)	Bchl *a, ok*	Coccus	0
T1f6	*Amoebobacter* sp.	TL México	Bchl *a, sp*	Coccus	2
DSM 1591	*Chr. purpuratum*	Sea sponge	Bchl *a, ok*	Rods	5
DSM 4395	*Chr. salexigens*	GS (Camargue France)	Bchl *a, sp, ly, rh*	Rods	7
DSM 3771	*Chr. tepidum*	YPN (USA)	Bchl *a, sp, ly, rh*	Rods	0
DSM 185	*Chr. vinosum*	Pond	Bchl *a, sp, ly, rh*	Rods	0
T9s60	*Chromatium* sp.	TL México	Bchl *a, sp*	Rods	2
T9s62	*Chromatium* sp.	TL México	Bchl *a, ly*	Rods	2
T9s64	*Chromatium* sp.	TL México	Bchl *a, ly*	Rods	2
T9s642	*Chromatium* sp.	TL México	Bchl *a, sp*	Rods	2
T7s9	*Chromatium* sp.	TL México	Bchl *a, ly*	Rods	2
T9s68	*Chromatium* sp.	TL México	Bchl *a, ly*	Rods	2
T11rosa	*Thiocapsa* sp.	TL México	Bchl *a, sp*	Coccus	2
T11s	*Thiocapsa* sp.	TL México	Bchl *a, sp*	Coccus	2
DSM 6210	*Tc. halophila*	GS (Camargue France)	Bchl *a, ok*	Coccus	7
DSM 5811	*Tc. marina*	LP France	Bchl *a, ok*	Rods	5
DSM 249	*Chlorobium limicola*	THS, USA	Bchl *c, cb*	Rods	0
DSM 266	*Cb. phaeobacteroides*	LB, Norway	Bchl *e, irt*	Rods	0
T11S	*Chlorobium* sp.	TL México	Bchl *c, cb*	Rods	2

Table 5. Purple photosynthetic sulfur bacteria strains used for fatty acid analysis.

DSM German Collection of Microorganisms and Cell Culture, *Amb. Amoebobacter, Cb. Chlorobium, Chr Chromatium, Tc Thiocapsa.* TL=Tampamachoco Lagoon, GS=Giraud Saltens, YNP=Yellowstone National Park. THS=Tassajara Hot Spring (USA), LB=Lake Blankvann, *Bchl* bacteriochlorophyll, *ok* okenone, *cb* chlorobactene, *sp* spirilloxanthin, *ly* lycopene, *rh* rhodopinal, *irt isorenieratene*, NaCl(%)=Optimal salinity requirement.

30.0 mL sodium bicarbonate solution (5% distilled water), 6.0 mL $Na_2S\cdot9H_2O$ (6% distilled water), and distilled water (ca. 950 mL) up to 1 L, and adjusted to pH 7.3. For *Chlorobium* species, the same medium was used but $Na_2S\cdot9H_2O$ was increased to 10.0 mL; solution SL10 and pH 6.8 were used. Test tubes were kept at room temperature under constant illumination by 60-W incandescent and fluorescent light bulbs placed at 20-25 cm, were used for growth of Chromatiaceae and Chlorobiaceae, respectively (van Niel, 1971). Agar shake technique on semisolid medium sealed with oil paraffin was used to isolate and purify photosynthetic purple and green sulfur bacteria (Pfennig & Trüper, 1981). Pure cultures were verified by optical microscope observations.

Table 5 contains the list of the cultures and collection strains used for fatty acid analysis, general characteristics (main pigments and cell forms), and optimal salinity requirements for each one. At the present document, all the bacteria strains are cited as it is in the original papers or as they were handled. This is because constantly photosynthetic bacteria are reclassified and their names also change.

3.2 Culture conditions

All strains were grown anaerobically by triplicate in 1500-mL bottles with metal screw caps and autoclavable rubber seals containing the specific culture medium described before; pH between 7.0-7.5 for Chromatiacea and 6.8 for green sulfur bacteria. Liquid cultures of purple sulfur bacteria were illuminated constantly by incandescent light bulbs (2000 lux) and green sulfur bacteria were exposed to fluorescent light bulbs. Incubation temperature was 23 °C except for the thermophilic strain *Chr. tepidum* (43 °C). Enough biomass of each culture was attained with a sodium sulfide solution neutralized with hydrochloric acid, added almost daily when sulfur was not present in cultures the latter was detected by the use of lead acetate paper (Merck).

After 15 days of incubation and without sulfur in the liquid cultures, the cells were concentrated by centrifugation (5000-10,000 rpm at 5 °C). Cultures of *Amoebobacter* and *Thiocapsa* were centrifuged at 10,000 rpm since it was ratter difficult to concentrate the cells. Once the cultures were concentrated, the cells were frozen at -20 °C, for further lyophilization.

3.3 Fatty acid analysis by gas chromatography

3.3.1 Sampling conditions

Temperature, age of the cells, as well as the composition of the medium, affect cells fatty acid composition and limits the comparison of the results among different laboratories and the application of these cellular components as a chemotaxonomic tool. Few reports deal with the importance of the size and sample conditions (dry or wet) for fatty acids analysis. It has been proposed that for heterotrophic bacteria, less than 40.0 mg of wet weight were enough (Sasser, 2001). It has also been indicated that 1.0 mg of dry weight (Brondz & Olzen, 1986) or a single colony could be enough for fatty acids analysis (Buyer, 2002). In the case of phototrophic sulfur bacteria, Tacks and Holt (1971) used 10-15 g of wet weight of *Thiocapsa floridiana* (*Thiocapsa roseopersicina*).

At the present work, and for the sake of reproducibility it was used cells of the same age (Janse, 1997; Sasser, 2001); grown in standard conditions (as previously described) and fatty acid analyses were done weighing 500 mg and 250 mg of dry cells, for purple and green photosynthetic sulfur bacteria, respectively.

3.3.2 Fatty acid analysis

For fatty acid analysis, fat present in the freeze-dried cells (three different samples of the same strain) was saponified, followed by diazomethane derivatization (Sigma N-nitroso-methylurea), and evaporation with nitrogen gas. Identification of fatty acids was accomplished by comparison of the retention times with standard mixtures of fatty acid methyl esters. Fatty acids final identification was done by gas chromatography and mass spectrometry (Muñoz et al., 1996; Núñez-Cardona et al., 2008). Details of these techniques are described below.

3.3.2.1 Extraction of lipids

Each sample was treated with 2.0 mL benzene and 8.0 mL of KOH solution (5.0 g/100 mL methanol) and containing 5% (w/v) potassium hydroxide in a screw-cap test tube (14X120 mm) fitted with a Teflon-lined cap and heated in a cover bath at 80 °C during 4 hours. The samples were cooled at room temperature and were acidified with a solution of H_2SO_4 (20% v/v) to pH 1, this was verified with pH paper test. Diethyl ether was added to extract lipids and it was added water to wash the extracts until neutral and the clean separated ether was evaporated using a nitrogen gas atmosphere.

3.3.2.2 Preparation of methyl esters

The lipids were methylated using diazomethane, each sample was prepared with 2.0 g N-Nitroso-N-methylurea (Sigma) dissolved in a precooled solution containing 30.0 mL of diethyl ether, 2.0 g KOH and 6.0 mL distilled water. The mixture was stirred for 5 to 10 minutes, it was removed the supernatant and placed in a new tube cooled in ice containing potassium hydroxide pellets. This solution must be conserved to -80 °C and handled in a hood because diazomethane is very toxic. At the end, 10 ml of diazomethane was added into each tube with the dried lipids and methylation was achieved within 10-20 minutes; after this time, the content was evaporated at 40 °C in a water bath.

3.3.2.3 Fatty acid methyl ester analysis by gas chromatography

N-hexane (0.04-1.0 mL) was added to the dried extracts and 2.0 µL were injected in a gas chromatograph supported by a silica capillary column (15m x 0.25 mm I.D.) with cross-linked methyl silicone (HP-1, Hewlett-Packard) as stationary phase. The column was inserted in an HP-5890A gas chromatograph equipped with a flame ionization detector. The column was programmed at 175 to 300 °C for 15 minutes. Injector and detector temperatures were 275 and 300 °C, respectively. The carrier gas was helium with a flow rate of approximately 1.0 mL/min and the split ratio was approximately 1:50. The chromatograms were integrated by using a HP3396 Series II Integrator. Identification of fatty acids was accomplished by comparison of the retention times with standard mixtures of fatty acid methyl esters.

3.3.2.4 Gas chromatography-mass spectrometry analysis

Fatty acids were subjected to gas chromatography-mass spectrometry, using an HP-5890 gas chromatograph attached to a HP5989X quadripole mass spectrometer with a dimethylpolysiloxane column TRB1 (30 m). Injector and detector temperatures were 225 °C, and two ramps were used, one from 10 °C/min to 240 °C and the other one from 40 °C/min to 270 °C. The injection mode was splitless, ca. 1:50.

3.3.2.5 Reproducibility of fatty acid pattern

With the aim of reaching reproducible data, the integration area of the fatty acid peaks was fixed to 1800 as minimal area and the extract of each sample was injected as many times to get a total area between 14×10^6 and 18×10^6 for purple sulfur and 7×10^6 for green photosynthetic sulfur bacteria.

To assess the reproducibility of the fatty acids pattern, the coefficient of variation (CV) was calculated; for this, it was used the peaks of each fatty acid generated in the chromatograms of the three independent samples of each strain. The standard deviation (of each peak that corresponds to one fatty acid) was calculated and divided by the mean and multiplied by 100, this was made following the recommendations of Bounsfield et al. (1983) and Rainey et al. (1994).

3.4 Results and discussion

With the technique applied for fatty acid analysis, it was possible to determine at least 90-99% of the total fatty acid composition from purple sulfur bacteria of the Chromatiaceae family. Data are the results of three different chromatograms from three different cultures of the same strain. The variation coefficient calculated for each fatty acid was less than 10%; therefore, the results here presented have a high degree of reproducibility. Major and minor fatty acids for photosynthetic sulfur bacteria are given in Table 6.

Strain	NI	C12:0	C13:0	C14:0	C16:1	C16:0	aC17:0	C17:0	C18:1(7)	C18:1(9)	C18:0	C20:1	X (%)
DSM 4197	1.13	0.53	0.42	0.31	28.17	11.66	1.80	0.53	41.53	6.61	0.21	0.60	93.50
T1f6	1.10	1.36	0.48	0.43	24.43	14.53	2.34	0.35	43.03	6.72	1.25	0.75	96.77
DSM 5811	1.86	0.49	0.43	0.29	27.64	12.05	0.18	0.04	41.72	6.14	1.84	0.70	93.38
DSM1591	1.78	0.47	0.00	0.19	30.18	6.59	0.00	0.37	45.10	4.62	3.37	0.83	93.50
DSM 4395	1.41	0.74	0.00	0.81	17.34	16.74	0.00	0.29	51.90	4.76	2.54	0.77	97.30
DSM 3771	1.11	0.33	0.00	0.41	16.17	25.86	0.00	0.55	36.27	7.11	4.81	2.44	95.06
DSM185	0.94	0.13	0.00	0.76	31.77	15.90	0.00	0.08	36.92	10.04	0.00	0.60	97.14
T9s60	2.34	0.93	0.00	0.25	26.38	14.23	0.00	0.30	44.00	6.01	1.74	0.68	96.86
T9s62	1.98	0.72	0.00	0.24	27.14	11.81	0.00	0.42	45.80	5.85	1.85	0.62	96.43
T9s64	1.99	0.71	0.00	0.20	27.83	12.39	0.00	0.41	46.94	6.08	1.65	0.79	98.99
T9s642	2.26	1.26	0.00	0.20	27.69	12.67	0.00	0.42	45.57	5.76	1.72	0.73	98.28
T7s9	1.99	0.51	0.00	1.27	32.26	10.46	0.00	0.87	41.27	8.61	0.23	0.42	97.89
T9s68	1.69	0.89	0.00	1.50	28.75	12.43	0.14	0.82	37.01	6.60	0.35	0.36	90.54
T11rosa	1.62	0.32	0.48	0.47	23.55	14.92	1.03	0.36	44.52	5.75	2.47	0.64	96.13
DSM 6210	1.57	1.26	0.43	0.49	25.61	22.73	0.11	0.28	35.86	5.92	4.00	0.63	98.89
T 11s	1.53	0.47	0.66	0.51	21.21	15.49	2.93	0.36	38.63	5.27	3.29	0.52	90.87

Table 6. Fatty acid composition of purple sulfur bacteria (Chromatiaceae) identified by Gas chromatography-Mass spectrometry. NI=no identified, X(%)=average (%).

As it is evident, the fatty acid profile of these bacteria not changed among the strains assayed. Similar total (%) fatty acid results have been reported for this physiological group by Haverkate et al. (1965); Tacks & Holt (1971); Frederickson et al. (1986); Imhoff (1988); Imhoff & Thiemann (1991); Macarrón (1998).

Results showed the presence of saturated and unsaturated fatty acids with chains from 12 to C20:1 long; C13:0 as well as aC17:0 were almost present only in phototrophic purple sulfur bacteria with spherical shapes, such as DSM 4197, DSM 5811, DSM 6210, and isolates T1f6, T11s and T11rosa. These results agree with those of Núñez-Cardona et al. (2008), who found the same fatty acids in *Amoebobacter* sp. (Bk18) and *Amoebobacter roseus* (DSM 235), which are also spherical. C13:0 fatty acid accounted for less than 1% (0.42-0.66) and aC17:0 for between 0.11 and 2.93% of the fatty acids present. Tindall et al. (2010) recommend that all components constituting 1% or more of the fatty acids must be reported, but according to these results, the identification of C13:0 and aC17:0 are specific components of two genera (*Amoebobacter* and *Thiocapsa*). The importance of reporting minor fatty acids for taxonomic purposes is evident even when the fatty acids are present in percentages below 1% because minor fatty acids point out the differences among genera and probably among species too.

Imhoff & Bias Imhoff (1996) expressed that photosynthetic purple sulfur bacteria have equilibrated proportions of C16:0, C16:1 and C18:1, pointing out C18:1 as the most abundant, but not as a major fatty acid. Table 6 shows that, in the conditions described above, C18:1 could account for more than 50% of the total fatty acids from strains like *Chr. salexigens* DSM 4395, *Tc. halophila* DSM 6210 and some *Chromatium* spp. from marine origin (e.g., T9s64, T9s62, T9s642, T9s69). The order of abundance of fatty acids from purple sulfur bacteria is as follows: C18:0, C16:1, and C16:0. With exception of *Chr. tepidum* DSM 3771, *Chr. salexigens* DSM 4395 and *Tc. halophila* DSM 6210 (extremophilic microorganisms) that present some variations in response to their own specific requirements for growth within the systems they inhabit.

Chr. tepidum DSM 3771 presents more C16:0 than C16:1; hence the fatty acid profile for this strain is: C18:1>C16:1>C16:0. In *Tc. halophila* DSM 6210, C18:1 is the major fatty acid, and the increase of C16:0 is evident. In *Chr. salexigens* DSM 4395, C16:0 and C16:1 are almost equals in abundance. In contrast, *Chr. purpuratum* DSM 1591 showed the lowest quantity of C16:0 (6.59%), although it conserves the pattern C18:1>C16:1>C16:0, common in phototrophic sulfur bacteria.

With the exception of the data reported by Imhoff and Thiemann (1991) for *Chr. purpuratum* BN5500, *Ec. halophila*, and *Ec. mobilis*, there is no available information on the influence of salinity on fatty acid composition in the Chromatiaceae family. Nevertheless, *Tc. halophila* DSM 6210, as in *Chr. salexigens* DSM 4395, grown in salt concentrations of 7 and 10%, respectively, showed the same behavior as *Chr. purpuratum* and *Ectothiorohospira* species; i.e., when there are higher salt concentrations in the culture media, these bacteria produce more saturated fatty acids, especially C16:0 and C18:0, and there is a consequent decrease of C18:1 (Imhoff & Thiemann, 1991).

Fatty acid C20:1 was detected in almost all strains of the Chromatiaceae family here analyzed; in general, it was not higher than 1% (0.3-0.83%) but in the thermophilic *Chr. tepidum*, C20:1 accounts for 2.44% of the total fatty acids. This cellular component was also identified in phospholipids of *Thiocapsa* spp. (5811 and BF6400) by Macarrón (1998) and in intact cells of *Chr. purpuratum* BN 5500 (Imhoff & Bias-Imhoff, 1995). It is important to mention that C20:1 was not recorded by Sucharita et al. (2010), using the MIS system, perhaps because MIS is designed to identify fatty acids among 9–20 carbons in length (Sasser, 2001), is not possible to detect the presence of this fatty acid with this system. It is also evident that the results of these authors showed the presence of main fatty acids in four strains of *Marichromatium*, but only in *Mch. indicum*, they detailed minor and major fatty acids.

The differences recorded here in the fatty acid compositions of purple sulfur bacteria are in agreement with molecular studies as have been reported by Imhoff & Caumette (2004). Analysis of 16S rDNA revealed genetic differences among marine, halophilic, and the freshwater photosynthetic bacteria. These results led to the taxonomic reclassification and description of new genera of purple photosynthetic sulfur bacteria (Imhoff et al., 1998).

Tree strains of green sulfur bacteria were used to validate the standardized technique for fatty acid analysis of photosynthetic sulfur bacteria and to know the fatty acid profile of another photosynthetic group; results of this analysis are shown in table 7.

Fatty acid	DSM 249	DSM 266	T11S
C12:0	1.29	0.97	1.96
C14:0	4.83	13.63	21.05
12-me-C14:0	0.52	0.42	0.31
C15:0	0.71	1.12	0.22
C16:1(9)	30.99	36.01	26.65
C16:0	12.29	11.96	22.12
aC17:0	0.92	0.61	1.37
15-me-C16:1 (11)	0.72	0.46	0.14
Cyc 17:0	3.42	1.78	0.81
C17:0	0.29	0.27	0.26
C18:1	10.33	2.97	2.9
C18:0	0.81	0.66	0.53
Total (%)	67.12	70.86	78.32

Table 7. Fatty acid detected in three strains of green sulfur bacteria.

As indicated before, for green sulfur bacteria only 250 mg of dried cells was used because of the difficulties to clean the samples. As with purple sulfur bacteria, each green sulfur bacterium was assayed using independent cultures in triplicate. Invariably, after fatty acid extraction and analysis by gas chromatography, the results showed the presence of 12 fatty acids with chains between 12 and 18 carbon atoms in all *Chlorobium* strains assayed. According to these results, a fatty acid profile for these strains could be C16:1>C16:0 and C14:0. It is also clear that fatty acid composition contains minor components including C12:0, 12-Me-C14:0 (aC15:0), C15:0, C14-Me-C16:0(aC17:0), 15-Me-C16:1, C17:0, C18:1, C18:0, and a cyclic fatty acid (Cyc17:0)

The two hydroxy-acids (C14:0 3OH and C16:0 3:OH) were not detected as reported by Knudsen et al. (1982). According to Boon et al. (1997), hydroxyl acids are part of the cellular wall of Gram-negative bacteria and bounded to lipid A and cannot be extracted with soft organic compounds; a special procedure for analysis and extraction is necessary. Nevertheless, in the *Marichromatium indicum* strain (Sucharita et al., 2010) it was possible to identify hydroxyl fatty acids using the MIS system

Unbranched fatty acids are characteristic in Gram-positive bacteria; although they have been detected in the *Clorobium* strains analyzed here. The fatty acids aC15:0 and an aC17:0 were detected previously linked to phospholipids in both *Chlorobium limicola* and *Cb. phaeobacteroides* in quantities less than 0.5% (Macarrón, 1998). Additionally, C15:0 was identified as a fatty acid component of the profile of *Chlorobium* sp. by Frederickson et al. (1986). 15-Me-C16:1 was the only fatty acid not reported before for green sulfur bacteria.

4. Conclusions

Information on fatty acids of photosynthetic purple and green sulfur bacteria is scarce, and the studies of these cellular components are limited to a very low number of collection strains; there is no detailed information about the techniques applied for fatty acid analysis and the genera most studied being *Chlorobium*, *Chromatium* and *Ectothiorhodospira*. With the technique assayed in members of the Chromatiaceae and Chlorobiaceae families it was possible to detect 12 fatty acids both major and minor components. There is clear evidence that extremophilic strains conserve the same fatty acid profile, but there are quantitative differences. The profile of fatty acids in sulfur and green photosynthetic sulfur bacteria is consistent among the strains analyzed, and almost all the fatty acids recorded in these bacteria were previously reported. Fatty acid analysis using standard conditions offers a valuable tool to utilize this information for chemotaxonomic purposes and minor components must be taken into account. Also, it is necessary intensify the study of fatty acids in photosynthetic purple and green sulfur bacteria.

5. Acknowledgements

Thanks to the Universitat Autonoma de Barcelona (UAB), Spain, where it was done all the experimental part of the present work with the academic and technical support from Doctors Marina Luquin, Jodi Mas, and Manuel Muñoz. Financial support from the Consejo Nacional de Ciencia y Tecnología (CONACyT, México) and the Universidad Autónoma Metropolitana-Xochimilco is gratefully acknowledged. Also thanks to Dr. Luis Raúl Tovar Galvez (Centro Interdisciplinario de Investigaciones y Estudios sobre Medio Ambiente y Desarrollo-IPN) for his valuable comments to this manuscript.

6. References

Abbas, C.A. & Card, G.L. (1980). The relationships between growth temperature, fatty acid composition and the physical state and fluidity of membrane lipids in *Yersinia enterocolitica*. *Biochimica et Biophysica Acta-Biomembranes*, Vol.602, No.3, (November 1980), pp. 469-476, ISSN 0005-2736

Abel, K.; Deschmertzing, H. & Peterson, J.I. (1963). Classification of Microorganisms by Analysis of Chemical Composition I. Flexibility of Utilizing Gas Chromatography. *Journal of Bacteriology*, Vol.85, (May 1963), pp. 1039-1044, ISSN 0021-9193

Asselineau, J. & Trüper H.G. 1982. Lipid composition of six species of phototrophic bacteria genus *Ectothiorhodospira*. *Biochimica et Biophysica Acta-Lipids and Lipid Metabolism*, Vol.712, No.1, (July 1982), pp. 111-116. ISSN ISSN 0006-3002

Boom, V.T. & Cronan Jr., J.E. (1989). Genetics and Regulation of Bacterial Lipid Metabolism. *Annual Reviews in Microbiology*, Vol.43, (October 1989), pp. 317-343, ISSN 0066-4227.

Boon, J.J.; de Leeuw, J.W.; Hoek, G.J. & Vosjan, J.H. (1977). Significance and Taxonomic Value of Iso and Anteiso Monoenoic Fatty Acids and Branched β-hydroxy acids in *Desulfovibrio desulfuricans*. *Journal of Bacteriology*, Vol.129, No.3, (March 1977), pp. 1183-1191, ISSN 0021-9193

Bounsfield, I.J.; Smith, G.L.; Dando, T.R. & Hobbs, G. (1983). Numerical Analysis of Total Fatty Acids Profiles in the Identification of Corenyform and Some Other Bacteria. *Journal of General Microbiology*, Vol.129, No.2, (February 1983), pp. 375-394, ISSN 0022-1287

Brondz, I. & Olsen, I. (1986). Chemotaxonomy of Selected Species of the *Actinobacillus Haemophilus-Pasteurella* Group by Means of Gas Chromatography-Mass Spectrometry and Bioenzymatic Methods. *Journal of Chromatography*, Vol.380, (July 1986), pp. 1-17, ISSN 1570-0232

Brown, J.L.; Ross T.; McMeekin, T.A. & Nichols, P.D. (1997). Acidic Habituation of *Escherichia coli* and the Potential Role of Cyclopropane Fatty Acids in Low pH Tolerance. *International Journal of Food Microbiology*, Vol.37, No.2-3, (July 1997), pp. 163-173, ISSN 0168-1605

Bryantseva, I.A.; Tourova, T.P.; Kovaleva, O.L.; Kostrikina, N.A. & Gorlenko, V.M. (2010). *Ectothiorhodospira magna* sp. nov., a New Large Alkaliphilic Purple Sulfur Bacterium. *Microbiology*, Vol.79, No.6, (December 2010), pp. 780-790, ISSN 1350-0872

Buyer, J.S. (2002). Identification of Bacteria from Single Colonies by Fatty Acid Analysis. *Journal of Microbiological Methods,* Vol.48, No.2-3, (February, 2002), pp. 259-265. ISSN 0167-7012

Cho, K.Y. & Salton, M.R.J. (1966). Fatty Acid Composition of Bacterial Membranes and Wall Lipids. *Biochimica et Biophysica Acta-Lipids and Metabolism*, Vol.116, No.1, (February 1966), pp. 73-79, ISSN 0006-3002.

Cronan J.E.; Reed J.R.; Taylor F.R. & Jackson, M.B. (1979). Properties and biosynthesis of cyclopropane fatty acids in *Escherichia coli*. *Journal of Bacteriology*, Vol.138, No.1, (April 1979) pp. 118-121, ISSN 0021-9193

De Gelder, J. (2008). Raman spectroscopy as a tool for studying bacterial cell compounds. In:http://hdl.handle.net/1854/11263,16.08.2011,   12.08.2011.   Available   from http://lib.ugent.be/fulltxt/RUG01/001/271/628/RUG01001271628_2010_0001_A C.pdf

Durand-Chastel, H. (1997). Production of *Spirulina* Rich in GLA and Sulfolipids. *International Symposium of Marine Cyanobacteria and Related Organisms*. Intitute Ocenographique, Paris. November 24-28, 1997.

Erwin, J. & Bloch, K. 1964. Biosynthesis of Unsaturated Fatty Acids in Microorganisms. *Science*, Vol.143, No.3610, (March 1964), pp. 1006-1012, ISSN 0036-8075

Fredrickson, H.L.; Cappenberg, T.E. & De Leeuw, W.J. (1986). Polar Lipid Ester-Linked Fatty Acid Composition of Lake Vechten Seston: an Ecological Application of Lipid Analysis. *FEMS Microbialogy Ecology*, Vol.38, No.6, (December 1986), pp. 381-396, ISSN 1574-6941

Grogan, D.W. & Cronan Jr., J.E. (1997). Cyclopropane Ring Formation in Membrane Lipids of Bacteria. *Microbiology and Molecular Biology Reviews*. Vol.61, No.4, (December 1997), pp. 429-441, ISSN 1092-2172

Guschina, I.A. & Harwood, L.J. (2008). Lipids: Chemical Diversity, In: *John Wiley & Sons Inc.* 10.08.2011, Available from http://onlinelibrary.wiley.com /book / 10.1002/ 9780470048672

Haack, S.K.; Garchow, H.; Odelsen, D.A.; Forney, L. & Klug, M.J. (1994). Accuracy, Reproducibility and Interpretation of Fatty Acid Methyl Ester Profiles of Model Bacterial Communities. *Applied and Environmental Microbiology*, Vol.50, No.7, (July 1994), pp. 2483-2493, ISNN 0099-2240

Haverkate, F.F.; Teulings, A.G. & Van Deenen, L.L.M. (1965). Studies on the Phospholipids of Photosynthetic Microorganisms. *Proceedings of the Koninklijke Nederlandse Akademie van Wetenschappen Series B*, Vol.68, pp.154-159, ISSN: 0920-2250

Imhoff, J.F. (1988). Lipids, Fatty Acids and Quinones in Taxonomy and Phylogeny of Anoxygenic Phototrophic Bacteria, In: *Green Photosynthetic Bacteria*, J.M Olson, J. Amez, E. Stackebrandt & H.G. Trüper (Eds.), 223-232. Plenum Press, ISBN 0306429209. New York, USA

Imhoff, J.F. (2008). Systematics of Anoxygenic Phototrophic Bacteria, In: *Sulfur Metabolism in Phototrophic Organisms*, H. Rüdiger, C. Dahl, D.B. Knaff & T. Leustek (Eds.), 269-287, Springer Verlag, ISBN-10 9048177421, New York, USA.

Imhoff, J.F. & Bias-Imhoff, U. (1995). Lipids, Quinones and Fatty acids of Anoxygenic Phototrophic Bacteria, In: *Anoxygenic Photosynthetic Bacteria*, R.E. Blakenship, M.T. Madigan & C.E. Bauer (Eds.), 179-205, Kluwer Academic Publishers, ISBN 0-7923-3681-X, The Netherlands

Imhoff, J.F. & Caumette, P. (2004). Recommended Standards for the Description of New Species of Anoxygenic Phototrophic Bacteria, *International Journal of Systematic and Evolutionary Microbiology*, Vol.54, No.4, (July 2004), pp. 1415-1421, ISSN 1466-5034

Imhoff, J.F.; Suling, J. & Petri, R. (1998). Phylogenetic Relationship Among the Chromatiaceae, their Taxonomic Reclassification and Description of the New Genera *Allochromatium, Halochromatium, Isochromatium, Marichromatium, Thiococcus, Thiohalocapsa* and *Thermochromatium*. *International Journal of Systematic Bacteriology*, Vol.48, No.4, (October 1998), pp. 1129-1143, ISSN 0020-7713

Imhoff, J.F. & Thiel, V. (2010). Phylogeny and Taxonomy of Chlorobiaceae. *Photosynthesis Research*, Vol.104, No.5-6, (November 2010), pp. 123-136, ISSN 0166-8595.

Imhoff, J.F. & Thiemann B. (1991). Influence of Salt Concentration and Temperature to the Fatty Acid Composition of *Ectothiorhodospira* and Other Halophilic Phototrophic Purple Bacteria. *Archives of Microbiology*, Vol.156, No.5, (October 1991), pp. 370-375, ISSN 0302-8933

Janse, J.D. (1997). Fatty Acid Analysis in the Identification, Taxonomy and Ecology of (Plant Pathogenic) Bacteria. In: *Diagnosis and identification of Plant Pathogens*, H.W. Dehne, G. Adam, M. Diekman M, F.J. Mauler-Machnik, P. van Halteren P. (Eds.), 63-70, Kluwer Academic Press Publishers, ISBN 0792347714, The Netherlands.

Jantzen, E. & Bryan, K. (1985). Whole Cell and Lipopolysaccharide Fatty Acids and Sugars of Gram-Negative Bacteria, In: *Chemical Methods in Bacterial Systematics*, M. Goodfellow & D.E. Minnikin (Eds.), 145-171, Academic Press, ISBN-10: 0122896750, London, UK

Kenyon, C.N. (1972). Fatty Acid Composition of Unicellular Strains of Blue-Green Algae. *Journal of Bacteriology*, Vol.109, No.2, (February 1972), pp. 827-834, ISSN 00 21-9193

Kenyon, C.N. & Gray, A.M. (1974). Preliminary Analysis of Lipids and Fatty Acids of Green Bacteria and *Chloroflexus aurantiacus*. *Journal of Bacteriology*, Vol.120, No.1, (October 1974), pp. 131-138, ISSN 0021-9193

Knudsen, E.; Jantzen, E.; Bryan, K.; Ormerod, J.G. & Sirevag, R. (1982). Quantitative and Structural Characteristics of Lipids in *Chlorobium* and *Chloroflexus*. *Archives of Microbiology*, Vol.132, No.2, (August 1982), pp. 149-154, ISSN 0302-8933

Kunistky, C.; Osterhout, G. & Sasser, M. (2006). Identification of Microorganisms Using Fatty Acids Methyl Ester (FAME) and the MIDI Sherlock Microbial Identification System,   20.08.2011. https://store.pda.org/bookstore/TableOfContents/ERMM_V3_Ch01.pdf

Lakshmi, K.V.N.S.; Sasikala, Ch.; Ramana V.V.; Ramaprasad, E.V.V. & Ramana, Ch.V. (2011). *Rhodovulum phaeolacus* sp. nov. a Phototrophic Alphaproteobacterium Isolated From a Brown Pond. *Journal of General and Applied Microbiology*, Vol.57, No.3, (July 2011), pp.145-151, ISSN 1349-8037

Lechevalier, H. & Lechevalier, M.P. (1988). Chemotaxonomic Use of Lipids-an Overview, In: *Microbial Lipids*, C. Ratledge & S.G. Wilkinson (Eds.), 869-902, Academic Press, ISBN 0125823043, London, UK

Lechevalier, M.P. & Moss, C.W. (1977). Lipids in Bacterial Taxonomy: a Taxonomist's View. *Critical Reviews in Microbiology*, Vol.5, No.2, (January 1977), pp. 109-210, ISSN 1040-841X

Li, Y.; Wu, S., Wang, L.; Li, Y. & Shi, F. (2010). Differentiation of Bacteria Using Fatty Acid Profiles from Gas Chromatography-Tandem Mass Spectrometry. *Journal of the Science of Food and Agriculture*. Vol.90, No.8, (June 2010), pp. 1380-1383, ISSN 1097-0010

Macarrón G.B. (1998). Utilización de Marcadores Lipídicos en el Estudio de la Biomasa, la Estructura y el Estado Nutricional de las Comunidades de los Tapices Microbianos del Delta del Ebro. Ph. D. Thesis, Universitat Autónoma de Barcelona, España. 250 pp.

Muñoz, M.; Julián, E.; García-Barceló, M.; Ausina V. & Luquin M. (1997). Easy Differentiation of *Mycobacterium mucogenicum*, from Other Species of the *Mycobacterium fortuitum* Complex by Thin Layer and Gas Chromatography of Fatty Esters and Alcohols. *Journal of Chromatography B*, Vol.689, No.2, (February 1997), pp. 341-347, ISSN 1570-0232

Newton, J.W. & Newton, N.G.A. (1957). Composition of the Photoactive Subcellular Particles from *Chromatium*. *Archives of Biochemistry and Biophysics*, Vol.71, No. 1, (September 1957), pp. 250-265, ISSN: 0003-9861

Nichols, D.S.; Nichols, P.D.; Russel, N.J.; Davies, N.W. & McMeekin, A.T. (1997). Polyunsaturated Fatty Acids in the Psychrophilic Bacteria *Shewanella gelidimarina* ACAM 456T: Molecular Species Analysis of Major Phospholipids and Biosynthesis of Eicosapentaenoic Acid. *Biochimica et Biohphysica Acta-Lipids and Metabolism*, Vol.1347, No.2-3 (August 1997), pp. 164-176, ISSN 0005-2760.

Núñez-Cardona, M.T.; Donato-Rondon, J.Ch.; Mas, J. & Reynolds, C. (2008). A Phototrophic Sulfur Bacterium from a High Altitude Lake in the Colombian Andes, *Journal of Biological Research-Thessaloniki*, Vol.9, (June 2008), pp. 17-24, ISSN 1790045X

Pfennig, N. & Trüper, H.G. (1981). Isolation of Members of the Families Chromatiaceae and Chlorobiaceae, In: *The Prokaryotes*, M. Starr, H. Stolp, H.G. Trüper, H.G. Balows, H. Schlegel, (Eds.), 279-289, Springer-Verlag, ISBN 0387-97258-7, New York, USA.

Purcaro, G.; Tranchida, P.Q.; Dugo, P.; La Camera, E.; Bisignano, G.; Conte, L. & Mondello, L. (2010). Characterization of Bacterial Lipid Profiles by Using Rapid Sample Preparation and Fast Comprehensive Two-Dimensional Gas Chromatography in Combination with Mass Spectrometry. *Journal of Separation Science*, Vol.33, No.15, (August 2010), pp. 2334- 2340, ISSN 1615-9306

Rabinowitch, H.D.; Sklan, D.; Chance, D.H.; Stevens, R.D. & Fridowich, I. (1993). *Escherichia coli* Produces Linoleic Acid During Late Stationary Phase. *Journal of Bacteriology*, Vol.175, No.17, (September 1993), pp. 5324-5328, ISSN 0021-9193

Rainey, P.B.; Thompson, I.B. & Palleroni, N.J. (1994). Genome and Fatty Acid Analysis of *Pseudomonas stutzeri*. *International Journal of Systematic Bacteriology*, Vol.44, No.1, (January 1994), pp. 54-61, ISSN 0020-77013

Ramana, V.V.; Sasikala, Ch.; Ramaprasad, E.V.V. & Ramana, Ch.V. (2010). Description of *Ecthiorhodospira salini* sp. nov. *Journal of General and Applied Microbiology*, Vol.56, No.4, (August 2010), pp. 313-319, ISSN 1349-8037

Ratledge, C. & Wilkinson, S.G. (1988). Fatty Acids, Related and Derived Lipids, In: *Microbial Lipids, Vol.1*, C. Ratledge & S.G Wilkinson, (Eds.), 23-52, Academic Press, ISBN 0125823045, London UK

Ryu, H.S.; Chung, B.S.; Park, M.; Lee, S.S. & Jeon, Ch.O. (2008). *Rheinheimera soli* sp. nov. a γ–Proteobacterium Isolated from Soil in Korea. *International Journal of Systematic and Evolutionary Microbiology*, Vol.58, No.10, (October 2008), pp. 2271-2274, ISSN 1466-5034

Russell, N.J. & Fukunaga, N. (1990). A Comparison of Thermal Adaptation of Membrane Lipids in Psychrophilic and Themophilic Bacteria. *FEMS Microbiology Reviews*, Vol.75, No.2-3, (June 1990), pp. 171-182, ISSN 1574-6968

Sasser, M. (2001). Identification of Bacteria by Gas Chromatography of Cellular Fatty Acids, In: MIDI Technical Note #101. Newark, USA, 08.08.2011. Available from http://www.midi.inc.com

Scheuerbrandt, G. & Bloch, K. (1962). Unsaturated Fatty Acids in Microorganisms. *Journal of Biological Chemistry*, Vol.237, No.7, (July 1962), pp. 2064-2068, ISSN 0021-9258

Stanier, R.Y. (1968). Biochemical and Immunological Studies on the Evolution of a Metabolic Pathway in Bacteria. In: *Chemotaxonomy and Serotaxonomy*, J.G. Hawks, (Ed.), 201-225, Academic Press, ISBN-10: 0123333504, London, UK.

Sucharita, K.; Kumar, E.S.; Sasikala, Ch.; Panda, B.B.; Takaichi, S. & Ramana, Ch.V. (2010). *Marichromatium fluminis* sp. nov., a Slightly Alkaliphilic, Phototrophic Gamma-proteobacterium Isolated from River Sediment. *International Journal of Systematic and Evolutionary Microbiology*, Vol.60, No.5, (May 2010), pp. 1103-1107, ISSN 1466-5034.

Tacks, B.J. & Holt, S.C. (1971). *Thiocapsa floridana*: a Cytological, Physical and Chemical Characterization. II. Physical and Chemical Characteristics of Isolated and Reconstituted Chromatophores. *Biochimica et Biophysica Acta-Biomembranes*, Vol.233, No.2, (April 1971), pp. 278-295, ISSN 0005-2736

Tindall, B.J.; Roselló-Mora, R.; Busse H-J.; Ludwing, W. & Kampfer, P. (2010). Notes on the Characterization of Prokaryote Strains for Taxonomic Purposes. *International*

Journal of Systematic and Evolutionary Microbiology, Vol.60, No.1, (August 2009), pp. 249-266, ISSN 1466-5034

Van Niel, C.B. (1971). Techniques for the Enrichment, Isolation and Maintenance of the Photosynthetic Bacteria. *Methods in Enzymology, Photosynthesis Part A (Vol. 23),* A. San Pietro (Ed.), 3-28. Academic Press, ISBN: 0121818861/0-12-181886-1, Exeter Dev. UK

Watanabe, K.; Ishikawa, Ch.; Ohtsuka, I.; Kamata, M; Tomita M.; Yazawa, K. & Muramats, H. (1997), Lipid and Fatty Acids Composition of Novel Docosahexaenoic Acid-Producing Marine Bacterium. *Lipids,* Vol.32, No.9, (September, 1997), pp. 975-978, ISSN: 0024-4201

Yakimov, M.M.; Giuliano, L.; Chernikova, T.N.; Gentile, G.; Abraham, W.-R., Lünsdorf, H.; Timmis, K.N. & Golyshin, P.N. (2001). *Alcalilimnicola halodurans* gen. nov., sp. nov., an Alkaliphilic, Moderately Halophilic and Extremely Halotolerant Bacterium, Isolated from Sediments of Soda-Depositing Lake Natron, East Africa Rift Valley. *International Journal of Systematic and Evolutionary Microbiology,* Vol.5, Part.6, (November 2001), pp. 2133-2143, ISSN 1466-5034

Yano, Y.; Nakayama, A.; Saito, H. & Ishiharo, K. (1994). Production of Docosahexaenoic Acid by Marine Bacteria Isolated From Deep Sea. *Lipids,* Vol.29, No.7, (July 1997), pp. 527-528, ISSN 0024-4201

Yawaza, K. (1996). Production of Eicosapentaenoic Acid From Marine Bacteria. *Lipids,* Vol.31, No.1, (March, 1996), pp. S297-S300, ISSN 0024-4201

Application of Gas Chromatography in a Study of Marine Biogeochemistry

Luisa-Fernanda Espinosa[1],
Silvio Pantoja[2] and Jürgen Rullkötter[3]
[1]INVEMAR, Cerro de Punta Betin, A.A., Santa Marta,
[2]Department of Oceanography and Center for Oceanographic Research in
the Eastern South Pacific, University of Concepcion, P.O., Concepción,
[3]Institute of Chemistry and Biology of the Marine Environment (ICBM),
Carl von Ossietzky University of Oldenburg, P.O. Oldenburg,
[1]Colombia
[2]Chile
[3]Germany

1. Introduction

A goal of marine biogeochemistry is to characterize the chemical composition of particulate organic matter (POM) in the oceanic water column in order to evaluate sources, reactivity and the potential for preservation of POM in the sedimentary environment, given its importance in biogeochemical cycles in the ocean. Organic matter transformation has traditionally been measured as changes in the concentration of organic matter (determined as organic carbon) at different depths (e.g., Gundersen et al., 1998; Wakeham, 1995). Nevertheless, studies of the transformation of specific compounds provide more precise information regarding degradation mechanisms (Abramson et al., 2010; Lee et al., 2004; Pantoja & Lee, 2003). Some of the specific compounds that have been investigated are amino acids (e.g., Abramson et al., 2010; Pantoja & Lee, 2003), carbohydrates (e.g., Çoban-Yildiz et al., 2000), and many lipid classes (Burns et al., 2003; Galois et al., 1996; Minor et al., 2003; Parrish et al., 2000; Sheridan et al., 2002; Sun et al., 1997; Treignier et al., 2006; Wakeham, 1995; Wakeham et al., 1997, 2002).

Lipids could be efficiently characterized with standard chromatographic techniques such as the gas chromatography-mass spectrometry techniques. Although lipids may be a small fraction of the organic matter in plankton (about 15 %) (Wakeham et al., 2000), their composition has been extensively studied to learn about the sources, fluxes, and alterations of organic matter in the water column and sediments (e.g., Grossi et al., 2001; Muri et al., 2004;; Treignier et al., 2006; Wakeham, 1995). Lipids are suitable for such studies because they can be source-specific and, as such, are appropriate "biomarkers" (Pantoja & Wakeham, 2000; Pazdro et al., 2001; Volkman & Tanoue, 2002).

In this chapter, we will present the results of our research carried out in the upwelling system off Antofagasta, northern Chile. The Antofagasta coastal zone ($\approx 23 \circ S$) is part of the

Humboldt Current System (HCS) and is characterized by high biological production (Daneri et al., 2000), which is followed by a high export of organic matter (González et al., 2000, 2009) and its preservation in the sediments. The quantity and quality of organic matter preserved in the sediments is controlled by production at the sea surface and decomposition processes during transport in the water column (Lee et al., 2004). In highly productive environments, such as HCS, we would expect to find a clear imprint of marine processes in the distribution of organic molecules in the water column, with a minor contribution of terrestrial biomarkers. Thus, we determined the distribution pattern of free alcohols, free fatty acids, and sterols with the goal of assessing the molecular distribution of lipids in the water column to evaluate sources and reactivity and to infer their potential for preservation in the sedimentary environment. This study of marine biogeochemistry represents an example of gas chromatography – mass spectrometry application.

2. Methods

The investigation was carried out off Antofagasta in northern Chile, at two stations: one coastal (23°16'S; 70°40.3'W) and one oceanic (23°06.9'S; 71°58'W) (Fig. 1). Sampling was carried out during the *FluMO* (*Fluxes of Organic Matter*) cruise on board the RV Abate Molina, from April 18 to April 25, 2001. Suspended particulate organic matter was collected from the oxygenated surface layer, the Oxygen Minimum Zone (OMZ), and the oxygenated deep layer using a rosette equipped with Niskin bottles and a CTDO. Between 120 and 150 L of water were filtered through a 90 mm diameter borosilicate filter (nominal pore size 0.7 µm) precombusted at 450°C for 4 hours. Filters were kept frozen (-20 °C) until analysis. Vertical profiles of dissolved O_2 were determined with the CTDO sensor, calibrated using the Winkler method, as published elsewhere (Pantoja et al., 2009).

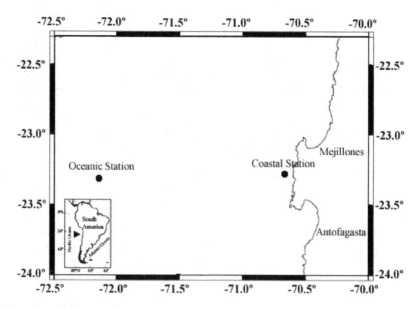

Fig. 1. Location of the sampling stations.

Filters were sub-sampled for determination of particulate organic carbon content, which was calculated as the difference between total carbon and inorganic carbon. Total carbon was analyzed by combustion in a Leco CS-444 instrument, and carbonate was determined by acidification using a UIC coulometer.

Filters were ultrasonically extracted with a solvent mixture of dichloromethane/methanol (2:1 v:v) for 30 minutes in centrifuge tubes (three times). After centrifugation, the combined supernatant was extracted three times with dichloromethane (30 mL each). The combined extracts were dried overnight with anhydrous sodium sulfate, concentrated in a rotary evaporator, and dried with nitrogen gas.

The total extract was dissolved in a small amount of dichloromethane, and 5α-androstan-3β-ol was added as an internal standard. The hexane-insoluble fraction (asphaltenes) was precipitated by adding an excess amount of n-hexane. After filtration, the asphaltene-free extract was separated into three fractions by liquid chromatography over an SiO_2 column (silica gel 60 Merck, activated with 5% water): Fraction 1 containing aliphatic and aromatic hydrocarbons was eluted with 7 mL hexane/dichloromethane (9:1); fraction 2 containing phthalates was eluted with 5 mL hexane/dichloromethane (1:1); and fraction 3 containing alcohols, sterols, stanols, and free fatty acids was eluted with dichloromethane/methanol (9:1). The third fraction was derivatized with N-methyl-N-trimethylsilyltrifluoroacetamide (MSTFA) to form trimethylsilyl (TMS) ethers of alcohols, sterols, and and TMS esters of fatty acids.

Compounds eluted in the third fraction were analyzed on a Hewlett-Packard 5890 Series II gas chromatograph equipped with a DB5-HT high temperature fused silica capillary column (30 m length; 0.25 mm ID; 0.1 µm film thickness), a flame ionization detector, and a Gerstel KAS3 cold injection system. Helium was used as the carrier gas. Separations were achieved using a temperature program of 4°C/min from 60°C to 380°C. Selected samples were analyzed by gas chromatography-mass spectrometry (GC-MS) using a Finnigan MAT SSQ 710B mass spectrometer equipped with the same type of GC column and using helium as the carrier gas. Mass spectra were acquired over the range of m/z 50-650 at a rate of 1 scan/s, with an ionization electron energy of 70 eV. The temperature was programmed from 60°C to 350°C at a rate of 3°C/min.

Most extracts were analyzed by GC-MS for compound identification based on the elution order and a comparison of the mass spectral pattern with published data. Lipid quantification was based on response factors derived from the internal standard (5α-Androstan-3β-ol).

A Principal Component Analysis (PCA) was performed in order to determine the distribution patterns of biomarkers in the water column and to identify lipid sources and lability. For this, we used 44 lipid compounds present in nearly all samples, combined with the oxygen concentration, organic carbon content, and water depth. A correlation matrix was set up with the composition of individual compounds relative to total lipids in each sample, normalized according to:

$$n_{ik} = \frac{\left(x_{ik} - x\right)}{S_i}$$

where x is the relative percent of compound (i) in sample (k), and S is the variance (Reemtsma & Ittekkot, 1992).

The PCA was performed using STATISTICA software v. 99. Principal components were calculated from the correlation matrix and treated by Varimax rotation in order to maximize the load of each variable on one factor (Reemtsma & Ittekkot, 1992).

3. Results and discussion

In this study, 44 lipid compounds were identified and quantified from the following classes: alcohols (C_{14} to C_{32}), fatty acids (C_{14} to C_{21}), and sterols (C_{26} to C_{30}); these represented between 0.8 and 0.04% of total organic carbon (Table 1). Alcohols constituted the greatest percentage of the three lipid classes analyzed at both stations (45% at the coastal station and 37% at the oceanic station). Phytol, which is released by the chlorophyll a molecule through enzymatic hydrolysis when zooplankton consumes phytoplankton, was the most abundant alcohol (Table 2). In the laboratory, phytol release was achieved through saponification (Wakeham et al., 2002). In the present study, phytol is used as a biomarker for the decay of chlorophyll a, as the samples were not saponified prior to GC-MS analysis.

Station	Depth (m)	C_{org} ($\mu g\ C_{org}/L$)	% C_{org} Total Lipids	% Total Lipids n-Alcohols	% Total Lipids Fatty acids	% Total Lipids Sterols
Oceanic	100	122	0.22	38	30	32
	220	66	0.22	43	31	26
	1000	91	0.04	31	46	21
Average				37	36	26
Coastal	20	177	0.78	55	15	30
	250	76	0.52	54	27	20
	300	68	0.29	42	34	24
	800	93	0.04	29	29	42
Average				45	26	29

Table 1. Organic carbon (C_{org}), and percent of lipid concentrations to C_{org} in particulate organic matter.

The total lipid concentration (alcohols + fatty acids + sterols) was five times greater at the coastal than at the oceanic station (Figs. 2a, b). At both stations, the concentration of total lipids was highest in the samples from shallowest water depth, reaching values of 1.97 µg/L at the coastal station (20 m) and 0.20 µg/L at the oceanic station (100 m). This agrees with the greater primary production measured at the coastal station (between 1.3 and 3.3 g C/m² d) as compared with that at the oceanic station (between 0.27 and 0.30 g C/m² d) (Daneri and Pantoja, unpublished results). The total lipid concentrations decreased with depth. At the coastal station, the decrease was more significant, reaching values two orders of magnitude lower (0.04 µg/L) in the deepest sample (800 m). At the oceanic station, the decrease was less significant, dropping only to one half (0.11 µg/L) at 1000 m depth (Figs. 2a, b).

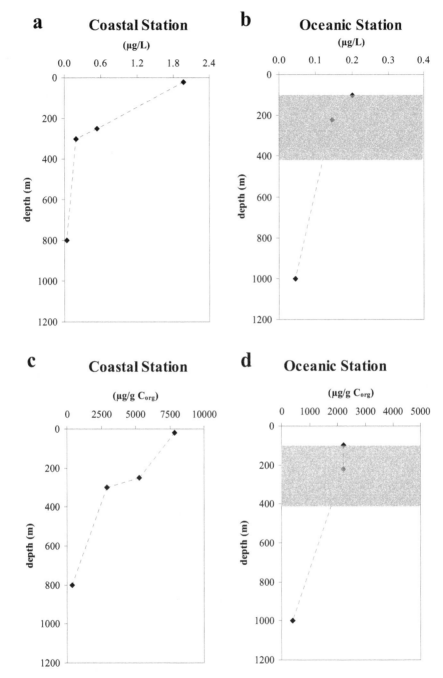

Fig. 2. Total lipid concentrations, and lipid/Corg ratios in particulate organic matter. Shaded areas represent the Oxygen Minimum Zone ([O2] < 22.5 μM).

Most of the particulate organic matter in the ocean is biosynthesized at the surface by photoautotrophic plankton via photosynthetic fixation of carbon (Lee et al., 2004). Although a lesser percentage could be biosynthesized by the microbial activity in the deeper ocean layers, the changes observed in the vertical concentrations (Table 2, Figs. 2a, b) mainly reflect the breakdown of the lipid molecules in transit to the seafloor. When the lipid concentration is normalized to the organic carbon content (Figs. 2c, d), it becomes obvious that the proportion of lipid components decreases in relation to the total organic carbon

Compound	Coastal Station				Oceanic Station		
	20 m	250 m	300 m	800 m	100 m	220 m	1000 m
Sterols							
A = 27-*nor*-24-Methylcholesta-5,22E-dien-3β-ol	3.43	2.24	2.07	nd	1.95	4.86	5.22
B = 27-*nor*-24-Methyl-5α-cholest-22E-en-3β-ol	4.5	nd	nd	nd	nd	nd	nd
C = Cholesta-5,22E-dien-3β-ol	4.58	0.93	nd	nd	1.06	1.28	1.93
D = 5α-Cholest-22E-en-3β-ol	4.07	0.62	nd				
E = Cholest-5-en-3β-ol	71.9	7.24	5.87	nd	nd	0.57	nd
F = 5α-Cholestan-3β-ol	8.70	2.45	4.01	7.24	5.86	6.76	5.64
G = 24-Methylcholesta-5,22(E)-dien-3β-ol	50.8	8.08	11.3	0.89	1.21	1.34	0.90
				nd	3.69	3.52	4.67
H = 24-Methyl-5α-cholest-en-3β-ol	4.63	1.26	0.55	nd	nd	0.82	1.28
I = 24-Methyl-5α-cholest-22E-dien-3β-ol	5.01						
		6.71	0.64	nd	2.49	0.78	nd
J = 24-Methylcholesta-5,24(28)-dien-3β-ol	107	14.4	3.41				
				0.75	8.00	nd	3.95
K = 24-Methyl-5α-cholest-24(28)-en-3β-ol	2.32	0.60	nd	nd	nd	nd	nd
L = 24-Methylcholest-5-en-3β-ol	59.4	6.84	0.88	0.93	nd	nd	nd
M = 23,24-Dimethylcholesta-5,22E-dien-3β-ol	6.29	0.92	nd	nd	0.79	nd	1.27
N = 24-Ethyl-5α-cholest-22E-en-3β-ol	16.1						
O = 23,24-Dimethylcholest-5-en-3β-ol		4.40	5.24	1.20	1.07	1.49	2.32
	2.37	1.32	0.84	nd	nd	nd	nd
P = 24-Ethylcholest-5-en-3β-ol	22.3						
Q = 24-Ethyl-5α-cholestan-3β-ol	4.12	3.98	2.62	nd	2.99	1.98	1.68
R = 4α,23,24-Trimethyl-5α-cholest-22E-en-3β-ol	31.8	4.55	4.34	5.03	0.910	nd	5.94
		9.04	1.52	nd	3.07	3.24	3.29
S = 4α,23,24-Trimethyl-5α-cholestan-3β-ol	3.68	1.01	0.91	nd	6.66	nd	3.64
Alcohols							
nC_{14}-OH	20.6	4.53	4.37	1.13	1.45	1.26	0.73
nC_{16}-OH	11.5	6.17	8.08	2.04	3.79	5.84	6.20
nC_{17}-OH	4.96	1.88	1.53	nd	nd	0.900	1.47
nC_{18}-OH	31.5	10.0	13.1	3.12	6.27	13.2	10.8
Phytol	580	143	10.5	0.70	5.69	1.65	2.95
nC_{19}-OH	3.16	2.13	1.65	nd	nd	0.79	0.98
nC_{20}-OH	13.5	3.93	6.19	1.94	1.80	3.66	2.49
nC_{21}-OH	6.74	1.59	0.87	nd	3.12	1.66	4.52
nC_{22}-OH	34.8	14.2	17.0	1.70	6.40	21.2	22.8
nC_{23}-OH	nd	nd	nd	nd	1.08	0.780	nd

nC_{24}-OH	19.0	6.31	5.91	0.580	2.17	3.69	5.06
nC_{26}-OH	18.5	4.77	5.80	nd	1.33	2.49	3.49
nC_{28}-OH	8.43	7.86	nd	nd	2.78	2.83	3.63
nC_{30}-OH	4.18	1.73	nd	nd	2.90	0.950	nd
nC_{32}-OH	4.71	2.50	0.80	nd	1.33	0.740	0.590
$(\Sigma C_{14}-C_{20})/(\Sigma C_{22}-C_{32})$	1.03	0.811	1.21	3.61	0.914	0.916	0.695
$(\Sigma oddC_{15}-C_{19})/(\Sigma evenC_{16}-C_{20})$	0.119	0.152	0.089	0.0	0.137	0.075	0.126
Fatty Acids							
$C_{14:0}$	24.6	23.3	16.8	2.14	4.22	5.87	9.16
$i\text{-}C_{15:0}$	3.92	1.80	1.97	nd	1.29	2.86	2.72
$ai\text{-}C_{15:0}$	4.72	3.30	2.78	nd	1.61	3.10	3.37
$C_{16:1}$	20.0	12.4	7.10	nd		1.33	2.93
$C_{16:0}$	82.4	38.5	14.2	4.89	12.4	17.4	38.9
$i\text{-}C_{17:0}$	4.80	2.35	nd	nd	1.82	nd	1.01
$ai\text{-}C_{17:0}$	3.30	nd	nd	nd		nd	nd
$C_{18:1}$	28.7	11.9	16.4	1.60	5.16	5.53	15.0
$C_{18:0}$	28.2	6.14	14.8	2.38	4.59	9.71	18.5
$C_{21:0}$	7.84	5.92	4.14	nd	1.60	nd	1.18

Table 2. Concentration of lipids (ng L^{-1}) identified in suspended particulate organic matter at the two sampling stations off Antofagasta. nd = not detected.

(C_{org}). This is interpreted to show that the lipid molecules analyzed in this study are more labile than other components representing the bulk of the organic matter.

At the coastal station, phytol and the sterols were the most abundant compounds at the surface (20 m) and had the greatest decrease in concentration with increasing depth (Table 2, Fig. 3a). At the oceanic station, the shallowest sample was collected at the oxycline (100 m), which is the transition zone between the well-oxygenated surface layer and the deep layer, where oxygen concentrations fall below 22.5 µM (suboxic conditions). At this depth, the most abundant compounds were the fatty acids; their concentration decreased to 50% at 1000 m depth (Fig. 3b).

The vertical distribution of the concentration of each lipid class normalized to C_{org} indicates, on the one hand, the relative importance of each class, and on the other, the preferential consumption of some compound types with respect to others (Lee et al., 2004). At the coastal station, the most abundant compounds at 20 m were phytol (3.3 x 10^3 µg/g C_{org}) and the sterols (2.3 x 10^3 µg/g C_{org}), but within the Oxygen Minimum Zone, the fatty acids (1.4 x 10^3

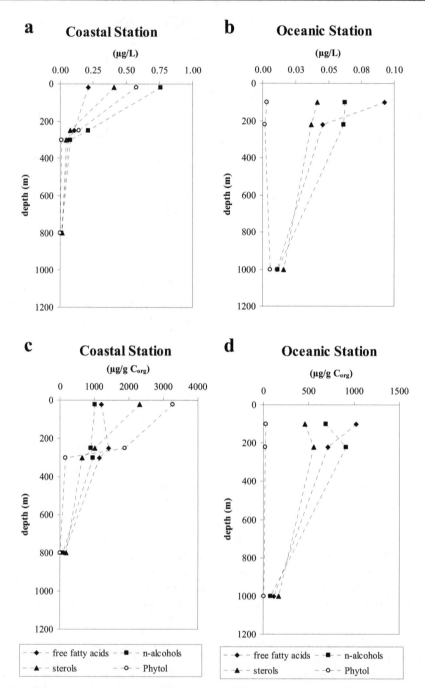

Fig. 3. Lipid class concentrations, and lipid classes/Corg ratios in particulate organic matter. Shaded areas represent the Oxygen Minimum Zone ([O2] < 22.5 µM).

μg/g C_{org}) and *n*- alkanols (0.95 x 10^3 μg/g C_{org}) were more abundant (Table 2, Fig. 3c). If we consider photosynthetic production to be the main source for the formation of organic matter at the ocean surface, our results show that phytol and sterols are preferentially consumed in the water column, as they disappear more quickly than C_{org} with increasing depth. At the coastal station, the fatty acids are the most abundant lipids at 100 m depth (1.0 x 10^3 μg/g C_{org}), but in the Oxygen Minimum Zone, the *n*- alkanols (0.9 x 10^3 μg/g C_{org}) and sterols (0.5 x 10^3 μg/g C_{org}) are more abundant than the fatty acids (Table 2, Fig. 3c). At the oceanic station, the compounds that disappear most quickly with depth are the fatty acids (Table 2, Fig. 3d), meaning that these are the most labile compounds in this environment.

3.1 Sterols

At the coastal station, the sterols represented 29% of all three lipid classes analyzed (Table 1). The surface sample (20 m) was dominated by 24-methylcholesta-5,24(28)-dien-3β-ol (24-methylenecholesterol), cholest-5-en-3β-ol (cholesterol), 24-methylcholest-5-en-3β-ol, and 24-methylcholesta-5,22(E)-dien-3β-ol (diatomsterol) (Table 2, Fig. 4a). The sterols 24-methylenecholesterol and diatomsterol are phytoplankton biomarkers (known as phytosterols) (Burns et al., 2003; Volkman et al., 1998; Wakeham et al., 1997), fundamentally diatom biomarkers (Burns et al., 2003; Parrish et al., 2000), and cholesterol is a biomarker of both phytoplankton and zooplankton (Burns et al., 2003). Although less abundant than the diatom biomarkers, some sterols (4α,23,24-trimethyl-5α-cholest-22E-en-3β-ol and 23,24-dimethylcholesta-5,22E-dien-3β-ol), biomarkers of dinoflagellates, were detected (Fig. 4a) (Burns et al., 2003; Volkman et al., 1998; Wakeham et al., 2002).

A general decrease in the relative abundance of phytosterols was observed within the Oxygen Minimum Zone (sampling depths 250 and 300 m). In this area, the phytosterols 24-methylenecholesterol, diatomsterol, cholesterol, 24-methylcholest-5-en-3β-ol, and 4α,23,24-trimethyl-5α-cholest-22E-en-3β-ol were equally predominant (Fig. 4a). An increase in the abundance of some stanols such as 5α-Cholestan-3β-ol, 24-methyl-5α(H)-cholest-22E-dien-3β-ol, and 24-ethyl-5α(H)-cholest-22E-en-3β-ol was also evident. Wakeham, 1989 sugested these compounds are produced by the bacterial hydrogenation of stenols under suboxic conditions, for which their presence has been recognized as the result of diagenetic transformation (Parrish et al., 2000; Volkman et al., 1998; Wakeham et al., 2007).

In the deepest sample (800 m), the relative abundance of cholesterol, 24-methylcholest-5-en-3β-ol, 24-ethyl-5α-cholestan-3β-ol, 24-ethyl-5α(H)-cholest-22E-en-3β-ol, and 5α-cholestan-3β-ol increased to different extents compared to the Oxygen Minimum Zone and partly also to the surface waters (Fig. 4a). At this depth, cholesterol was the most common sterol and, although common in phytoplankton, the principal source of this compound is zooplankton.

At the oceanic station, sterols represented 26% of the lipids on average (Table 1). The relative abundance pattern was different from that at the coastal station (Fig. 4b). At 100 m depth (oxycline), the most abundant sterols were 24-methylcholesta-5,24(28)-dien-3β-ol, cholesterol, and diatomsterol. At this station, a high proportion of the A-ring stanols 24-methyl-5α-cholest-24(28)-en-3β-ol, and 24-ethyl-5α-cholestan-3β-ol was also observed; these are used as indicators of bacterial decay of organic matter under suboxic conditions (Wakeham & Ertel, 1987).

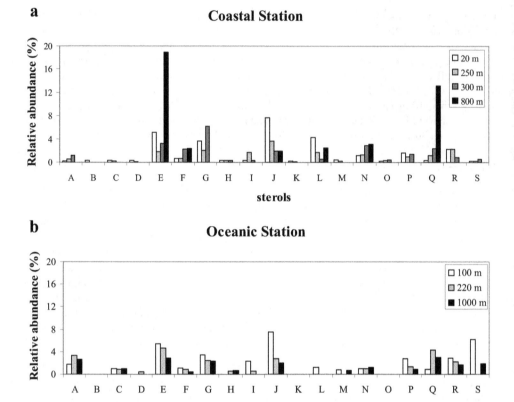

Fig. 4. Distribution of sterols in particulate organic matter. Letter codes are defined in Table 2.

Within the Oxygen Minimum Zone (220 m), the relative abundance of the majority of the sterols identified decreased (Fig. 4b), except for 27-*nor*-24-methylcholesta-5,22E-dien-3β-ol and 24-ethyl-5α-cholestan-3β-ol. The sterol 27-*nor*-24-methylcholesta-5,22E-dien-3β-ol, although commonly reported in sediments, has also been found in some dinoflagellates. Unlike at the coastal station, 24-ethyl-5α-cholestan-3β-ol was the only stanol that increased within the Oxygen Minimum Zone (Fig. 4b).

In spite of their different profiles in terms of relative sterol abundance, biomarkers for the diatoms, i.e. 24-methylenecholesterol and diatomsterol, were predominant at both stations (Fig. 4a, b). This result is not surprising, since studies performed along the northern coast of Chile off Antofagasta have shown that the phytoplankton assemblages in this area are more dominated by diatoms than by dinoflagellates (González et al., 2000; Iriarte y González, 2004; Iriarte et al., 2000).

3.2 Alcohols

The *n*-alkanols are formed by hydrolysis of esterified alcohols and are derived from a large variety of marine organisms (phytoplankton, zooplankton, bacteria) or higher plants

(Treignier et al., 2006; Jeng et al., 2003). In the Antofagasta area, the identified n-alkanols represented, on average, 45% of the total lipids at the coastal station and 37% at the oceanic station (Table 1). At both stations, 16 n-alkanols were identified in the range between $nC_{14}OH$ and $nC_{32}OH$ (Table 2), with a predominance of even-carbon-numbered chains, i.e. $nC_{16}OH$, $nC_{18}OH$, $nC_{20}OH$, and $nC_{22}OH$; of these, $nC_{22}OH$ was the most abundant alkanol (Fig. 5). The short-chain n-alkanols ($\leq C_{20}$) have a marine origin (phytoplankton, zooplankton, bacteria), whereas the long-chain alkanols ($\geq C_{22}$) are attributed to terrestrial plants (Treignier et al., 2006; Wakeham, 2000; Jeng et al., 2003). Nonetheless, $nC_{22}OH$ is also attributed to occur in cyanobacteria. For example, studies carried out in the Baltic Sea found that $nC_{22}OH$ was the most abundant n-alkanol in samples containing principally cyanobacteria (Volkman et al., 1998). Given the very low contribution from terrestrial sources in the arid zone off Antofagasta, we attribute the presence of $nC_{22}OH$ to cyanobacteria.

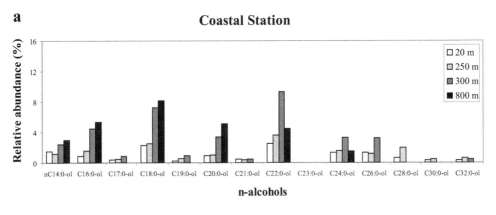

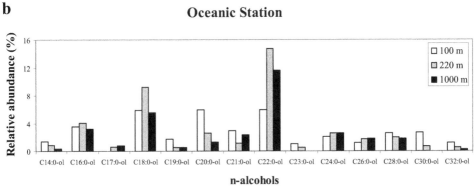

Fig. 5. Distribution of n-alkanols in particulate organic matter.

In order to determine the sources of the n-alkanols in aquatic environments, Fukushima and Ishiwatari developed a parameter that is based on the ratio of the sum of short-chain n-alkanols (Σ [$nC_{14}OH$–$nC_{20}OH$]) over the sum of long-chain n-alkanols (Σ [$nC_{22}OH$–$nC_{30}OH$])

(Treignier et al., 2006). Values >1 reveal a predominance of short-chain compounds and values <1 a predominance of long-chain compounds (Treignier et al., 2006). The values calculated for our samples at each depth level show that ratios >1 occur in the coastal zone (except at 250 m depth) and in the oceanic zone, indicating that the n-alkanols are primarily planktonic in origin (Table 2). We feel that the dominance of the $nC_{22}OH$ alcohol found at all depths at the oceanic station (Fig. 5b) should be attributed to cyanobacteria, as found for other ecosystems (Volkman et al., 1998).

The isoprenoid alcohol phytol (3,7,11,15-tetramethylhexadec-2-en-1-ol) was also found in all samples. This compound occurred at high concentrations (0.58 µg/g) in the coastal zone at 20 m depth and its concentration decreased rapidly with increasing water depth until values three orders of magnitude lower (0.001 µg/g) were reached at 800 m depth (Fig. 3a). At the oceanic station, the phytol concentration was greatest at 100 m (0.006 µg/g) and, as at the coastal station, decreased with depth until, at 1000 m, the compound was no longer detected (Fig. 3b). Phytol is a biomarker of phytoplankton consumed by zooplankton. It is released from chlorophyll *a* through enzymatic hydrolysis when zooplankton grazes on phytoplankton (Sheridan et al., 2002). As explained earlier, primary production is higher at the coastal than at the oceanic station, so that we expect greater grazing activity in the former environment as well. The abundance of metazoans (or zooplankton) was three times greater (2189 ± 1888 ind/m³) in the surface layer (0-50 m) at the coastal station than at the oceanic station (625 ± 184 ind/m³) (Zenteno-Devaud, unpublished results), which is supporting evidence for the phytol profiles observed. On the other hand, phytol is a fairly labile molecule, since it practically disappears in the deepest zones at both stations (800 and 1000 m) (Figs. 3a, b).

The isoprenoid alcohol phytol (3,7,11,15-tetramethylhexadec-2-en-1-ol) was also found in all samples. This compound occurred at high concentrations (0.58 µg/g) in the coastal zone at 20 m depth and its concentration decreased rapidly with increasing water depth until values three orders of magnitude lower (0.001 µg/g) were reached at 800 m depth (Fig. 3a). At the oceanic station, the phytol concentration was greatest at 100 m (0.006 µg/g) and, as at the coastal station, decreased with depth until, at 1000 m, the compound was no longer detected (Fig. 3b). Phytol is a biomarker of phytoplankton consumed by zooplankton. It is released from chlorophyll *a* through enzymatic hydrolysis when zooplankton grazes on phytoplankton (Sheridan et al., 2002). As explained earlier, primary production is higher at the coastal than at the oceanic station, so that we expect greater grazing activity in the former environment as well. The abundance of metazoans (or zooplankton) was three times greater (2189 ± 1888 ind/m³) in the surface layer (0-50 m) at the coastal station than at the oceanic station (625 ± 184 ind/m³) (Zenteno-Devaud, unpublished results), which is supporting evidence for the phytol profiles observed. On the other hand, phytol is a fairly labile molecule, since it practically disappears in the deepest zones at both stations (800 and 1000 m) (Figs. 3a, b).

3.3 Fatty acids

The free fatty acids represented 26% of the total lipids at the coastal station and 36% at the oceanic station (Table 1). As with the alcohols, the most abundant fatty acids were straight-chain compounds, which represented more than 65% of fatty acids at all depths, followed in abundance by unsaturated and branched (especially iso + anteiso) fatty acids, respectively

(Fig. 6). The straight-chain fatty acids C_{14}, C_{16}, and C_{18} come from mixed planktonic sources of phytoplankton, zooplankton (Canuel & Zimmerman, 1999; Wakeham , 1995), and bacteria (Gong & Hollander, 1997). In cyanobacteria, fatty acids are dominated by the C_{16} straight-chain acid (Wakeham, 1995), whereas diatoms biosynthesize large amounts of the 16:1ω7 acid (Wakeham et al., 2007). The absence of long-chain fatty acids (>24 carbon atoms) is indicative of the very low continental contribution in the northern zone of Chile (absence of important rivers), since the long-chain fatty acids are terrestrial higher plant markers (Derieux et al., 1998; Parrish et al., 2000).

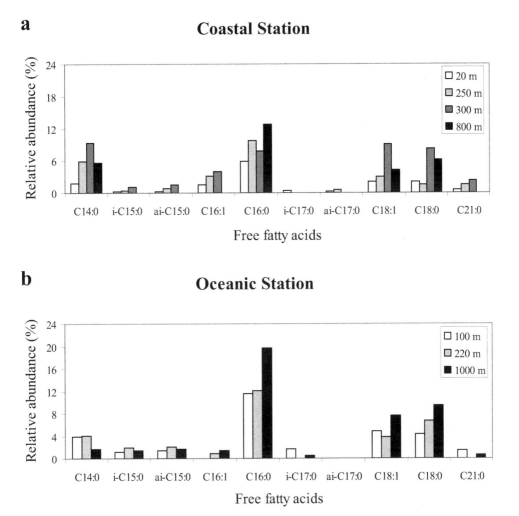

Fig. 6. Distribution of fatty acids in particulate organic matter.

In natural systems, fatty acids are found free or associated with other compounds through ester bonds. The associated fatty acids are more abundant (S.D. Killops & V.J. Killops, 1994; Pazdro et al., 2001) and mainly occur as constituents of phospholipids which are integral constituents of the membranes of all living organisms (Petsch et al., 2003, Wakeham et al., 2007). The relative abundance of free fatty acids analyzed in this work can be interpreted as the result of reworking of the organic matter of decayed organisms in the water column, more than as specific indicators of organism abundance (Pazdro et al., 2001; Wakeham, 1995). Thus, the greater relative abundance of saturated fatty acids over monounsaturated fatty acids (Fig. 6) and the absence of polyunsaturated fatty acids can be interpreted as the result of the increasing metabolization of organic matter with depth since the monounsaturated and polyunsaturated fatty acids are more susceptible to breakdown than the saturated fatty acids (Derieux et al., 1998; Galois et al., 1996; Pazdro et al., 2001).

The relative abundance of *iso-* and *anteiso-*$C_{15:0}$ was greater within the Oxygen Minimum Zone, as compared to surface and deep samples at both the coastal (250 – 300 m) and oceanic (220 m) stations (Figs. 6a, b). Considering that i and ai-$C_{15:0}$ fatty acids are markers of bacteria and that they are commonly used as indicators of intense reworking of organic matter (Pazdro et al., 2001; Wakeham, 2007), their occurrence within the Oxygen Minimum Zone can be interpreted as pointing to significant bacterial activity on the organic matter in this layer.

3.4 Principal Component Analysis

Principal component analysis (PCA) has become a powerful tool for reducing the quantity of variables and uncovering tendencies in data series having numerous entries. This analysis allows a simple graphic representation of the degree of likeness existing within a data group, thereby establishing the main characteristics of the group of variables in a series of samples, reducing it to a smaller set of derived variables (principal components). More information can often be extracted from these principal components than from the original variables.

Application of PCA to our data set showed that the three first components explained a large proportion of the total variance (75 %). Component I explained 32% of the variance, Component II 26%, and Component III 17% (Table 3). Figure 7 shows the plane that represents the first two components. For Factor I, to which the variables were most positively loaded (Table 3), comprised the sterols 24-ethyl-5α-cholest-22E-en-3β-ol, 24-ethyl-5α-cholestan-3β-ol, and 5α-cholestan-3β-ol; the C_{20}, C_{14}, and C_{16} n-alkanols; and the C_{14} fatty acid (in the hatched oval, Fig. 7), which suggests that these variables are related to each other. The n-alkanols, fatty acids, and two of the stenols, precursors of the stanols 24-ethylcolesta-5,22E-dien-3β-ol and cholest-5-en-3β-ol, which loaded positively to Component I, were derived from phytoplanktonic organisms. On the other hand, free alcohols and free fatty acids are considered lipid decay indicators (since most of these compounds are found linked to other molecules (Galois et al., 1996; Gong & Hollander, 1997; Parrish et al., 2000); the same applies to the increase of the relative abundance of straight-chain fatty acids at the expense of their unsaturated counterparts (Wakeham, 1995) and the presence of stanols that are produced by the microbial reduction of their stenol analogues under suboxic conditions (Minor et al., 2003; Wakeham, 1995). These results lead us to conclude that the substances in Component I represented highly degraded organic matter of planktonic origin.

Variable	Factor 1	Factor 2	Factor 3
N = 24-Ethyl-5α-cholest-22E-en-3β-ol	**0.905**	0.093	-0.385
$nC_{20}OH$	**0.833**	-0.102	-0.216
Q = 24-Ethyl-5α-cholestan-3β-ol	**0.782**	-0.036	0.168
F = 5α-Cholestan-3β-ol	**0.750**	-0.001	-0.545
$nC_{14}OH$	**0.624**	0.476	-0.547
$nC_{16}OH$	**0.614**	-0.547	-0.273
$C_{14:0}$	**0.600**	-0.317	-0.612
E = Cholest-5-en-3β-ol	**0.600**	0.378	0.002
R = 4α,23,24-Trimethyl-5α-cholest-22E-en-3β-ol	**-0.964**	0.018	0.097
P = 24-Ethylcholest-5-en-3β-ol	**-0.823**	-0.061	-0.442
$nC_{28}OH$	**-0.807**	-0.374	0.130
$nC_{30}OH$	**-0.710**	0.026	-0.517
$\Sigma\ i+ai(C_{15}\text{-}C_{17})$	**-0.737**	-0.504	-0.3598
J = 24-Methylcholesta-5,24(28)-dien-3β-ol	**-0.758**	0.553	-0.102
M = 23,24-Dimethylcholesta-5,22E-dien-3β-ol	**-0.761**	-0.007	0.014
$nC_{32}OH$	**-0.641**	0.191	-0.719
L = 24-Methylcholest-5-en-3β-ol	-0.029	**0.964**	0.115
Phytol	-0.305	**0.823**	0.135
K = 24-Methyl-5α-cholest-24(28)-en-3β-ol	-0.285	**0.805**	0.223
Oxygen	-0.165	**0.783**	0.117
B = 27-*nor*-24-Methyl-5α-cholest-22E-en-3β-ol	-0.243	**0.718**	0.316
$nC_{22}OH$	0.053	**-0.904**	0.237
A = 27-*nor*-24-Methylcholesta-5,22E-dien-3β-ol	-0.347	**-0.865**	0.272
$C_{18:0}$	0.446	**-0.812**	0.023
$nC_{24}OH$	0.172	**-0.785**	-0.301
H = 24-Methyl-5α-cholest-en-3β-ol	-0.100	-0.460	**0.616**
$C_{21:0}$	-0.155	-0.006	**-0.838**
$C_{16:1}$	-0.645	-0.109	**-0.713**
Eigenvalues	**13.337**	**11.057**	**7.115**
% total variance	**32**	**26**	**17**

Table 3. Loading scores for the three factors explaining 75% of the variance in particulate organic matter.

On the other hand, the variables that were negatively loaded to Component I (in the dark grey oval, Fig. 7) were the sterols 4□,23,24-trimethyl-5α-cholest-22E-en-3β-ol, 24-ethylcholest-5-en-3β-ol, 24-methylcholesta-5,24(28)-dien-3β-ol, and 23,24-dimethylcholesta-5,22E-dien-3β-ol; the C_{28}, C_{30}, and C_{32} *n*-alkanols; and the *i*- and *ai*-C_{15} and -C_{17} ($\Sigma\ i+ai(C_{15}$-C_{17})) bacterial fatty acid markers (Table 3). The long-chain *n*-alkanols and 24-ethylcholest-5-en-3β-ol are usually considered biomarkers of terrestrial higher plants (Parrish et al., 2000; Treignier et al., 2006; Hinrichs et al., 1999), but 24-ethylcholest-5-en-3β-ol in highly productive marine ecosystems can also arise from planktonic sources. Nonetheless, since the PCA showed this sterol to be highly related to terrestrial biomarkers such as long-chain *n*-alcohols, this compound in the investigated marine environment appears to be more terrestrial than planktonic in origin.

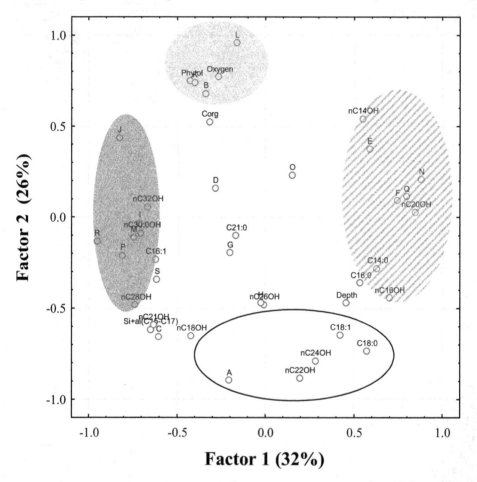

Factor 1 (32%)

Fig. 7. Compound loadings on Factor 1 vs. Factor 2 for the principal component analysis of lipids in particulate organic matter. The hatched oval shows components with positive loadings on Factor I, the dark grey oval shows components with negative loadings on Factor I, the light gray oval shows components with positive loadings on Factor II, and the white oval shows components with negative loadings on Factor II.

For Factor II, the positively loaded variables were 24-methylcholest-5-en-3β-ol, 24-methyl-5α-cholest-24(28)-en-3β-ol, and 27-*nor*-24-methyl-5α-cholest-22E-en-3β-ol, phytol and the oxygen concentration (shown in the light gray oval, Fig. 7; Table 3). This Component is represented by diatom biomarkers. Phytol is the indirect indicator of phytoplankton, being released from the chlorophyll molecule, and the three sterols are bioindicators of diatoms.

These molecules, in turn, are directly related to the oxygen concentration, that is, as the oxygen concentration drops in the water mass, the concentration of these compounds decreases, which means that they are effectively degraded under suboxic conditions; this is supported by the disappearance of these biomarkers below the OMZ (Figs. 3 and 4).

The variables that were negatively loaded for Factor II (in the oval, Fig. 7) were the C_{22} and C_{24} n-alkanols; the $C_{18:0}$ fatty acid; and the sterol 27-nor-24-methylcholesta-5,22E-dien-3β-ol (Table 3). Both the C_{18} fatty acid and 27-nor-24-methylcholesta-5,22E-dien-3β-ol are planktonic in origin, the sterol was reported to be a bioindicator of some flagellates (Wakeham, 1995). The alkanols >C_{22} have been reported to be biomarkers of higher plants; nonetheless, there is uncertainty with respect to nC_{22}-OH, since it can also be derived from cyanobacteria (Volkman et al., 1998). In Figure 7, this alcohol is observed to be more related to planktonic biomarkers than to compounds of terrestrial origin, so we consider n-alkanol in the Antofagasta area to be derived from cyanobacteria.

PCA allowed the separation of planktonic biomarkers (positively charged in Component I) from terrestrial biomarkers (negatively charged in Component I). Moreover, it showed phytol, a biomarker of grazing activity, to be closely correlated with the oxygen concentration in the water column, as it was more abundant at the surface where the oxygen concentration was higher.

4. Conclusions

The analysis of individual compounds within each of several lipid classes (alcohols, fatty acids, and sterols) allowed us to determine that phytoplankton, mainly diatoms, are the main sources of lipids in the suspended particulate organic matter in the Antofagasta area. Continental contributions are scarce, as seen in the low concentrations of terrestrial biomarkers, e.g., the sterol 24-ethylcholest-5-en-3β-ol and the long-chain n-alkanols (>C_{22}), as well as in the absence of long-chain fatty acids.

On the other hand, the analysis of these lipid classes showed that, as depth increased, the particles were intensely degraded. The rapid disappearance of labile molecules such as phytol, the increase in n-alkanols and straight-chain fatty acids, the decrease of unsaturated fatty acids, and the increase of bacterial biomarkers all indicate increased microbial activity on the lipid molecules.

5. Acknowledgment

We thank S. Wakeham (Skidaway Institute of Oceanography) for his analytical suggestions, and G. Daneri, H. Gonzalez, J. Sepúlveda, R. Castro, P. Rossel, and G. Lorca for their collaboration during FluMO cruise sampling. We are grateful to T. Stiehl, J. Köster, B. Scholz-Böttcher, J. Maurer and K. Adolph from the University of Oldenburg for their help with analyses. LE thanks the German Academic Exchange Service (DAAD) and the MECESUP-UCO0002 grant (Ministry of Education-Chile) for supporting her Ph.D. studies at University of Concepción. DAAD is also acknowledged for funding LE's research visit in Professor Rullkötter's. This work was supported by FONDECYT Grant 1000366 and the FONDAP-COPAS Center (Grant 150100007).

6. References

Abramson, L.; Lee, C.; Liu, Z.; Wakeham, S.G. & Szloseka, J. (2010). Exchange between suspended and sinking particles in the northwest Mediterranean as inferred from the organic composition of in situ pump and sediment trap samples. *Limnology and Oceanography*, Vol. 55, No. 2, (March 2010), pp. 725–739, ISSN 0024-3590

Burns, K.A.; Volkman, J.K.; Cavanagh, J.A. & Brinkman D. (2003). Lipids as Biomarkers for Carbon Cycling on the Northwest Shelf of Australia: Results from a Sediment Trap Study. *Marine Chemistry*, Vol. 80, (January 2003), pp. 103-128, ISSN 0304-4203

Canuel, E.A. & Zimmerman, A.R. (1999). Composition of Particulate Organic Matter in the Southern Chesapeake Bay: Sources and Reactivity. *Estuaries*, Vol. 22, (December 1999), pp. 980-994, ISSN 01608347

Çoban-Yildiz, Y.; Chiavari, G.; Fabbri, D.; Gaines, A.F.; Galletti, G. & Tugrul, S. (2000). The Chemical Composition of Black Sea Suspended Particulate Organic Matter: Pyrolysis-GC/MS as a Complementary Tool to Traditional Oceanographic Analysis. *Marine Chemistry*, Vol. 69, (March 2000), pp. 55-67, ISSN 0304-4203

Daneri, G.; Dellarossa, V.; Quiñones, R.; Jacob, B.; Montero, P. & Ulloa, O. (2000). Primary Production and Community Respiration in the Humboldt Current System off Chile and Associated Oceanic Areas. *Marine Ecology Progress Series*, Vol. 197, (May 2000), pp. 41–49, ISSN 0171-8630

Derieux, S.; Fillaux, J. & Saliot, A. (1998). Lipid Class and Fatty Acid Distributions in Particulate and Dissolved Fractions in the North Adriatic Sea. *Organic Geochemistry*, Vol. 29, (November 1998), pp. 1609-1621, ISSN 0146-6380

Galois, R.; Richard, P. & Fricourt, B. (1996). Seasonal Variations in Suspended Particulate Matter in the Marennes-Oléron Bay, France, Using Lipids as Biomarkers, *Estuarine, Coastal and Shelf Science*, Vol. 43, (September 1996), pp. 335-357, ISSN 0272-7714

Gong, C. & Hollander, D.J. (1997). Differential Contribution of Bacteria to Sedimentary Organic Matter in Oxic and Anoxic Environments, Santa Monica Basin, California. *Organic Geochemistry*, Vol. 26, (May-June 1997), pp. 545-563, ISSN 0146-6380

González, H.E.; Ortiz, V.C. & Sobrazo, M. (2000). The role of Fecal Material in the Particulate Organic Carbon Flux in the Northern Humboldt Current, Chile (23°S), Before and During the 1997–1998 El Niño. *Journal of Plankton Research*, Vol. 22, (March 2000), pp. 499-529, ISSN 0142-7873

González, H.E.; Daneri, G; Iriarte, J.L.; Yannicelli, B.; Menschel, E.; Barría, C.; Pantoja, S. & Lizárraga, L. (2009). Carbon fluxes within the epipelagic zone of the Humboldt Current System off Chile: The significance of euphausiids and diatoms as key functional groups for the biological pump. *Progress in Oceanography*. Vol. 83, (December 2009), pp. 217–227, ISSN 0079-6611

Grossi, V.; Blokker, P. & Sinninghe-Damsté, J.S. (2001). Anaerobic Biodegradation of Lipids of the Marine Microalga *Nannochloropsis salina*. *Organic Geochemistry*, Vol. 32, (June 2001), pp. 795-808, ISSN 0146-6380

Gundersen, J.S.; Gardner, W.D.; Richardson, M.J. & Walsh, I.D. (1998). Effects of Monsoons on the Seasonal and Spatial Distribution of POC and Chlorophyll in the Arabian Sea. *Deep-Sea Research II*, Vol. 45, (August 1998), pp. 2103-2132, ISSN 0967-0645

Hinrichs, K.U.; Schneider, R.R.; Müller, P.J. & Rullkötter, J. (1999). A Biomarker Perspective on Paleoproductivity Variations in Two Late Quaternary Sediment Sections from the Southeast Atlantic Ocean. *Organic Geochemistry*, Vol. 30, (May 1999), pp. 341-366, ISSN 0146-6380

Iriarte, J.L.; Pizarro, G.; Troncoso, V.A. & Sobarzo, M. (2000). Primary Production and Biomass Size-Fractioned Phytoplankton off Antofagasta, Chile (23–241S) During Pre-El Niño and El Niño 1997. *Journal of Marine Systems*, Vol. 26, (September 2000), pp. 37–51, ISSN 0924-7963

Iriarte, .J.L. & González, H.E. (2004). Phytoplankton Size Structure During and After the 1997/98 El Niño in a Coastal Upwelling Area of the Northern Humboldt Current System. *Marine Ecology Progress Series*, Vol. 269, (March 2005), pp. 83-90, ISSN 0171-8630

Jeng, W.L.; Lin, S. & Kao, S.J. (2003). Distribution of Terrigenous Lipids in Marine Sediments off Northeastern Taiwan. *Deep-Sea Research II*, Vol. 50, (March-April 2003), pp. 1179-1201, ISSN 0967-0645

Killops, S.D. & Killops, V.J. (1994). *An Introduction to Organic Geochemistry*. Longman Scientific & Technical, ISBN 0-582-08040-1, New York, USA

Lee, C.; S. Wakeham & C. Arnosti. (2004). Particulate Organic Matter in the Sea: The Composition Conundrum. *Ambio*, Vol. 33, No. 8, (December 2004), pp. 565-+575, ISSN 0044-7447

Minor, E.C.; Wakeham, S.G. & Lee, C. (2003). Changes in the Molecular-Level Characteristics of Sinking Marine Particles with Water Column Depth. *Geochimica et Cosmochimica Acta*, Vol. 67, (November 2003), pp. 4277-4288, ISSN 0016-7037

Muri, G.; Wakeham, S.G.; Pease, T.K. & Faganeli, J. (2004). Evaluation of Lipid Biomarkers as Indicators of Changes in Organic Matter Delivery to Sediments from Lake Planina, a Remote Mountain Lake in NW Slovenia. *Organic Geochemistry*, Vol. 35, (October 2004), pp. 1083-1093, ISSN 0146-6380

Pantoja, S. & Wakeham, S.G. (2000). Marine Organic Geochemistry: A general overview. In: *Chemical processes in marine environments*, A. Gianguzza; E. Pelizzetti & S. Sammartano (Eds.), pp. 43-74, Springer-Verlag, ISBN, Berlin, Germany

Pantoja, S. & Lee, C. (2003). Amino Acid Remineralization and Organic Matter Lability in Chilean Coastal Sediments. *Organic Geochemistry*, Vol. 34, (August 2003), pp. 1047-1056, ISSN 0146-6380

Pantoja, S.; Rossel, P.; Castro, R.; Cuevas, L.A.; Daneri, G.; Córdova, C. (2009). Microbial Degradation Rates of Small Peptides and Amino Acids in the Oxygen Minimum Zone of Chilean Coastal Waters. *Deep-Sea Res. II*, Vol. 56, No. 16, (July 2009), pp. 1055-1062, ISSN

Parrish. C.C.; Abrajano, T.A.; Budge, S.M.; Helleur, R.J.; Hudson, E.D.; Pulchan, K. & Ramos, C. (2000). Lipid and Phenolic Biomarkers in Marine Ecosystems: Analysis and Applications. In: *The handbook of environmental chemistry. Vol. 5, Part D. Marine Chemistry*, P. Wangersky (Ed.), Springer-Verlag, pp. 193-223, ISBN, Berlin, Germany

Pazdro, K.; Staniszewski, A.; Bełdowski, J.; Emeis, K.C.; Leipe, T. & Pempkowiak, J. (2001). Variations in Organic Matter Bound in Fluffy Layer Suspended Matter from the Pomeranian Bay (Baltic Sea). *Oceanologia*, Vol. 43, (2001), pp. 405-420, ISSN 0078-3234

Petsch, S.T.; Edwards, K.J. & Eglinton, T.I. (2003). Abundance, Distribution and $\delta^{13}C$ Analysis of Microbial Phospholipid-Derived Fatty Acids in a Black Shale Weathering Profile. *Organic Geochemistry*, Vol. 34, (June 2003), pp. 731-743, ISSN 0146-6380

Reemtsma, T. & Ittekkot, V. (1992). Determinations of Factors Controlling the Fatty Acid Composition of Settling Particles in the Water Column by Principal-Component Analysis and their Quantitative Assessment by Multiple Regression. *Organic Geochemistry*, Vol. 18, (January 1992), pp. 121-129, ISSN 0146-6380

Sheridan, C.C.; Lee, C.; Wakeham, S.G. & Bishop, J.K.B. (2002). Suspended Particle Organic
    Composition and Cycling in Surface and Midwaters of the Equatorial Pacific
    Ocean. *Deep-Sea Research I*, Vol. 49, (November 2002), pp. 1983-2008, ISSN 0967-
    0637

Sun, M.Y.; Wakeham, S.G. & Lee, C. (1997). Rates and Mechanisms of Fatty Acid
    Degradation in Oxic and Anoxic Coastal Marine Sediments of Long Island Sound,
    New York, USA. *Geochimica et Cosmochimica Acta*, Vol. 61, (January 1997), pp. 341-
    355, ISSN 0016-7037

Treignier, C.; Derenne, S. & Saliot, A. (2006). Terrestrial and Marine n-Alcohol Inputs and
    Degradation Processes Relating to a Sudden Turbidity Current in the Zaire
    Canyon. *Organic Geochemistry*, Vol. 37, (September 2006), pp. 1170–1184, ISSN 0146-
    6380

Volkman, J.K.; Barrett, S.M.; Blackburn, S.I.; Mansour, M.P.; Sikes, E.L. & Gelin, F. (1998).
    Microalgal Biomarkers, a Review of Recent Research Developments. *Organic
    Geochemistry*, Vol. 29, (November 1998), pp. 1163–1179, ISSN 0146-6380

Volkman, J.K. & Tanoue, E. (2002). Chemical and Biological Studies of Particulate Organic
    Matter in the Ocean. *Journal of Oceanography*, Vol. 58, No. 2 (March 2002), pp. 265-
    279, ISSN 0916-8370

Wakeham, S.G. & Ertel, J.R. (1987). Diagenesis of Organic Matter in Suspended Particles and
    Sediments in the Cariaco Trench. *Organic Geochemistry*, Vol. 13, (1988), pp. 815-82,
    ISSN 0146-6380

Wakeham; S.G. & Lee, C. (1989). Organic Geochemistry of Particulate Matter in the Ocean:
    The Role of Particles in Oceanic Sedimentary Cycles. *Organic Geochemistry*, Vol. 14,
    (1989), pp. 83-96, ISSN 0146-6380

Wakeham, S.G. (1995). Lipid Biomarkers for Heterotrophic Alteration of Suspended
    Particulate Organic Matter in Oxygenated and Anoxic Water Columns of the
    Ocean. *Deep-Sea Research Part I*, Vol. 42, (October 1995), pp. 1749-1771, ISSN 0967-
    0637

Wakeham, S.G.; Hedges, J.I.; Lee, C.; Peterson, M.L. & Hernes, P.J. (1997). Composition and
    Transport of Lipid Biomarkers Through the Water Column and Surficial Sediments
    of Equatorial Pacific Ocean. *Deep-Sea Research II*, Vol. 44, (1997), pp. 2131-2162, ISSN
    0967-0645

Wakeham, S.G. (2000). Organic Matter Preservation in the Ocean: Lipid Behavior from
    Plankton to Sediments. In: *Chemical Processes in Marine Environments*, A. Gianguzza;
    E. Pelizzetti & S. Sammartano (Eds.), pp. 127-139, Springer-Verlag, ISBN, Berlin,
    Germany

Wakeham, S.G.; Lee C. & Hedges, H.I. (2000): Fluxes of Major Biochemicals in the Equatorial
    Pacific Ocean. p. 117– 140. In: *Dynamics and Characterization of Marine Organic
    Matter*, H. Handa; E. Tanoue & T. Hama, (Ed.) Terra Sci. Publ., ISBN 978-
    0792362937, Tokyo, Japan

Wakeham, S.G.; Peterson, M.L.; Hedges, J.I. & Lee, C. (2002). Lipid Biomarker Fluxes in the
    Arabian Sea, with a Comparison to the Equatorial Pacific Ocean. *Deep Sea Research
    II*, Vol. 49, (2002), pp. 2265-2301, ISSN 0967-0645

Wakeham, S.G.; Amann, R.; Freeman, K.H.; Hopmans, E.C.; Jørgensen, B.B.; Putnam, I.F.;
    Schouten, S.; Sinninghe Damsté, J.S.; Talbot, H.M. & Woebken, D. (2007). Microbial
    Ecology of the Stratified Water Column of the Black Sea as Revealed by a
    Comprehensive Biomarker Study. *Organic Geochemistry*, Vol. 38, No. 12, (December
    2007), pp. 2070–2097, ISSN 0146-6380

7

Gas Chromatography Results Interpretation: Absolute Amounts Versus Relative Percentages

G.M. Hon[1], S. Abel[2], C.M. Smuts[3], P. van Jaarsveld[2], M.S. Hassan[1],
S.J. van Rensburg[4], R.T. Erasmus[5] and T. Matsha[1]

[1]Cape Peninsula University of Technology,
[2]South African Medical Research Council,
[3]North-West University,
[4]National Health Laboratory Services,
[5]University of Stellenbosch,
South Africa

1. Introduction

1.1 Fatty acid functions

Impaired fatty acid metabolism has been implied in a number of diseased states such as multiple sclerosis, sudden cardiac death, insulin resistance, atherosclerosis and hypertension (Blaak, 2003; Oram & Bornfeldt, 2004; Pilz et al., 2007; Sarafidis & Bakris, 2007; Hon et al., 2009a, 2009b). The fatty acid status may be assessed in plasma, platelets, red blood cells, white blood cells and brain tissue (Zamaria, 2004, Hon et al., 2009a). However, the plasma fatty acid profile may vary considerably, whereas red blood cell membrane fatty acids are said to reflect dietary fat intake in relation to the biological half life of cells (Romon et al., 1995; Zamaria, 2004). Furthermore, red blood cell membranes, in contrast to other cells, lack the desaturase enzymes and must get their fatty acids preformed from plasma (Allen et al., 2006). Fatty acids may be saturated, monounsaturated (the n-7 and n-9 fatty acid series) or polyunsaturated (the n-3 and n-6 fatty acid series) and membrane fatty acids may further be non-esterified or esterified into glycerophospholipids (Koay & Walmsley, 1999).

1.1.1 Cell membrane fluidity – Phospholipids

Cell membrane phospholipids differ in their head-groups and hydrocarbon chains (fatty acids) (Hazel & Williams, 1990; Williams, 1998; Barenholz, 2002). Their polar head-groups can be choline, ethanolamine, serine, inositol, inositol phosphates or glycerol (Koay & Walmsley, 1999). Cell membrane phospholipids as well the type of fatty acids contained within the different phospholipid fractions are responsible for a variety of cellular functions (Manzoli, 1970; Caret et al. 1997; Williams, 1998). Cell membrane phospholipids determine membrane structure and fluidity. Phosphatidylethanolamine phospholipids are ordered-crystalline-phase lipids and can pack closely in membranes, while phosphatidylcholine phospholipids are liquid-crystalline-phase lipids, and do not pack close in the membrane

(Harlos & Eibl, 1981; Williams, 1998; Hamai et al. 2006). A combination of ordered-crystalline-phase lipids and liquid-crystalline-phase lipids are needed in cell membranes to regulate membrane fluidity.

1.1.2 Cell membrane fluidity – Fatty acids

Polyunsaturated fatty acids influence membrane fluidity, are involved with neurotransmission and prostaglandin formation, and are involved in enzyme and receptor expression (Horrobin & Manku, 1990; Horrobin, 1999; De Pablo & De Cienfuegos, 2000; Nakamura & Nara, 2004; Zamaria, 2004). Similar to the role of cell membrane phospholipids, a balance in the saturated and unsaturated fatty acids composition is needed for optimal membrane fluidity (Allen et al. 2006). Saturated fatty acids contain single carbon-carbon bonds, while mono- and polyunsaturated fatty acids contain double carbon-carbon bonds which allow for greater flexibility of these chains (Horrobin, 1999; Allen et al. 2006). Membrane fluidity is an important function of cell membranes and organisms adjust the fluidity of their cellular membranes in response to changes in their physiochemical environment (Hazel & Williams, 1990; Williams, 1998; Barenholz, 2002; Allen et al. 2006). In this regard, phagocytosis is an important mechanism in many cells for the elimination of micro-organisms or foreign particles, and membrane fluidity plays an important role in this process (De Pablo & De Cienfuegos, 2000). Although the underlying cause of multiple sclerosis remains unknown, both an auto-immune and viral aetiology has been hypothesized as contributing factors to the disease (Stinissen et al. 1997; Brown, 2001; Hunter & Hafler, 2000).

1.1.3 Fatty acids in inflammation and infection

Polyunsaturated fatty acids are also precursors for eicosanoid production. Eiosanoids are inflammatory mediators of inflammation and their functions depend on each precursor fatty acid (Horrobin & Manku, 1990; De Pablo & De Cienfuegos, 2000; Zamaria, 2004). The effects of monounsaturated fatty acids on eicosanoid metabolism are less clear than that of the polyunsaturated fatty acids (Harwood & Yaqoob, 2002). However, studies suggest that diets rich in monounsaturated fatty acids may decrease the expression of adhesion molecules on peripheral blood mononuclear cell membranes, and may therefore have specific anti-inflammatory effects (Yaqoob et al. 1998; Harwood & Yaqoob, 2002). Saturated fatty acids have been shown to display anti-viral and -bacterial properties (Sands, 1977; Cordo et al. 1999; Narasimhan et al. 2006).

1.1.4 Fatty acid supplementation

It is important to be able to separate and quantify the individual fatty acids because they differ in structure and function. Altered lipid metabolism in the cell membrane is believed to contribute to central nervous system injury (Adibhatla & Hatcher, 2007) and influences the function of immune cells (Calder, 2007). Fatty acids have been implicated in the pathogenesis and treatment of multiple sclerosis (Zamaria, 2004; Harbige & Sharief, 2007). However, although some studies reported an improvement in symptoms during fatty acid supplementation (Bates et al. 1978; Nordvik et al. 2000; Weinstock-Guttman et al. 2005; Harbige & Sharief, 2007) others found no improvement in the disease progression of patients (Paty et al. 1978; Farinotti et al. 2007). Therefore, it is important to investigate fatty acid metabolic abnormalities and possible treatment with fatty acid supplements in a disease such as multiple sclerosis carefully and comprehensively.

1.2 Gas chromatography applications of fatty acids

The ability of gas chromatography to separate and quantify fatty acids makes it possible to measure fatty acids in human cells. Generally, the results obtained from gas chromatography analysis are expressed either as absolute amounts or relative percentages. The preferred method for the quantification of fatty acids is in absolute amounts, using calibration methods and internal standards. However, quantification in relative percentages may in some cases supply additional information and this method should therefore not be rejected without evaluating its potential. This paper discusses the discrepancies arising when reporting results quantified in either absolute amounts or relative percentages when investigating the fatty acid profile in a diseased state.

1.3 Fatty acids quantified in relative percentages and in absolute amounts

It is important to know whether fatty acids are present in optimum amounts in relation to one another in a diseased state. For example, fatty acids from the n-6 and n-3 fatty acid series fulfil different functions as precursors for pro- and anti-inflammatory mediators respectively. Prostaglandin E2, which is derived from the metabolism of C20:4n-6 (see Table 1 for common names) is highly pro-inflammatory and prostaglandin E1 derived from the metabolism of C20:3n-6 has intermediate properties (Horrobin & Manku, 1990; Simopoulos, 2002; Bagga et al., 2003; Haag, 2003). In contrast, prostaglandin E3 derived from the metabolism of C20:5n-3 and docosanoids derived from the metabolism of C22:6n-3 have anti-inflammatory properties (Bagga et al., 2003; Haag, 2003; Zamaria, 2004; Farooqui et al., 2007; Chen et al., 2008). The n-3 fatty acids, C20:5n-3 and C22:6n-3 compete with the n-6 fatty acid, C20:4n-6 for enzymatic metabolism (Simopoulos, 2002; Bagga et al., 2003; Culp et al., 1979; Calder, 2007) and in a disease such as multiple sclerosis which is inflammatory in nature, it could be useful to know whether the n-6 and n-3 fatty acids are present in relative percentages comparable to that from control subjects.

The long-chain n-6 and n-3 polyunsaturated fatty acids are also required in large amounts in the brain, which requires four times the amount of C20:4n-6 than C22:6n-3 4 times the amount of C20:4n-6 than C22:6n-3 on a daily basis (Harbige & Sharief, 2007; Rapoport et al., 2007). Furthermore, the composition of membrane phospholipids and their fatty acids determines the degree of membrane fluidity and the degree of fatty acid unsaturation plays an important role in maintaining required physiological levels of fluidity (Mouritsen & Jorgensen, 1998; Zamaria, 2004). In a diseased state it may be useful to know their relative percentages as measured against that of total fatty acids present as compared to that of control subjects. However, using relative percentages as a measure could bias results when comparisons are made between subjects. Comparing similar percentages of different absolute values could become misleading when these absolute values differ substantially and it has been suggested that relative percentages do not represent a true reflection of the different fatty acid compounds present in biological fluids/tissue. Absolute values are also an indication of the movement of fatty acids between phospholipid classes within the cell membrane and could therefore reflect their possible availability for metabolite synthesis, particularly inflammatory signalling.

1.4 Fatty acids in multiple sclerosis

Reports on fatty acid composition in blood and brain tissue from patients with multiple sclerosis may vary substantially (Harbige & Sharief, 2007; Hon et al., 2009a, 2009b). Cultural and ethnic differences as well as dietary variability have been implicated in these differences (Harbige & Sharief, 2007). However, a further cause for these differences could possibly be related to the quantification method used. Measurement of the fatty acid composition in absolute amounts and relative percentages could both be informative, but could well differ in the information given. The absolute amounts of fatty acids are identified in relation to the phospholipids and membrane content whereas the relative percentage composition is a measure of each fatty acid present in relation to the total fatty acids identified, that is changes in one fatty acid will affect the percentage of all other fatty acids. Therefore, ideally, fatty acid analysis and interpretation should be evaluated according to outcome required. In this chapter we evaluate whether the two methods of measurement of fatty acids in control subjects would show sufficient agreement and to evaluate possible deviations from these relationships in patients with multiple sclerosis. Furthermore, to evaluate the two methods of measurement when correlated/associated with inflammation as measured by C-reactive protein as well as disease outcome as measured by the Kurtzke Expanded Disability Status Scale (Kurtzke, 1983). C-reactive protein is a recognised marker of inflammation and has been shown to correlate with infectious episodes and clinical relapses in patients with multiple sclerosis (Giovannoni et al., 1996; Giovannoni et al., 2001; Sellner et al., 2008). Thin-layer chromatography and gas chromatography were used for the extraction and separation of lipid classes and quantification of fatty acids in plasma, red blood cell and peripheral blood mononuclear cell membranes from control subjects and patients with multiple sclerosis (Folch et al., 1957; Gilfillan et al., 1983; Van Jaarsveld et al., 2000; Hon et al., 2009b; Hon et al., 2011a).

2. Methodology

2.1 Ethics statement

Ethics approval for the study was obtained from the Health and Applied Sciences Research Ethics Committee of the Cape Peninsula University of Technology. Patients with multiple sclerosis were contacted and recruited through the Multiple Sclerosis Society, Western Cape Branch, South Africa. The study population consisted of 31 female patients with multiple sclerosis and 30 age- race- and gender-matched control subjects. Informed written consent was obtained from all study participants. All results were treated confidentially.

2.2 Study population

The diagnoses of the recruited patients were verified by a neurologist based on clinical, laboratory and magnetic resonance imaging findings. Six of the patients had active disease, 11 had a relapse 5-12 months prior to recruitment and 14 did not relapse for more than a year. The number of years since the patients were diagnosed was 7 (11) years (median and quartile range). Ten patients were using non-steroidal anti-inflammatory drugs and five patients were using immunosuppressive medication. Exclusion criteria included the use of interferon, steroids, and fatty acid supplements.

2.3 Blood sampling and processing

Venous blood was collected after an overnight fast into anti-coagulant ethylenediaminetetraacetic acid tubes (Beckman Coulter, Cape Town, South Africa). Blood was separated into its different components using histopaque-1077 separation medium as per manufacturer's instructions (Sigma-Aldrich, Cape Town, South Africa). Blood was layered onto histopague in a ratio of 15 mL blood per 12 mL histopague and centrifuged at 400 x g for 20 min at room temperature. The plasma layer was kept, spun twice at 1500 x g for 5 min to remove platelet contaminants, and frozen in 1 mL aliquots. The peripheral blood mononuclear cell interface was recovered, washed twice with 0.85 % saline solution, resuspended in 1 mL of a 0.85 % saline solution and frozen. Three mL red blood cells were washed twice with 0.85 % saline solution and frozen as packed cells without added saline. The samples were frozen at -80°C immediately after separation of blood fractions. A 0.85 % saline solution was used in this study instead of the prescribed balanced phosphate buffered saline solution as additional tests such as membrane phosphate determination were to be carried out at a later stage.

Solvent mixtures and thin-layer chromatography was used for the extraction and separation of fatty acids respectively, with further conversion to fatty acid methyl esters for quantification by gas chromatography as described by Folch et al. (1957), Gilfillan et al. (1983), Van Jaarsveld et al. (2000), Hon et al. (2009b), Hon et al. (2011a). The following sections summarises the methodology used for quantification of the fatty acids in human tissue, plasma, red blood cell and peripheral blood mononuclear cell membranes.

2.4 Thin – Layer chromatography

The plasma, red blood cell and peripheral blood mononuclear cell membrane lipids were extracted according to a modified method of Folch et al. (1957), Van Jaarsveld et al. (2000), and described in Hon et al. (2009b), Hon et al. (2011a). All the extraction solvents were of high-performance liquid chromatography-grade and contained 0.01 % butylated hydroxytoluene (Sigma-Aldrich, Cape Town, South Africa) as an antioxidant. The red blood cell samples were extracted and the dried lipid residue resuspended in 80 µl chloroform/methanol (ratio 2:1 v/v) for thin-layer chromatography. Of this 20 µl was used for esterified fatty acid analysis and 40 µl for non-esterified fatty acid separation on thin-layer chromatography. The peripheral blood mononuclear cell samples were resuspended in 70 µl chloroform/methanol; 20 µl was used for esterified fatty acid analysis and 30 µl for non-esterified fatty acid analysis. Plasma samples were resuspended in 80 µl chloroform/methanol of which 20 µl was used for esterified fatty acid analysis and 30 µl for non-esterified fatty acid analysis.

Neutral lipids (non-esterified fatty acids, mono- di- and triacylglyceryl and cholesteryl fatty acid esters) were separated from the total phospholipid fraction by thin-layer chromatography on pre-coated silica gel plates (10 x 10 cm), using the solvent system petroleum benzene (boiling point 40-60°C)/diethyl ether (peroxide free)/acetic acid (90:30:1; v/v/v) (Sigma-Aldrich, Cape Town, South Africa) as previously described (Van Jaarsveld et al., 2000). Individual phospholipid classes were separated by thin-layer chromatography on pre-coated silica gel plates (10 x 10 cm) using chloroform/petroleum benzene/methanol/acetic acid/boric acid (40:30:20:10:1.8; v/v/v/v/w) as solvent (Gilfillan et al., 1983).

The lipid bands containing phosphatidylcholine (PC), phosphatidylethanolamine (PE), phosphatidylserine (PS) and sphingomyelin (SM) were identified by comparison to a known mixed phospholipid standard run in parallel to the samples and visualized with long wave ultraviolet light after spraying the plates with chloroform/methanol (ratio 1:1; v/v) containing 10 mg per 100 mL 2,5-bis-(5´-tert-butylbenzoxazolyl-[2´])thiophene (Sigma Chemical Company, Cape Town, South Africa). These bands were scraped off the thin-layer chromatography plates and used for further fatty acid analysis.

Fatty acids	
Abbreviated chemical formulae	Common name
C18:3n-3	Alpha-linolenic acid
C18:4n-3	Stearidonic acid
C20:4n-3	Eicosatetraenoic
C20:5n-3	Eicosapentaenoic acid
C22:5n-3	Docosapentaenoic acid
C22:6n-3	Docosahexaenoic acid
C18:2n-6	Linoleic acid
C18:3n-6	Gamma-linolenic acid
C20:2n-6	Eicosadienoic acid
C20:3n-6	Dihomo-gamma-linolenic acid
C20:4n-6	Arachidonic acid
C22:4n-6	Adrenic acid
C22:5n-6	Docosapentaenoic acid
C16:1n-7	Palmitoleic acid
C18:1n-7	Vaccenic acid
C18:1n-9	Oleic acid
C20:1n-9	Gadoleic acid
C14:0	Myristic acid
C16:0	Palmitic acid
C18:0	Stearic acid
C20:0	Arachidic acid
C22:0	Behenic acid
C24:0	Lignoceric acid

Table 1. Abbreviated chemical formulae and common names of major plasma and blood cell membrane fatty acids. References: Nightingale et al., 1990; Horrobin, 1999; Pereira et al., 2003.

2.5 Gas chromatography

The lipid fractions scraped off the thin-layer chromatography plates were transmethylated, using 5 % sulphuric acid/methanol at 70°C for 2 hours (sphingomyelin fraction: 18 hours) (Van Jaarsveld et al., 2000; Hon et al., 2009b; Hon et al., 2011a). After cooling, the resulting fatty acid methyl esters were extracted with 1 mL of distilled water and 2 mL of n-hexane. The top hexane layer was removed and evaporated to dryness in a waterbath at 37°C under nitrogen gas, re-dissolved in carbon disulphide and analyzed by gas chromatography

(Finnigan Focus Gas Chromatography, Thermo Electron Corporation, USA, equipped with flame ionization detection), using a 30 meter BPX 70 capillary columns of 0.32 mm internal diameter (SGE International Pty Ltd, Australia). Gas flow rates were as follows: nitrogen, 25 mL per min; air, 250 mL per min; and hydrogen (carrier gas), 25 mL per min and a split ratio of 20:1. Temperature programming was linear at 5°C per min, with initial temperature of 140°C, final temperature 220°C, injector temperature 240°C, and detector temperature 250°C. The fatty acid methyl esters were identified by comparison of the retention times to those of a standard fatty acid methyl ester mixture (Nu-Chek-Prep Inc., Elysian, Minnesota). The individual fatty acid methyl esters were quantified against an internal standard with known concentration (C17:0, Sigma-Aldrich, South Africa). Plasma fatty acids were quantified in absolute values in µg fatty acids per mL plasma analyzed. Red blood cell membrane fatty acids were quantified in absolute values in µg fatty acid per mL packed red blood cells analyzed. Peripheral blood mononuclear cell membrane protein was measured and fatty acids were quantified against membrane protein present in µg fatty acids per mg protein (see protein assay below). All individual fatty acids were also reported as relative percentage of the total fatty acids identified.

2.6 Protein analysis

The bicinchoninic acid protein determination assay was used to determine the protein content of peripheral blood mononuclear cell membranes (Kaushal & Barnes, 1986) for quantification of membrane fatty acids. Peripheral blood mononuclear cell membrane lipids were quantified against membrane protein present because of the high variation normally found in white blood cell counts. Of the starting material prepared in saline, a 200 µl sample aliquot was diluted with 2 % sodium dodecyl sulphate solution (Fluka, Sigma-Aldrich, Cape Town, South Africa) to denaturate protein prior to protein determination and assayed for protein content in triplicate. Bovine serum albumin (Sigma-Aldrich, Cape Town, South Africa) was used to prepare a standard curve. Sample and standard optical density was read on a spectrophotometer at a wavelength of 562 nm. Sample values were obtained from the bovine serum albumin linear standard curve. Peripheral blood mononuclear cell membrane fatty acids were quantified against membrane proteins and expressed in µg fatty acids per mg protein.

2.7 Phospholipid determination

Red blood cell and peripheral blood mononuclear cell membrane phospholipids were determined using a colorimetric assay with malachite green dye, as previously described (Itaya & Ui, 1966; Smuts et al., 1994). Phospholipid classes, identified and recovered as described above, were reduced to inorganic phosphates and quantified according to their phosphorous (Pi) content.

The different factors used for conversion of phosphorous to phospholipids were: for phosphatidylcholine: 25.4; phosphatidylethanolamine: 23.22; phosphatidylserine: 25.4; and sphingomyelin: 24.21. Red blood cell membrane phospholipid quantification was expressed in µg phospholipid per mL packed cells and peripheral blood mononuclear cell membrane phospholipid in µg phospholipid per mg protein (see protein assay).

2.8 C-reactive protein determination

Plasma C-reactive protein concentrations were determined in a routine Chemical Pathology laboratory using a Beckman nephelometer auto-analyser using reagents from Beckman, Cape Town, South Africa. A positive diagnostic value is considered as a C-reactive protein value equal to or greater than 5 µg per mL plasma.

2.9 Statistical analysis

A statistics programme, STATISTICA 9 {StatSoft, Inc. (2009). STATISTICA} was used to perform all statistical analyses. Correlation studies using Spearman's Rank correlation coefficient were used to evaluate the measure of agreement between the two methods (absolute amounts and relative percentages) in the plasma, red blood cell and peripheral blood mononuclear cell membranes from control subjects and patients with multiple sclerosis (Table 2). Linear regression analyses were used to measure strength of association between fatty acids quantified in absolute amounts and relative percentages, and C-reactive protein as well as the Kurtzke Expanded Disability Status Scale (Tables 3 and 4). The results were considered significant if P-values were less than 0.05.

3. Results

3.1 Method of measurement comparison

correlation studies between fatty acids quantified in absolute amounts and relative percentages are summarized in Table 2, Figure 1 and Results: sections 3.1.1, 3.1.2 and 3.1.3: Overall, the absolute amounts and relative percentage fatty acid composition results were moderately and mostly positively associated with each other in both study groups, although a number of fatty acids correlated not significantly in both groups (data not shown). However, it should be noted that though these correlations were significant, they were not strong as most of the R values were less than 0.5. Differences in the strength of association between the two study groups are listed below.

3.1.1 Plasma fatty acids

SM C20:0 showed significant correlations in controls only, and non-esterified fatty acid C18:2n-6 and PC C16:0 showed significance in patients only. Further, SM C18:0 showed weaker correlations in plasma from patients than in controls. Non-esterified fatty acids C20:4n-6, C16:0 and C18:0 showed no correlation in either of the two study groups.

3.1.2 Red blood cell membrane fatty acids

PS C22:4n-6, SM C16:0 and PS C18:0 showed significant correlations in controls only and PS C20:4n-6, PS C22:6n-3, PC C18:1n-9 and non-esterified fatty acid C22:0 were significantly associated in patients only. PE C18:0 showed a stronger association between absolute amounts and relative percentage fatty acids in red blood cells from patients with multiple sclerosis than in control subjects. Furthermore, PE C20:4n-6, PC C16:0, PE C16:0, PC C18:0, PE C18:0 and non-esterified fatty acid C16:0 showed no correlation in either of the two study groups.

	Total group		Controls		Patients	
	R	P-value	R	P-value	R	P-value
Plasma fatty acids						
Polyunsaturated fatty acids						
NEFA C18:2n-6	0.42	0.0008	0.31	0.0961	0.48	0.0060
Saturated fatty acids						
PC C16:0	0.33	0.0095	0.35	0.0601	0.36	0.0469
SM C18:0	0.64	< 0.0001	0.80	< 0.0001	0.50	0.0039
SM C20:0	0.49	0.0001	0.65	0.0001	0.34	0.0644
Red blood cell membrane fatty acids						
Polyunsaturated fatty acids						
PE C20:4n-6	0.36	0.0041	0.35	0.0583	0.28	0.1245
PS C20:4n-6	0.40	0.0012	0.17	0.3699	0.50	0.0043
PS C22:4n-6	0.53	< 0.0001	0.74	< 0.0001	0.18	0.3413
PS C22:6n-3	0.45	0.0002	0.31	0.0951	0.57	0.0008
Monounsaturated fatty acids						
PC C18:1n-9	0.40	0.0012	0.31	0.0992	0.47	0.0072
Saturated fatty acids						
NEFA C22:0	0.41	0.0012	0.27	0.1528	0.45	0.0113
PC C18:0	0.23	0.0737	0.33	0.0780	0.18	0.3363
PE C16:0	0.32	0.0118	0.34	0.0622	0.27	0.1373
PE C18:0	0.34	0.0075	0.30	0.1099	0.35	0.0518
PS C18:0	-0.34	0.0079	-0.37	0.0461	-0.26	0.1557
SM C16:0	0.41	0.0011	0.58	0.0009	0.36	0.0509
Peripheral blood mononuclear cell membrane fatty acids						
Polyunsaturated fatty acids						
PC C18:2n-6	0.65	< 0.0001	0.32	0.1156	0.81	< 0.0001
PE C20:4n-6	0.56	< 0.0001	0.38	0.0599	0.69	0.0001
Monounsaturated fatty acids						
NEFA C18:1n-9	0.69	< 0.0001	0.89	< 0.0000	0.39	0.0463
PC C18:1n-9	0.62	< 0.0001	0.42	0.0351	0.78	< 0.0001
Saturated fatty acids						
NEFA C18:0	-0.10	0.4700	-0.43	0.0337	0.08	0.6903
PE C18:0	0.39	0.0051	0.49	0.0121	0.31	0.13153
PS C18:0	0.38	0.0064	0.40	0.0448	0.34	0.0924
SM C16:0	0.51	0.0001	0.24	0.2432	0.69	0.0001
SM C18:0	0.51	0.0002	0.22	0.3168	0.74	< 0.0001

Table 2. Method of measurement comparison: correlation studies between fatty acids quantified in absolute amounts and relative percentages. Key: NEFAs: Non-esterified fatty acids, PC: Phosphatidylcholine, SM: Sphingomyelin, PE: Phosphatidylethanolamine, PS: Phosphatidylserine.

3.1.3 Peripheral blood mononuclear cell membrane fatty acids

PE C18:0, PS C18:0 and non-esterified fatty acid C18:0 showed significant correlations in controls only and PC C18:2n-6, PE C20:4n-6, SM C16:0 and SM C18:0 were significant in patients only. PC C18:1n-9 showed stronger correlations between absolute amounts and relative percentage fatty acids in peripheral blood mononuclear cells from patients with multiple sclerosis, while non-esterified fatty acid C18:1n-9 showed stronger correlations in controls.

3.2 Differences between plasma, red blood cell and peripheral blood mononuclear cell membrane fatty acids from control subjects and patients with multiple sclerosis

Measures of central tendencies have been published previously. Polyunsaturated fatty acids measured in plasma, red blood cell and peripheral blood mononuclear cell membranes from control subjects and patients with multiple sclerosis showed similar results when measured in absolute values or relative percentages. They include the following: PC C20:4n-6 quantified in absolute values (Hon et al., 2011a) and relative percentages (data not shown) was decreased in plasma from patients. PC C20:4n-6 quantified in absolute values (Hon et al., 2009a) and relative percentages (data not shown) was decreased in red blood cell membranes from patients. PE C22:4n-6 and PS C22:4n-6 quantified in absolute values as well as relative percentages were decreased in peripheral blood mononuclear cell membranes from patients with multiple sclerosis (Hon et al., 2009b).

3.3 Association studies between C-reactive protein and PC fatty acids are summarized in Table 3

3.3.1 Plasma fatty acids

C-reactive protein showed no association with any of the fatty acids quantified in absolute amounts in the PC phospholipid fraction in plasma from controls, but a positive association with PC C20:3n-6 in plasma from patients. It showed a weak association with PC C20:3n-6 quantified in relative percentages in plasma from controls and a strong association in plasma from patients similar to that found with the fatty acid quantified in absolute amounts. It further showed a positive association with PC C16:1n-7 quantified in relative percentages in plasma from controls and an inverse association with PC C18:1n-9 in plasma from patients.

3.3.2 Red blood cell membrane fatty acids

The C-reactive protein showed a positive association with PC C16:1n-7 (Hon et al., 2010) quantified in absolute amounts in the red blood cells from controls and a positive association with PC C20:3n-6 in red blood cells from patients. Similar to results for that of plasma PC fatty acids, it showed a weak association with PC C20:3n-6 quantified in relative percentages in red blood cells from controls and a strong association in red blood cells from patients. Also, similar to results for that of plasma PC fatty acids, the C-reactive protein showed a positive association with PC C16:1n-7 quantified in relative percentages in red blood cells from controls and an inverse association with PC C18:1n-9 in red blood cells from patients, but this inverse correlation was also found in red blood cells from controls (Hon et al., 2010).

	Controls				Patients			
	Absolute amounts		Percentages		Absolute amounts		Percentages	
	b*	P-value	b*	P-value	b*	P-value	b*	P-value
Plasma fatty acids								
Polyunsaturated fatty acids								
PC C20:3n-6	0.30	0.1242	0.33	0.0973	**0.39**	**0.0378**	**0.49**	**0.0071**
Monounsaturated fatty acids								
PC C16:1n-7	0.32	0.1152	**0.40**	**0.0444**	0.15	0.4485	0.12	0.5416
PC C18:1n-9	-0.01	0.9653	-0.12	0.5585	-0.16	0.4044	**-0.46**	**0.0126**
Saturated fatty acids								
PC C18:0	-0.01	0.9570	-0.19	0.3390	0.33	0.0840	0.21	0.2694
Red blood cell membrane fatty acids								
Polyunsaturated fatty acids								
PC C20:3n-6	0.20	0.3059	0.36	0.0625	0.34	0.0679	**0.52**	**0.0041**
PC C22:4n-6	0.18	0.3826	0.37	0.0601	-0.26	0.1759	-0.24	0.2113
Monounsaturated fatty acids								
PC C16:1n-7	**+ Association***		**0.39**	**0.0468**	No association*		0.14	0.4788
PC C18:1n-9	No association*		-0.34	0.0817	No association*		**-0.41**	**0.0264**
Peripheral blood mononuclear cell membrane fatty acids								
Polyunsaturated fatty acids								
PC C20:3n-6	0.22	0.3255	0.24	0.2818	0.16	0.4482	0.37	0.0714
PC C22:4n-6	0.27	0.2262	0.38	0.0773	-0.14	0.5093	0.07	0.7534
PC C22:5n-6	**0.45**	**0.0358**	**0.46**	**0.0326**	0.07	0.7629	0.23	0.2806
PC C20:5n-3	-0.05	0.8133	-0.06	0.7823	-0.36	0.0842	-0.36	0.0944
Monounsaturated fatty acids								
PC C16:1n-7	**+ Association***		**0.54**	**0.0095**	No association*		0.12	0.5970
PC C18:1n-7	**+ Association***		0.42	0.0537	No association*		**0.46**	**0.0230**
PC C18:1n-9	No association*		**-0.45**	**0.0353**	No association*		-0.02	0.9398
Saturated fatty acids								
PC C14:0	0.30	0.1772	0.29	0.1953	-0.34	0.0992	-0.25	0.2426
PC C18:0	0.04	0.8744	-0.03	0.8887	**-0.47**	**0.0204**	-0.24	0.2694
PC C20:0	-0.07	0.7563	-0.14	0.5429	-0.36	0.0866	-0.14	0.5003
PC C22:0	-0.39	0.0762	-0.40	0.0680	-0.16	0.4487	0.04	0.8700

Table 3. Association studies between C-reactive protein and PC fatty acids in plasma, red blood and peripheral blood mononuclear cell membranes. * Results: Hon et al. (2010)

3.3.3 Peripheral blood mononuclear cell membrane fatty acids

The C-reactive protein showed a weak association with PC C20:3n-6 quantified in relative percentages in peripheral blood mononuclear cell membranes from patients. It showed positive associations with PC C22:5n-6, PC C16:1n-7 (Hon et al., 2010) and PC C18:1n-7 (Hon et al., 2010) quantified in both absolute amounts and relative percentages in peripheral blood mononuclear cell membranes from controls, while only PC C18:1n-7 quantified in relative percentages showed a positive association in peripheral blood mononuclear cell membranes from patients. It also showed an inverse association with PC C18:1n-9 quantified in relative percentages in peripheral blood mononuclear cell membranes from controls, in contrast to results in plasma and red blood cells where this association was found in patients.

3.4 Association studies between the Kurtzke Expanded Disability Status Scale and fatty acids from patients with multiple sclerosis are summarized in Table 4

3.4.1 Plasma fatty acids

The Kurtzke Expanded Disability Status Scale showed near-significant inverse associations with non-esterified fatty acid C20:4n-6 and SM C20:0 quantified in relative percentages, but not with absolute amounts in plasma from patients. It also showed a significant inverse correlation with non-esterified fatty acid C14:0 quantified in relative percentages, but a near-significant positive association with non-esterified fatty acid C14:0 quantified in absolute amounts.

| | Patients with multiple sclerosis | | | | |
| | Absolute amounts | | Relative percentages | | |
	b*	P-value	b*	P-value	R^2
Plasma fatty acids					
Polyunsaturated fatty acids					
NEFA C20:4n-6	-0.04	0.8188	-0.34	0.0716	0.05
Saturated fatty acids					
NEFA C14:0	0.42	0.0550	**-0.48**	**0.0307**	0.11
NEFA C16:0	0.26	0.1532	-0.31	0.0869	0.08
SM C20:0	-0.01	0.9683	-0.34	0.0828	0.05
Red blood cell membrane fatty acids					
Polyunsaturated fatty acids					
PC C20:2n-6	-0.42	0.2154	**0.72**	**0.0397**	0.12
PE C20:3n-6	-0.30	0.2806	**0.61**	**0.0322**	0.12
PS C22:5n-3	-0.06	0.7511	**-0.41**	**0.0320**	0.14
Saturated fatty acids					
PC C16:0	-0.31	0.0771	-0.31	0.077	0.13
PC C18:0	-0.13	0.4928	**0.41**	**0.0372**	0.09
PE C16:0	-0.06	0.7650	0.34	0.0835	0.04
Peripheral blood mononuclear cell membrane fatty acids					
Polyunsaturated fatty acids					
PE C20:5n-3	0.95	0.1296	-1.10	0.0791	0.06
Monounsaturated fatty acids					
PC C24:1n-9	2.14	0.0708	-1.99	0.0927	0.07

Table 4. Association studies between the Kurtzke Expanded Disability Status Scale and absolute amounts and relative percentage fatty acids in plasma, red blood cell and peripheral blood mononuclear cell membranes from patients with multiple sclerosis

3.4.2 Red blood cell membrane fatty acids

The Kurtzke Expanded Disability Status Scale showed significant positive associations with PC C20:2n-6, PE C20:3n-6 and PC C18:0 as well as a near-significant positive association with PE C16:0 quantified in relative percentages, but not with absolute amounts in red blood cell membranes from patients. In contrast, it showed a significant inverse correlation with PS C22:5n-3 quantified in relative percentages in red blood cell membranes from patients.

3.4.3 Peripheral blood mononuclear cell membrane fatty acids

The Kurtzke Expanded Disability Status Scale showed no significant associations with the fatty acids in peripheral blood mononuclear cell membranes from patients, but a near-significant inverse correlation with PE C20:5n-3 quantified in relative percentages.

4. Discussion

The objective was to establish whether using absolute amounts in quantifying fatty acids in biological tissue would provide sufficient information to investigate abnormalities in a diseased state, in this case investigating the fatty acid profile in patients with multiple sclerosis. Results showed that correlation strengths between fatty acids quantified in absolute amounts and relative percentages varied in some instances in plasma, red blood cell as well as peripheral blood mononuclear cell membranes between study groups (Table 2). Furthermore, even when a correlation was observed between absolute amounts and relative percentages the R value was not always close to the value one as would have been expected if these theoretically represent each other. Of note is that some of these correlations were lost or gained in diseased subjects. For example, PC C18:1n-9 showed stronger correlations in peripheral blood mononuclear cells from patients, than in control subjects. In contrast, PS C22:4n-6 showed significance in red blood cell membranes from controls, but not in patients with multiple sclerosis (Figure 1).

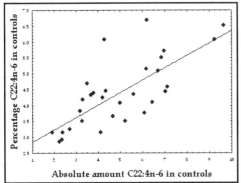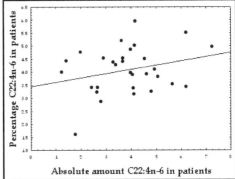

Fig. 1. Measurement comparison of PS C22:4n-6 quantified in absolute amounts and relative percentages in red blood cell membranes. Controls, R = 0.74; P = < 0.0000. In patients with multiple sclerosis the correlation was not significant, R = 0.18; P = 0.3413.

Multiple sclerosis is an inflammatory disease of the central nervous system in which an abnormal fatty acid profile has been reported, but with inconclusive findings (Cherayil, 1984; Navarro & Segura, 1989; Harbige & Sharief, 2007; Hon et al., 2009a, 2009b). The n-6 and n-3 polyunsaturated fatty acids, C18:2n-6, C20:4n-6 and C20:5n-3 have been reported to be decreased in plasma and blood cell membranes from these patients (Baker et al., 1964; Cherayil, 1984; Holman et al., 1989; Nightingale et al., 1990; Hon et al., 2009a, 2009b), but other research groups have found no differences compared to control subjects (Cumings et al., 1965; Fisher et al., 1987; Evans & Dodd, 1989; Koch et al., 2006). Disturbed metabolic relationships between C18:2n-6 and C20:3n-6 (Harbige & Sharief, 2007), as well as between C20:3n-6 and C20:4n-6 (Harbige & Sharief, 2007; Hon et al., 2009b) in the peripheral blood mononuclear cell membranes from patients with multiple sclerosis have also been reported. Homa et al. (1980) showed the relationship between C18:2n-6 and C20:4n-6 in red blood cells from patients to be disturbed, while Hon et al. (2009a) reported disturbances between PC C18:2n-6 and PC C20:3n-6, PC C18:2n-6 and PC C20:4n-6, PC C20:3n-6 and PC C20:4n-6 in red blood cells from patients. The lack of consensus on the role of fatty acids in the pathogenesis of multiple sclerosis, may be due to the variable results observed in previous studies which may be attributed to the different forms of assessments used (absolute amounts and relative percentages).

4.1 Quantification of polyunsaturated fatty acid C20: 4n-6

The n-6 polyunsaturated fatty acid PC C20:4n-6 (data not shown) showed significant correlations between absolute amounts and relative percentage composition in plasma from both the control group and patients with multiple sclerosis, but showed a wide variation in red blood cell membranes. PE C20:4n-6 showed a weak correlation in controls only, while PS C20:4n-6 in contrast showed a strong correlation in red blood cells from patients. PE C20:4n-6 showed a strong correlation in peripheral blood mononuclear cell membranes from patients and a weak correlation in controls. In contrast to the differences in correlations between absolute amounts and relative percentage composition of C20:4n-6 in red blood cell and peripheral blood mononuclear cell membranes from patients with that of the control group, the n-3 polyunsaturated fatty acids in general showed similar strong associations between absolute amounts and relative percentage in plasma, red blood cell and peripheral blood mononuclear cell membranes from controls and patients (Table 2). These results have demonstrated that in control subjects there is a wide range of association strengths between the n-6 fatty acids quantified in absolute amounts and relative percentages with these associations being further disturbed in patients with multiple sclerosis. On the other hand, the n-3 polyunsaturated fatty acids showed close associations between absolute amounts and relative percentages and that this was maintained in patients with multiple sclerosis.

It is unclear why the n-6 polyunsaturated fatty acid C20:4n-6 should show such a wide variation in association between absolute amounts and relative percentages (from very weak to strong) in plasma, red blood cell and peripheral blood mononuclear cell membranes from controls. Furthermore, it is unclear why C20:4n-6 should have stronger associations between absolute amounts and relative percentages specifically in peripheral blood mononuclear cell membranes from patients, unless the reversal of association strength in these immune cells are part of the pathogenesis of multiple sclerosis. C20:4n-6 is released from cell membranes during inflammation as precursor for pro-inflammatory prostaglandin E2 production (Horrobin & Manku, 1990; Simopoulos, 2002; Bagga et al., 2003; Haag, 2003), while its precursor fatty acid in the n-6 fatty acid series, C20:3n-6 is the precursor fatty acid for the synthesis of prostaglandin E1, a less inflammatory mediator of inflammation, than prostaglandin E2. In contrast, prostaglandin E3 derived from the metabolism of C20:5n-3 and docosanoids derived from the metabolism of C22:6n-3 has anti-inflammatory properties (Bagga et al., 2003; Haag, 2003; Zamaria, 2004; Farooqui et al., 2007; Chen et al., 2008). In a disease such as multiple sclerosis, which is inflammatory in nature, it is useful to know whether the n-6 and n-3 polyunsaturated fatty acids show any metabolic abnormalities and these results suggested abnormalities within the n-6 polyunsaturated fatty acid series rather than in the n-3 polyunsaturated fatty acids.

Results from previous studies (Hon et al., 2009a, 2009b; Hon et al., 2011a) and this study showed similar decreases in polyunsaturated fatty acid C20:4n-6 and its elongation product C22:4n-6 when quantified in absolute amounts or relative percentages in plasma, red blood cell and/or peripheral blood mononuclear cell membranes from patients with multiple sclerosis as compared to controls. Therefore, investigating measures of central tendencies in fatty acids from control subjects and patients with multiple sclerosis using both quantifications methods does not explain the differences in reports of C20:4n-6 composition in patients with multiple sclerosis as compared to control subjects. However, the variations in association strengths between C20:4n-6 quantified in absolute amounts and relative percentages in patients compared to control subjects (Table 2) showed that other factors must contribute to the change in association. In this regard, the C-reactive protein has been shown to correlate inversely with PE C20:4n-6 and PE C22:4n-6 quantified in absolute values in red blood cell membranes from patients with multiple sclerosis (Hon et al., 2009a) and also inversely with PC C20:4n-6, PS C20:4n-6, PS C22:4n-6 and PI C22:4n-6 quantified in absolute values in peripheral blood mononuclear cell membranes from patients with multiple sclerosis (Hon et al., 2009b). It is possible therefore to expect C-reactive protein concentrations to show associations with blood fatty acids depending on the inflammatory status of the patients.

4.2 Association between C-reactive protein and polyunsaturated fatty acids

In this study, results (Table 3) showed that the C-reactive protein showed very few associations with fatty acids quantified in absolute amounts in plasma, red blood cell or peripheral blood mononuclear cell membranes from either control subjects or patients with multiple sclerosis. In contrast, it showed a number of associations with fatty acids quantified in relative percentages in both controls and patients. Furthermore, these results indicated that the association between the C-reactive protein and certain fatty acids quantified in relative percentages are valid, because the same associations were found in plasma and red blood cell membrane fatty acids. For example, the C-reactive protein showed positive associations with PC C20:3n-6 quantified in absolute amounts as well as in relative percentages in plasma from patients but an association in plasma from controls only with this fatty acid quantified in relative percentages. The same association was found between the C-reactive protein and PC C20:3n-6 quantified in relative percentages in red blood cells from controls and patients but not in peripheral blood mononuclear cell membranes from patients or controls. These and previous results showed that association and correlation studies between the C-reactive protein and fatty acids as well as using absolute amounts and relative percentages of fatty acids in evaluating their role in inflammation in patients with multiple sclerosis gave different outcomes, and that relative percentages gave additional information. Specifically, these results showed that correlation studies highlighted the inverse correlations between the C-reactive protein and C20:4n-6 as well as C22:4n-6 quantified in absolute amounts in patients, while association studies showed a positive association between C-reactive protein and C20:3n-6 and that this association was found primarily with the relative percentages.

These results are important because C20:4n-6 and C20:3n-6 are precursors for prostaglandins E2 and E1 respectively and of these E1 is the less inflammatory mediator, and clearly, the C-reactive protein showed opposite correlations/associations with these two fatty acids. It is not clear from the results of this study whether the positive association between the C-reactive protein and C20:3n-6 was due to the membrane content of the fatty acid or whether it was due to the amount made available for prostaglandin E2 production, that is whether a higher membrane content suggested lower levels of the less inflammatory prostaglandin. However, fatty acid metabolic abnormalities have been reported between C20:3n-6 and C20:4n-6 in the peripheral blood mononuclear cell (Harbige & Sharief, 2007; Hon et al., 2009b) and red blood cell (Hon et al., (2009a) membranes from patients.

4.3 Association between C-reactive protein and monounsaturated fatty acids

Furthermore, the C-reactive protein showed positive associations with PC C16:1n-7 quantified in relative percentages in plasma from controls and with PC C16:1n-7 quantified in both absolute amounts (Hon et al., 2010) and relative percentages in both red blood cell and peripheral blood mononuclear cell membranes from controls (Table 3). In contrast, it showed no association with PC C16:1n-7 in any of these blood compartments from patients. The associations between the C-reactive protein and monounsaturated fatty acid C16:1n-7 in control subjects suggested a specific role for this fatty acid in the immune cells in the inflammatory process. The absence of this association in patients suggested an irregularity in the immune process in patients with multiple sclerosis. The C-reactive protein did show a positive association with PC C18:1n-7 in peripheral blood mononuclear cell membranes from patients but only with the percentage composition.

In contrast to the positive associations with the n-7 fatty acids, the C-reactive protein showed an inverse association with PC C18:1n-9 quantified in relative percentages in plasma and red blood cells from patients, but in peripheral blood mononuclear cell membranes from controls only. These results again showed the importance of including relative percentages when evaluating the relationship between inflammation and fatty acids in patients with multiple sclerosis. Monounsaturated fatty acids may have anti-inflammatory effects, although their role in inflammation is not as well defined as that of the polyunsaturated fatty acids (Yaqoob et al., 1998; Harwood & Yaqoob, 2002). The results from this and previous studies showed that the altered associations between the C-reactive protein and the n-7 and n-9 monounsaturated fatty acids may have had an effect on their anti-inflammatory functions and may therefore have contributed to the pathogenesis of the disease.

4.4 The relationship between polyunsaturated fatty acids and the Kurtzke Expanded Disability Status Scale

Previous studies have shown an inverse correlation between the Kurtzke Expanded Disability Status Scale and PC C20:4n-6 quantified in absolute values in red blood cell membranes from patients (Hon et al., 2009a), but a positive correlation with PC C20:3n-6 in peripheral blood mononuclear cell membranes from patients with multiple sclerosis (Hon et al., 2009b). Results from this study showed no association between the Kurtzke Expanded Disability Status Scale and C20:4n-6, but it did show positive associations with PC C20:2n-6 and PE C20:3n-6 quantified in relative percentages in red blood cell from patients (Table 4). These findings are important because the C-reactive protein has been shown in previous studies to show inverse correlations with C20:4n-6 quantified in absolute amounts in patients, while a positive association was shown in this study with PC C20:3n-6 quantified in relative percentages in peripheral blood mononuclear cell membranes from patients. Furthermore, previous studies have shown that the C-reactive protein also showed a positive correlation with the Kurtzke Expanded Disability Status Scale in patients with multiple sclerosis (Hon et al., 2009b).

Although these results would need further investigation, their associations/correlations would suggest that with an increase in the C-reactive protein, there is a decrease in blood cell membrane C20:4n-6 and an increase in C20:3n-6, with an accompanying increase in the Kurtzke Expanded Disability Status Scale in patients with multiple sclerosis. It is possible to hypothesize that the decrease in membrane C20:4n-6 could have been due to the release of this fatty acid for eicosanoid production because this fatty acid is the major fatty acid to be released for eicosanoid production during the inflammatory processes, as measured by the C-reactive protein (Calder, 2007). In addition, these results further suggested that an imbalance in C20:4n-6 and C20:3n-6 release for prostaglandin E2 and E1 production respectively could be associated with the inflammatory condition experienced by patients with multiple sclerosis and that this effect could possibly be a constant condition, hence the significant correlation/association with the Kurtzke Expanded Disability Status Scale respectively.

4.5 Membrane phospholipids

Membrane lipids consist of phospholipids, into which saturated and unsaturated fatty acids are incorporated, as well as cholesterol. Choline containing phospholipids, phosphatidylcholine and sphingomyelin are mainly on the outer leaflet of plasma membranes, while phosphatidylethanolamine and phosphatidylserine are located on the inner leaflet. The properties of the different phospholipid molecules depend on their different head-groups as well as on the type of fatty acids they contain (Williams, 1998; Horrobin, 1999) and their composition in cell membranes determines the degree of fluidity of the membrane structures. Phospholipids from both red blood cell (Hon et al., 2009c) and peripheral blood mononuclear cell (Hon et al., 2011b) membranes have been reported in detail and specifically their function in determining cell membrane fluidity and their role in disease outcome in patients with multiple sclerosis.

5. Conclusions

The validity of using fatty acid composition in relative percentages and/or absolute amounts has been discussed with regards to measurements of these in the blood components of patients with multiple sclerosis, an inflammatory disease of the central nervous system, in which various fatty acid imbalances have been described, but without any firm conclusions as to cause or consequences. The results of this study showed that using a combination of fatty acids quantified in both absolute amounts and relative percentages could be relevant, especially in a disease state such as multiple sclerosis. This was further confirmed by the association between C-reactive protein, a marker of inflammation and the fatty acids. Therefore, it is recommended that fatty acids should be analysed and interpreted using both absolute amounts and relative percentages.

6. References

Adibhatla, R.M. & Hatcher, J.F. (2007). Role of lipids in brain injury and diseases. *Future Lipidology*, Vol.2, pp. 403-422

Allen, H.G.; Allen, J.C.; Boyd, L.C.; Alston-Mills, B.P. & Fenner, G.P. (2006). Determination of membrane lipid differences in insulin resistant diabetes mellitus type 2 in whites and blacks. *Nutrition*, Vol.22, pp. 1096-1102

Bagga, D.; Wang, L.; Farias-Eisner, R.; Glaspy, J.A. & Reddy, S.T. (2003). Differential effects of prostaglandin derived from omega-6 and omega-3 polyunsaturated fatty acids on COX-2 expression and IL-6 secretion. *Proceedings of the National Academy of Sciences USA*, Vol.100, pp. 1751 - 1756

Baker, R.W.R.; Thompson R.H.S. & Zilkha K.J. (1964). Serum fatty acids in multiple sclerosis. *Journal of Neurology, Neurosurgery, and Psychiatry*, Vol.27, pp. 408-414

Barenholz, Y. (2002). Cholesterol and other membrane active sterols: from membrane evolution to "rafts". *Progress in Lipid Research*, Vol.41, pp. 1-5

Barenholz, Y. & Thompson, T.E. (1999). Sphingomyelin: biophysical aspects. *Chemistry and Physics of Lipids*, Vol.102, pp. 29-34

Bates, D.; Fawcett, P.R.W.; Shaw, D.A. & Weightman, D. (1978). Polyunsaturated fatty acids in treatment of acute remitting multiple sclerosis. *British Medical Journal*, Vol.2, pp. 1390-1391

Blaak, E.E. (2003). Fatty acid metabolism in obesity and type 2 diabetes mellitus. *Proceedings of the Nutrition Society*, Vol.62, pp. 753-60

Brown, K.A. (2001). Factors modifying the migration of lymphocytes across the blood-brain barrier. *International Immunopharmacology*, Vol.1, pp. 2043-2062

Calder, P.C. (2007). Immunomodulation by omega-3 fatty acids. *Prostaglandins, Leukotrienes and Essential Fatty Acids*, Vol.77, pp. 327-335

Caret, R.L.; Denniston, K.J. & Topping, J.J. (1997). Lipids and their functions in biochemical systems, In: *Principles and applications of inorganic, organic and biological chemistry*, C.H. Wheatley, (Ed.), 379-407, McGraw-Hill, Boston

Chen, C.T.; Green, J.T.; Orr, S.K. & Bazinet, R.P. (2008). Regulation of brain polyunsaturated fatty acid uptake and turnover. *Prostaglandins, Leukotrienes and Essential Fatty Acids*, Vol.79, pp. 85-91

Cherayil, G.D. (1984). Sialic acid and fatty acid concentrations in lymphocytes, red blood cells and plasma from patients with multiple sclerosis. *Journal of the Neurological Sciences*, Vol.63, pp. 1-10

Cordo, S.M.; Candurra, N.A. & Damonte, E.B. (1999). Myristic acid analogs are inhibitors of Junin virus replication. *Microbes and Infection*, Vol.1, pp. 609-614

Culp, B.R.; Titus, B.G. & Lands, W.E. (1979). Inhibition of prostaglandins by eicosapentaenoic acid. *Prostaglandins and Medicine*, Vol.3, pp. 269-278

Cumings, J.N.; Shortman, R.C. & Skrbic, T. (1965). Lipid studies in the blood and brain in multiple sclerosis and motor neurone disease. *Journal of Clinical Pathology*, Vol.18, pp. 641-644

De Pablo, M.A. & De Cienfuegos, G.Á. (2000). Modulatory effects of dietary lipids on immune system functions. *Immunology and Cell Biology*, Vol.78, pp. 31-39

Evans, P. & Dodd, G. (1989). Erythrocyte fatty acids in multiple sclerosis. *Acta Neurologica Scandinavica*, Vol.80, pp. 501-503

Farinotti, M.; Simi, S.; Di Pietrantonj, C.; McDowell, N.; Brait, L.; Lupo, D. & Filippini, G. (2007). Dietary interventions for multiple sclerosis. *Cochrane Database Syst Rev*, (1) CD004192

Farooqui, A.A.; Horrocks, L.A. & Farooqui, T. (2007). Modulation of inflammation in brain: a matter of fat. *Journal_of_Neurochemistry*, Vol.101, pp. 577-599

Fisher, M.; Johnson, M.H.; Natale, A.M.; & Levine, P.H. (1987). Linoleic acid levels in white blood cells, platelets, and serum of multiple sclerosis patients. *Acta Neurologica Scandinavica*, Vol.76, pp. 241-245

Folch, J.; Lees, M. & Sloane-Stanley, G.H. (1957). A simple method for the isolation and purification of total lipids from animal tissues. *Journal of Biological Chemistry*, Vol.226, pp. 497-509

Gilfillan, A.M.; Chu, A.J.; Smart, D.A. & Rooney, S.A. (1983). Single plate separation of lung phospholipids including disaturated phosphatidylcholine. *Journal of Lipid Research*, Vol.24, pp. 1651-1656

Giovannoni, G.; Miller, D.H.; Losseff, N.A.; Sailer, M.; Lewellyn-Smith, N.; Thompson, A.J. & Thompson, E.J. (2001). Serum inflammatory markers and clinical/MRI markers of disease progression in multiple sclerosis. *Journal of Neurology*, Vol.248, pp. 487-495

Giovannoni, G.; Thorpe, J.W.; Kidd, D.; Kendall, B.E.; Moseley, I.F.; Thompson, A.J.; Keir, G.; Miller, D.H.; Feldmann, M. & Thompson, E.J. (1996). Soluble E-selectin in multiple sclerosis: raised concentrations in patients with primary progressive disease. *Journal of Neurology, Neurosurgery, and Psychiatry*, Vol.60, pp. 20-26

Haag, M. (2003). Essential fatty acids and the brain. *Canadian Journal of Psychiatry*, Vol.48, pp. 195-203

Hamai, C.; Yang, T.; Kataoka, S.; Cremer, P.S. & Musser, S.M. (2006). Effect of average phospholipid curvature on supported bilayer formation on glass by vesicle fusion. *Biophysical Journal*, Vol.90, pp.1241-1248

Harbige, L.S. & Sharief, M.K. (2007). Polyunsaturated fatty acids in the pathogenesis and treatment of multiple sclerosis. *British Journal of Nutrition*, Vol.98, pp. S46-S53

Harlos, K. & Aibl, H. (1981). Hexagonal phases in phospholipids with saturated chains: Phosphatidylethanolamines and phosphatidic acids. *Biochemistry*, Vol.20, pp. 2888-2892

Harwood, J.L. & Yaqoob, P. (2002). Nutritional and health aspects of olive oil. *The European Journal of Lipid Science and Technology*, Vol.104, pp. 685-697

Hazel, J.R. & Williams, E.E. (1990). The role of alterations in membrane lipid composition in enabling physiological adaptation of organisms to their physical environment. *Progress in Lipid Research*, Vol.29, pp. 167-227

Holman, R.T.; Johnson, S.B. & Kokmen, E. (1989).Deficiencies of polyunsaturated fatty acids and replacement by nonessential fatty acids in plasma lipids in multiple sclerosis. *Proceedings of the National Academy of Sciences USA*, Vol.86, pp. 4720-4724

Homa, S.T.; Belin, J.; Smith, A.D.; Monro, J.A. & Zilkha, (1980). K.J. Levels of linoleate and arachidonate in red blood cells of healthy individuals and patients with multiple sclerosis. *Journal of Neurology, Neurosurgery and Psychiatry*, Vol.43, pp. 106-110

Hon, G.M.; Hassan, M.S.; Van Rensburg, S.J.; Abel, S.; Marais, D.W.; Van Jaarsveld, P.; Smuts, C.M.; Henning, F.; Erasmus, R.T. & Matsha, T. (2009a). Erythrocyte membrane fatty acids in patients with multiple sclerosis. *Multiple Sclerosis*, Vol.15, pp. 759-762

Hon, G.M.; Hassan, M.S.; Van Rensburg, S.J.; Abel, S.; Marais, D.W.; Van Jaarsveld, P.; Smuts, C.; Henning, F.; Erasmus, R. & Matsha, T. (2009b). Immune cell membrane fatty acids and inflammatory marker, C-reactive protein, in patients with multiple sclerosis. *British Journal of Nutrition*, Vol.102, pp. 1334-1340

Hon, G.M.; Hassan, M.S.; Van Rensburg, S.J.; Abel, S.; Van Jaarsveld, P.; Erasmus, R.T. & Matsha, T. (2009c). Red blood cell membrane fluidity in the etiology of multiple sclerosis. *Journal of Membrane Biology*, Vol.232, pp. 25-34

Hon, G.M.; Hassan, M.S.; Van Rensburg, S.J.; Abel, S.; Erasmus, R.T. & Matsha, T. (2010). Monounsaturated fatty acids in blood cell membranes from patients with multiple sclerosis. *Inflammation*, 2010 Dec 1. PMID: 21120595 [Epub ahead of print]

Hon, G.M.; Hassan, M.S.; Janse van Rensburg, S.; Abel , S.; Erasmus, R.T. & Matsha, T. (2011a). Plasma Non-esterified Fatty Acids in Patients with Multiple Sclerosis. *Neurology Asia*, In Press.

Hon, G.M.; Hassan, M.S.; Van Rensburg, S.J.; Abel, S.; Erasmus, R.T. & Matsha, T. (2011b). Peripheral Blood Mononuclear Cell Membrane Fluidity and Disease Outcome in Patients with Multiple Sclerosis. *Indian Journal of Hematology and Blood Transfusion*, (DOI) 10.1007/s12288-011-0087-x.

Horrobin, D.F. & Manku, M.S. (1990). Clinical biochemistry of essential fatty acids, In: *Omega-6 essential fatty acids. Pathophysiology and roles in clinical medicine*, D.F. Horrobin, (Ed.), 21-49, Alan R. Liss Inc, New York

Horrobin, D.F. (1999). The phospholipid concept of psychiatric disorders and its relationship to the neurodevelopmental concept of schizophrenia, In: *Phospholipid spectrum disorder in psychiatry*, M. Peet, I. Glen & D.F. Horrobin, (Eds.), 3-16, Marius Press, Lancashire, UK

Hunter, S.F. & Hafler, D.A. (2000). Ubiquitous pathogens. Links between infection and autoimmunity in MS? *Neurology*, Vol.55, pp. 164-165

Itaya, K. & Ui, M. (1966). A new micromethod for the colorimetric determination of inorganic phosphate. *Clinica Chimica Acta*, Vol.14, pp. 361-366

Kaushal, V. & Barnes, L.D. (1986). Effect of zwitterionic buffers on measurement of small masses of protein with bicinchoninic acid. *Analytical Biochemistry*, Vol.157, pp. 291-294

Koay, E.S.C. & Walmsley, N. (1999). Plasma lipids and lipoproteins, In: *A primer of Chemical Pathology*, E.S.C. Koay & N. Walmsley, (Eds.), 191-211, World Scientific, Singapore

Koch, M.; Ramsaransing, G.S.M.; Fokkema, M.R.; Heersema, D.J. & De Keyser, J. (2006). Erythrocyte membrane fatty acids in benign and progressive forms of multiple sclerosis. *Journal of the Neurological Sciences*, Vol.244, pp. 123-126

Kurtzke, J.F. (1983). Rating neurologic impairment in multiple sclerosis: an expanded disability status scale (EDSS). *Neurology,* Vol.33, pp. 1444-1452

Manzoli, F.A.; Stefoni, S.; Manzoli-Guidotti, L. & Barbieri, M. (1970). The fatty acids of myelin phospholipids. *Federation of European Biochemical Societies,* Vol.10, pp. 317-320

Mouritsen, O.G. & Jørgensen, K. (1998). A new look at lipid-membrane structure in relation to drug research. *Pharmaceutical Research,* Vol.15, pp. 1507-1519

Nakamura, M.T. & Nara, T.Y. (2004). Structure, function, and dietary regulation of Δ6, Δ5, and Δ9 desaturases. *The Annual Review of Nutrition,* Vol.24, pp.345-376

Narasimhan, B.; Mourya, V. & Dhake, A. (2006). Design, synthesis, antibacterial, and QSAR studies of myristic acid derivatives. *Bioorganic & Medicinal Chemistry Letters,* Vol.16, pp. 3023-3029

Navarro, X. & Segura, R. (1989). Red blood cell fatty acids in multiple sclerosis. *Acta Neurologica Scandinavica,* Vol.79, pp. 32-37

Nightingale, S.; Woo, E.; Smith, A.D.; French, J.M.; Gale, M.M.; Sinclair, H.M.; Bates, D. & Shaw, D.A. (1990). Red blood cell and adipose tissue fatty acids in mild inactive multiple sclerosis. *Acta Neurologica Scandinavica,* Vol.82, pp. 43-50

Nordvik, I.; Myhr, K.M.; Nyland, H. & Bjerve, K.S. (2000). Effect of dietary advice and n-3 supplementation in newly diagnosed MS patients. *Acta Neurologica Scandinavica,* Vol.102, pp. 143-149

Oram, J.F. & Bornfeldt, K.E. (2004). Direct effects of long-chain non-esterified fatty acids on vascular cells and their relevance to macrovascular complications of diabetes. *Frontiers in bioscience,* Vol.9, pp. 1240-1253

Paty, D.W.; Cousin, H.K.; Read, S. & Adlakha, K. (1978). Linoleic acid in multiple sclerosis: Failure to show any therapeutic benefit. *Acta Neurologica Scandinavica,* Vol.58, pp. 53-58

Pereira, S.L.; Leonard, A.E. & Mukerji, P. (2003). Recent advances in the study of fatty acid desaturases from animals and lower eukaryotes. *Prostaglandins, Leukotrienes and Essential Fatty Acids,* Vol.68, pp. 97-106

Pilz, S.; Scharnagl, H.; Tiran, B.; Wellnitz, B.; Seelhorst, U.; Boehm B.O. & März, W. (2007). Elevated plasma free fatty acids predict sudden cardiac death: a 6.85-year follow-up of 3315 patients after coronary angiography. *European Heart Journal,* Vol.28, pp. 2763-2769

Rapoport, S.I.; Rao, J.S. & Igarashi, M. (2007). Brain metabolism of nutritionally essential polyunsaturated fatty acids depends on both the diet and the liver. *Prostaglandins, Leukotrienes and Essential Fatty Acids,* Vol.77, pp. 251-261

Romon, M.; Nuttens, M.C.; Theret, N.; Delbart, C.; Lecerf, J.M.; Fruchart, J.C. & Salomez, J.L. (1995). Comparison between fat intake assessed by a 3-day food record and phospholipid fatty acid composition of red blood cells. *Metabolism - Clinical and Experimental,* Vol.44, pp. 1139-1145

Sands, J.A. (1977). Inactivation and inhibition of replication of the enveloped bacteriophage φ6 by fatty acids. *Antimicrobial Agents and Chemotherapy,* Vol.12, pp. 523-528

Sarafidis, P.A. & Bakris, G.L. (2007). Non-esterified fatty acids and blood pressure elevation: a mechanism for hypertension in subjects with obesity/insulin resistance? *Journal of Human Hypertension,* Vol.21, pp. 12-19

Sellner, J.; Greeve, I. & Mattle, H.P. (2008). Atorvastatin decreases high-sensitivity C-reactive protein in multiple sclerosis. *Multiple Sclerosis,* Vol.14, pp. 981-984

Simopoulos, A.P. (2002). Omega-3 fatty acids in inflammation and autoimmune diseases. *Journal of American College of Nutrition,* Vol.2, pp. 495-505

Smuts, C.M.; Weich, H.F.H.; Weight, M.J.; Faber, M.; Kruger, M.; Lombard, C.J. & Benade, A.J.S. (1994). Free cholesterol concentrations in the high-density lipoprotein subfraction-3 as a risk indicator in patients with angiographically documented coronary artery disease. *Coronary Artery Disease,* Vol.5, pp. 331-338

Stinissen, P.; Raus, J. & Zhang J. (1997). Autoimmune pathogenesis of multiple sclerosis: role of autoreactive T lymphocytes and new immunotherapeutic strategies. *Critical Reviews in Immunology,* Vo.17, pp. 33-75

Van Jaarsveld, P.J.; Smuts, C.M.; Tichelaar, H.Y.; Kruger, M. & Benadé, A.J.S. (2000). Effect of palm oil on plasma lipoprotein concentrations and plasma low-density lipoprotein composition in non-human primates. *International Journal of Food Sciences & Nutrition,* Vol.51, pp. S21-S30

Weinstock-Guttman, B.; Baier, M.; Park, Y.; Feichter, J.; Lee-Kwen, P.; Gallagher, E.; Venkatraman, J.; Meksawan, K.; Deinehert, S.; Pendergast, D.; Awad, A.B.; Ramanathan, M.; Munschauer, F. & Rudick, R. (2005). Low fat dietary intervention with ω-3 fatty acid supplementation in multiple sclerosis patients. *Prostaglandins, Leukotrienes and Essential Fatty Acids,* Vol.73, pp. 397-404

Williams, E.E. (1998). Membrane lipids: What membrane physical properties are conserved during physiochemically-induced membrane restructuring? *American Zoologist,* Vol.38, pp. 280-290

Yaqoob, P.; Knapper, J.M.E.; Webb, D.H.; Williams, E.; Newsholme, A. & Calder, P.C. (1998). The effects of olive oil consumption on immune functions in middle-aged men. *American Journal of Clinical Nutrition.* Vol.67, pp. 129-135

Zamaria, N. (2004). Alteration of polyunsaturated fatty acid status and metabolism in health and disease. *Reproduction Nutrition Development,* Vol.44, pp. 273-282

Part 2

Narcotics

8

Gas Chromatography in Forensic Chemistry: Cannabinoids Content in Marijuana Leaves (*Cannabis sativa* L.) from Colombia

N.M. Florian-Ramírez[1], W.F. Garzón-Méndez[2] and F. Parada-Alfonso[3]

[1]Scientific Lab, Criminalistic Group,
Departamento Administrativo de Seguridad-DAS, Bogotá,
[2]Criminalistic Chemistry Lab, Fiscalía General de la Nación, Bogotá,
[3]Chemistry Department, Science Faculty-Universidad Nacional de Colombia, Bogotá,
Colombia

1. Introduction

Despite that the narcotics had generated a regrettable stigma for Colombia, our country does not have a deep knowledge about the narcotic chemistry. In this study it is show the analysis of cannabinoids in the Colombian illicit crops of *Cannabis sativa* L. using Gas Chromatography (GC), Figure 1. The main important cannabinoids from *C. sativa* were identified by Gas Chromatography-Mass Spectrometry (GC-MS). The conditions for extraction and quantification of these cannabinoids (cannabidiol (CBD), Δ9-tetrahydrocannabinol (THC) and cannabinol (CBN)) from *C. sativa* leaves were optimized by the response surface analysis-RSA, using as variables the extraction time and the extraction numbers. Then, the better sample amount and the best detection method were determined, after the validation of the GC analytic method was made. Additionally, the cannabinoids of samples of different Colombian geographic regions were analyzed, determining that the THC content vs THC/CBN content can serve as a discriminating factor to distinguish the geographic origin of the samples studied. Finally, it was showed that the GC is a good analytic tool for forensic chemistry in Colombia.

The study of extraction process was carried out using RSA, the best response and the less variability for the cannabinoids was obtained with three extraction steps, each one of 7.5 minutes with ethanol as solvent. The quantitative analysis was made by Gas Chromatography-Flame Ionization Detector (GC-FID) with diphenhydramine (DPH) as surrogate standard. In the GC method for quantification of CBD, THC and CBN were assessed their chromatographic parameters; selectivity, linearity, precision, accuracy, Limit of Detection (LOD) and Limit of Quantification (LOQ).

The cannabinoids samples from four different Colombian regions (North region (NR), at Magdalena department, South region (SR) at Cauca department, East region (ER) at Meta department and Central region (CR), at Caldas department) were determined. The content average THC in samples of the NR was 2.81% ± 1.72, in the SR was 10.98% ± 6.70, in the ER was 15.74% ± 2.92 and at the CR was 1.87% ± 1.25. The higher content of THC at cannabis plants of the ER and SR could be indicative of the use of improved varieties, which creates a major concern in potential effects around of the health of drug consumers.

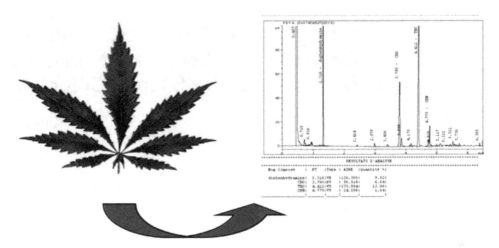

Fig. 1. Cannabis leave sample and a sample gas chromatogram of its extract.

It was established that the value of the Waller phenotype (Turner et al., 1979, 1980), equation 1, did not serve as a differentiation criterion on the geographical origin of the Colombian samples, due to overlapping ranges for three of the regions studied (NR, SR and ER). In order to establish a criterion to discriminate the geographic origin of the Colombian marijuana, different types of correlation among the CBD, THC and CBN were tested, finding that for the four groups of studied samples the ratio THC vs THC/CBN could be used.

$$\text{Waller phenotype} = (THC + CBN) / CBD \tag{1}$$

2. Forensic chemistry of cannabinoids

GC analysis is a primary analytical technique. It is used in every forensic lab. GC is widely used by forensic scientists, from analysis of body fluids for the presence of illegal substances, for testing of fiber and blood from a crime scene, and to detect residue from explosives (Müller et al., 1999). In the next parts of the chapter cannabinoids works would be reference as well as some aspects of Colombian cannabis.

2.1 Cannabinoids analysis by chromatography techniques

Regarding on the cannabinoids analysis by chromatographic methods, it is worth to mention the following works:

Davis & Farmilo performed an analysis of cannabinoids in leaves and flowering of cannabis by thin layer chromatography and GC. From this analysis it was proposed that the relationship of the THC and CBN content based on the CBD content allowed cannabis differentiation of samples by geographical origin (1963).

Hemphill et al. analyzed the cannabinoids content in different organs of cannabis plants selected from geographical regions by gas-liquid chromatography. The results showed quantitative variations in cannabinoid content in different parts of the plants belonging to the selected varieties. However, each part of the plant presented a profile of cannabinoids that characterized the phenotype of which it comes from chemical, in addition, the accumulation of a specific cannabinoid variety was not found and it was observed that factors such as degree of maturity of body, sex plant, the location of the organ in the plant and the sampling procedures influence the concentration of cannabinoids. Finally, cannabinoids were not detected in pollen grains and seeds (1980).

Gambaro et al. determined THC, CBD and CBN in cannabis preparations by GC-FID and High Performance Liquid Chromatography-Ultraviolet (HPLC-UV) (2002). De Meijer et al. studied the inheritance of chemical phenotype in cannabis varieties, for it they made four crosses between plants of C. sativa with CBDs and THCs pure chemotypes (2003).

Mechtler et al. analyzed the variation in the distribution of a population of THC in hemp, for it a number of plants from five well-known populations of hemp grown in regions of Austria and Hungary were sampled (2004). ElSohly & Slade reported the composition of the complex mixture of natural cannabinoids from cannabis (2005).

Pellegrini et al. developed a simple and rapid procedure for the determination of cannabinoids in hemp food products derived by GC-MS (2005). Lewis et al. determined the relative amounts of the three most abundant cannabinoids (THC, CBN and CBD) and other abundant compounds present in six compressed bars of cannabis resin by GC-MS analysis (2005).

A global scientific community has recently expressed concern over the detection in the illicit market in cannabis varieties with high content of THC. According to the National Coordination of Drug Policy, which is headquartered in Sweden, in 1961 the levels of THC, the more important psychoactive component, did not exceed 5%, today this number is above 20%.

Recently, De Backer et al. developed an High Performance Liquid Chromatography-Diode Array Detector (HPLC-DAD) method for the qualitative and quantitative determination of major cannabinoids in plant material of cannabis (2009). Then, the solubilities of some cannabinoids in supercritical carbon dioxide were determined (Perrotin-Brunel et al. 2010).

Milman et al. reported simultaneous quantification of cannabinoids in oral fluid by two dimensional GC-MS (2010). Tipparat et al. studied the cannabinoids composition of cannabis plants grown in Northern Thailand and its forensic application (2011).

2.2 Cannabis in Colombia

In Colombia, the introduction of marijuana in the Sierra Nevada de Santa Marta was conducted in the early seventeenth century, the use of hemp was quickly displaced by the twine. It is recognizes the uses as intoxicants of this plant since 1927, used by laborers and sailors from the Colombian Caribbean Coast. However, it was until 1945 went it was saw the first illegal cultivation of marijuana in this region.

With the completion of the Vietnam War in 1975, there was an increased cultivation of marijuana encouraged by the intoxicating veterans. Moreover marijuana was used until the mid-twentieth century by the popular classes to relieve the pain associated with arthritis. It was common that the local police take a jar of alcohol and added seized marijuana to produce a medicinal poultice. Then the habit of smoking marijuana as a pain medication was spread, but this was a practice of alternative medicine without scientific studies and without institutional control (Fernandez-Alonso et al, 2007).

Currently the forensic study over confiscated samples of *C. sativa* L., is based on the qualitative determination of THC, present in extracts obtained from plant samples, without performing an quantitative level analysis of that component and some other cannabinoids present in seized samples. As a result of the above at present there is no systematic information to determine the content of THC present in plant samples of *C. sativa* illicitly traded in Colombia, there is just an essential information for monitoring the potential levels of THC present in plant samples who smoke marijuana consumers.

The Dirección Nacional de Estupefacientes (DNE) of Colombian government has reported about the results achieved in Colombia in the fight against drug trafficking (2002, 2007). Florian et al. reported higher content of THC in vegetable samples of Colombian cannabis (2009).

This paper shows the GC as an important tool in the analysis of forensic samples, specifically in the determination of cannabinoids. Leaf samples of *C. sativa* L. in four regions of Colombia were analyzed by GC. The analyzed region were (i) Northern region (NR) at Magdalena department, (ii) Southern region (SR) at Cauca department, (iii) Eastern region (ER) at Meta department and (iv) Central region (CR) at Caldas department, see Figure 2. The analysis revealed differences in the cannabinoid content of the samples under study according to their origin region.

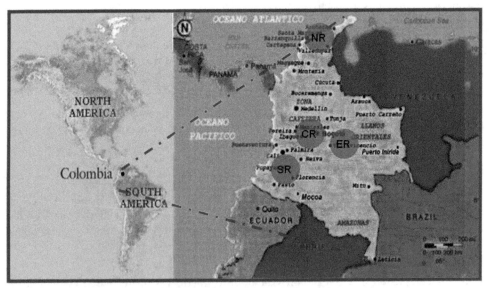

Fig. 2. Geographic origin of the *C. sativa* L. samples.

3. Optimization of the cannabinoids analysis conditions

To optimize the analytic method cannabis leaves from NR were used. The first objective was the optimization of the extraction method and the selection of the analysis method. Then, the sample quantity and quantification method were defined. After, the analytic method was validated.

3.1 Optimization of the extraction method and the selection of the analysis method

In order to obtain extracts of cannabinoids from plant material samples, De Meijer et al. method was taken as a based because of its quickness, efficiency and reproducibility, this method use extraction with ethylalcohol (EtOH), ultrasound and centrifugation at 4000 rpm (2003). The extraction time (t= 5, 10 and 15 min) and the number of extractions with EtOH (n= 1, 3 and 5) were considered as factors; for each extraction 750 µL were used and graduated to 5.00 mL. Using a 3^2 factorial model, cannabis leaves were subjected to each one of the extraction conditions, in triplicate, and at the central point (t= 10 min, n= 3) there were done five triplicates. The extracts obtained were analyzed by GC with different forms of detection: GC-FID, GC-MS scan mode and GC-MS Selected Ion Monitoring (SIM) mode. In all cases the response variable was the chromatographic area corresponding to each of the cannabinoids detected.

By comparing the coefficients of variation (C.V.) of the response surfaces for each analyte, based on the detection methods considered, we observed that the detection by GC-FID showed the lowest variation, see Table 1. Therefore, it was decided to use this method for quantitative analysis of cannabinoids, the GC-MS analysis was used for the qualitative identification of cannabinoids in the extracts.

Method	CBD C.V. (%)	THC C.V. (%)
GC-FID	9.36	7.89
GC-MS Scan	27.33	17.14
GC-MS SIM	15.51	11.58

Table 1. Coefficient of variation of the response surface for CBD and THC.

Based on the response obtained by GC-FID, the ANOVA for the quadratic regression model for each analyte, indicated that the difference was significant with a low probability value to a level of 0.05 (for the CBD pmodel = 0.0003 < α and the THC pmodel = 0.0105 < α). In that way, the models generated by the equations 2 and 3 explained satisfactorily the response in terms of the considered factors (time (t), number of extractions (n)), n being the most significant factor for the models.

Figure 3a shows the generated response surface to optimize the extraction conditions of THC, in it is observed that the response increased with the extraction time and the number of extractions. On the other hand, Figure 3b shows that the standard error associated with the analysis model, increased in the extreme values.

$$CBD\ area = 9050 + 169*t + 1944*n \tag{2}$$

$$THC\ area = 31518 - 2200*t + 10177*n + 123*t^2 - 1008*n^2 - 35*t*n \tag{3}$$

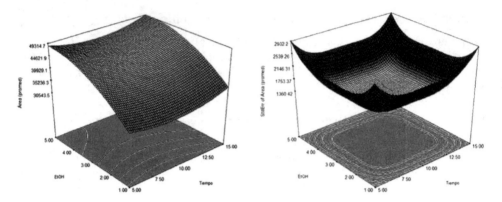

(a) Optimizing the extraction of THC,   (b) Standard error associated with the THC RSA.

Fig. 3. Response surface analysis (RSA)

To maximize the response of each analyte with the lowest associated error, the model suggested ten solutions, which found that the optimum conditions for extraction suggested by the quadratic model based on response time (t) and the number of extractions (n) for the analysis of CBD and THC by GC-FID were three extractions with ethanol for a period of 7.5 minutes, see Table 2. These conditions suggested by the model were experimentally verified for ensuring that a fourth extraction does not provides betters detectable levels of CBD and THC.

Solutions	CBD		THC	
	t	n	t	n
1	7.53	2.83	6.42	4.16
2	6.21	1.71	7.48	3.01
3	13.45	1.76	9.25	2.99
4	9.83	1.99	12.68	2.18
5	12.29	3.52	14.94	1.93
6	10.05	3.46	8.03	4.41
7	5.50	2.90	5.60	1.91
8	12.83	3.24	7.26	4.27
9	11.11	1.08	8.43	3.26
10	5.55	1.71	9.38	1.03

Table 2. Solutions suggested by the model for the CBD and THC extractions.

3.2 Sample quantity and quantification method

Once the method of analysis, time and number of extractions were defined, we proceeded to determine the mass of sample and the quantification method that would be used. For this there was prepared by triplicate, two extract sets: i) internal standard and ii) surrogate standard, using in each case 25.0, 50.0, 100.0 and 200.0 mg of vegetal material. For this series

Gas Chromatography in Forensic Chemistry: Cannabinoids Content in Marijuana Leaves
(Cannabis sativa L.) from Colombia
169

of experiments DPH at 100 ppm was used as internal and surrogate standards. In the method of internal standard DPH was added to the extract obtained, while in the surrogate standard method DPH was added to the vegetal sample.

The results were analyzed by ANOVA, evaluating for each of the quantifying methods the confidence interval and the response factor for each of the cannabinoids depending on the amount of sample, the results obtained for CBD and THC are presented in Figure 4.

The linear fit showed that there was a greater linearity in the results obtained by the method of surrogate standard, Figures 4b and 4d (R^2_{CBD} = 0.9934; R^2_{THC} = 0.9941) compared against the internal standard method, Figures 4a and 4c (R^2_{CBD} = 0.9843; R^2_{THC} = 0.8562). In addition, the surrogate standard method for both CBD and THC showed less dispersion in the response.

From this it could be concluded that the method of surrogate standard allows more accurate answers because it offers a lower confidence interval. Regarding the amount of sample to be extracted, it was found that using 100 mg of vegetal material minimized the error in the response for CBD and THC, as is evidenced in Figures 4b and 4d.

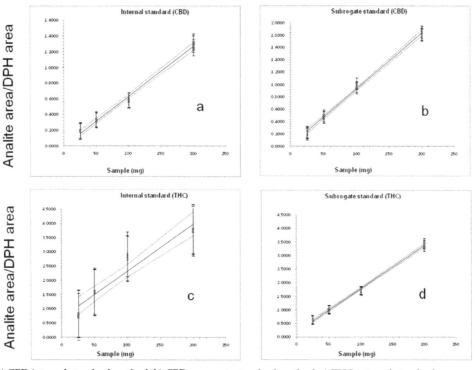

a) CBD internal standard method, b) CBD surrogate standard method, c) THC internal standard method, d) THC surrogate standard method.

Fig. 4. Linear fit for the chromatographic response depending on the quantity method of sample

Based on these arguments it was determined that the amount of sample to be used was of 100 mg and for quantification of cannabinoids by GC-FID was convenient to use the method of surrogate standard.

3.3 Validation of analytic method

Regarding the validation of CBD, THC and CBN determination method by GC-FID: (i) it was determined the repeatability of the retention times of the analytes and the standard, (ii) there was calculated the chromatographic parameters (iii) the linearity was established for each analyte, (iv) precision, (v) accuracy, (vi), Limit of Detection (LOD) and Limit of Quantification (LOQ).

Retention times (tr) of DPH (compound used as surrogate standard) and the cannabinoids CBD, THC and CBN, showed an excellent repeatability, the tr of these had C.V. less than 1%, see Table 3.

Compound	Replicas of tr (min)						Avg	SD	C.V. (%)
	1	2	3	4	5	6			
DPH	2.04	2.02	2.04	2.05	2.03	2.04	2.04	0.010	0.507
CBD	3.93	3.91	3.90	3.95	3.91	3.93	3.92	0.018	0.468
THC	4.60	4.59	4.62	4.60	4.61	4.63	4.61	0.015	0.319
CBN	5.25	5.25	5.23	5.24	5.26	5.25	5.25	0.010	0.197

Table 3. Characterization of the retention time for DPH, CBD, THC and CBN injected into the mixture.

Table 4 shows the chromatographic parameters of the cannabinoids by GC-FID. The relative tr (trr) were calculated regarding on the tr of THC.

The CBN has a response factor (RF) higher compared against the CBD and THC, compounds that exhibit relatively similar RF. Resolution (Rs) values obtained are higher than the acceptance criterion of 1.5, so the method can be considered as selective for the analytes considered. The CBN is the compound most congenial with the stationary phase and therefore it is more retained in this (K= 4.25) and the DPH has a lower retention (K= 1.04), being the compound with less affinity with the stationary phase. The highest number of theoretical plates (N) required for the separation correspond to CBN, followed in order by CBD, THC and DPH.

Compound	tr	Area	RF Area/g	Wb (s)	trr	Rs	α	K	N
DPH	2.04	543.6	28810.0	0.112	0.44	14.06	1.54	1.04	5290.8
CBD	3.92	576.2	11019.0	0.156	0.85	3.89	2.60	2.92	10111.4
THC	4.61	347.1	12178.9	0.197	1.00	3.35	3.30	3.61	8755.4
CBN	5.25	897.8	29055.0	0.184	1.14			4.25	13009.2

Table 4. Validation of chromatographic parameters for DPH, CBD, THC y CBN.

To assess the linearity in the quantification of cannabinoids of interest, ten calibration curves in the linear range were prepared (approximately 15 to 300 ppm), each point on the line with a concentration of 100.15 ppm of DPH. With linear regression the curves were evaluated to obtain the correlation coefficient, the slope and the intercept.

For each one of the cannabinoids the slopes or the intercepts does not differ significantly, so the equations of the curves for CBD, THC and CBN were calculated from average values in Table 5. To eliminate the possibility of systematic errors the slopes and correlation factors were compare with those obtained by forcing the calibration curve through the origin, in both cases we obtain values of slope and correlation factor without significant differences.

To determine the accuracy or the dispersion of measurements around their average value a series of ten measurements in intermediate precision conditions were made (varying the scientist) on solutions of cannabinoids of interest with DPH within the linear ranges established. For the three analytes considered, at all levels, data from the relative standard deviation are lower than those predicted by the Horwitz equation (PRSD (R) = $2C^{-0.15}$), indicating that the dispersions obtained experimentally are within acceptable ranges for each concentration. Furthermore the values obtained for HORRAT coefficients vary from 0.26 to 0.68 for CBD, between 0.29 to 1.08 for THC and from 0.32 to 0.72 for CBN, for the three cannabinoids these ratios are within the recommended range for intra-laboratory trials; values between 0.3 and 1.3. With this it can be concluded that the method of analysis has good accuracy for each of the analytes.

	Slope (m)	SDm	Intercept (y_b)	SDb	Error y SDxy	R	Error Xo SDxo
CBD	0.7480	0.0040	-0.1080	0.0075	0.0318	0.9981	0.1722
THC	0.6000	0.0050	-0.0246	0.0081	0.0376	0.9954	4.2504
CBN	1.0990	0.0053	-0.0577	0.0098	0.0422	0.9984	0.5899

Table 5. Parameters for the linear calibration curves estimation for CBD, THC and CBN using DPH as surrogate standard.

For the three cannabinoids considered when applying the t-student test, comparing the data obtained with the tabulated critical value for nine degrees of freedom at a confidence level of 95%, we can say that the accuracy is acceptable because the three levels of concentration does not exceeded the tabulated value, as summarized in Table 6.

Analyte		Level		
		High	Medium	Low
CBD	texp	1.973	2.046	2.184
	t tab 95%	2.262	2.262	2.262
THC	texp	1.304	2.198	1.723
	t tab 95%	2.262	2.262	2.262
CBN	texp	1.113	1.773	0.392
	t tab 95%	2.262	2.262	2.262

Table 6. Comparison t-student values for assessment of accuracy for the analysis of CBD, THC y CBN.

Moreover, the recoveries percentages and the relative deviation for the analytes, the low, medium and high, are within the ranges permitted for a nominal component that is greater than 10% in a recovery interval from 98 to 102%.These two tests allow us to conclude that the method has the accuracy required for low, medium and high ranges , for the analysis of these analytes, without evidence of systematic errors.

For each analyte, the linear range, the calibration curve, the LOD and the LOQ are presented in Table 7. The detection limits have very similar values for the three cannabinoids, however the lower sensitivity for the detection of THC is due because this compound has a higher limit of quantification in relation to the CBD and CBN.

Regarding this part of the work we can conclude: (i) the optimal conditions for extraction suggested by the quadratic model based on response time (t) and the number of extractions (n) for the extraction of CBD and THC from a Cannabis plant sample; (ii) the method for quantification of CBD and THC, using DPH as a surrogate standard, showed an excellent repeatability (coefficient of variation for the retention time of all substances was less than 1%); (iii) in addition, linear behavior was observed for the quantification of CBD, THC and CBN using DPH as standard in the following ranges: CBD between 19.6 to 315.2 ppm, THC between 13.2 to 287.6 ppm and for CBN 16.4 to 313.7 ppm.

Parameter	CBD	THC	CBN
Lineal Range ppm	19.6 – 315.2	13.2 – 287.6	16.4 – 313.7
Number of levels	7	7	7
Calibration curve	y=0.7480x-0.1080	y=0.6000x-0.0246	y=1.0990x-0.0577
LOD (ng)	1.02	1.06	1.06
LOQ (ng)	1.68	2.44	1.76

Table 7. Detection and quantification limits for CBD, THC y CBN.

The results experimentally obtained allow to conclude that the analysis method for CBD, THC and CBN had a very good accuracy and the required exactitude at different concentration levels, with no systematic error evidence.

Otherwise, the detection limits using DPH as surrogate standard were of 1,02 ng for CBD, 1,06 ng for THC and 1,06 ng for CBN. The quantification limits were of 1,68 ng for CBD, 2,44 ng for THC y 1,76 ng for CBN.

4. Profile analysis of cannabinoids in samples collected in four areas of illicit cultivation of cannabis

In order to determine the profile of cannabinoids in samples from four different regions of Colombia, were sampled four regions, in each region thirteen samples were taken, each was analyzed by triplicate. The areas sampled were: (i) Northern region (NR), (ii) Southern region (SR) , (iii) Eastern region (ER) and (iv) Central region (CR), see Figure 2.

In all samples of the four regions of the cannabinoid THC was the majority (NR 1.89-4.00% average 2.81%, SR 6.72-15.48% avg 10.98%, ER 13.37-17.63% avg 15.74%, CR 1.11-3.17% avg 1.87%). In regions NR, SR and ER the minor cannabinoid was CBN (NR 0.04-0.06% avg 0.05%, SR 0.05-0.17% avg 0.11%, ER 0.56-0.80% 0.68% avg), with the CBD as the intermediate (NR 0.29-2.40% avg 0.91%, SR 0.60-4.86% avg 2.52%, ER 0.92-4.14% 1.86% avg). Unlike the three previous regions, in CR CBD was minority (0.01-0.06% avg 0.02%) and CBN was the intermediate cannabinoid (0.83-1.94% 1.36% avg), see Figure 5. In the Table 8 cannabinoid content averages and phenotype averages of each studied region are shown.

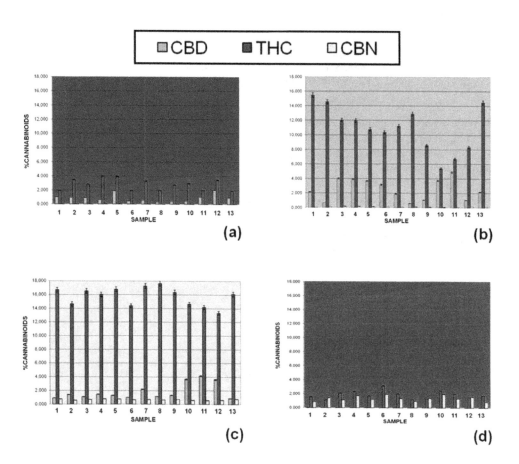

Fig. 5. Cannabinoids content in samples collected in different areas of Colombia a) NR, b) SR, c) ER, d) CR.

	NR			SR			ER			CR		
Cannabinoid	Avg.	SD	C.V.	Avg.	SD	C.V.	Avg.	SD	C.V.	Avg.	SD	C.V.
CBD	0.91	0.550	60.6	2.52	1.441	57.2	1.86	1.166	62.8	0.02	0.020	108.9
THC	2.81	0.792	28.2	10.98	3.075	28.0	15.74	1.340	8.5	1.87	0.574	30.7
CBN	0.05	0.005	10.3	0.11	0.040	36.2	0.68	0.080	11.8	1.36	0.366	27.0
Phenotype	4.14	2.233	53.9	7.41	7.035	94.9	11.82	5.504	46.5	414.52	354.087	85.4

Table 8. Cannabinoids content in samples collected in different areas of Colombia.

With respect to THC, the substance which presents psychoactive effects and potentially can generate dependence in consumers, samples from ER and SR showed the highest content (15.74% and 10.98%, respectively), much higher than those reported for THC in samples from Colombia (Baker et al., 1980). The NR (2.81%) and CR (1.87%) showed lower values of THC.

The high content of THC in samples ER and SR regions could indicate the use of improved varieties of cannabis, whereas worldwide are certified varieties without genetic manipulation with less than 7% THC (Montero 2008), which comes great concern about the higher drug potential effects among consumers (Licata et al., 2005).

Phenotype values calculated according to Waller for each group of samples were found within a given range for different regions, see Figure 6. These results are consistent with data published by Turner et al. (1979), who noted that the variability in the content of cannabinoids in plant samples difficult phenotypic classification based on the result of a single analysis, and additionally Mechtler et al. (2004) found that THC may have an erratic and in some plants with hemp can be outside of the characteristic values for the population farmed.

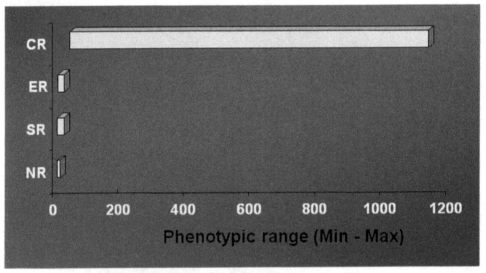

Fig. 6. Variation ranges of the phenotype for the analyzed samples of North region (NR), South region (SR), East region (ER) and Central region (CR).

The results of the minimum and maximum phenotype values calculated according to Waller relationship, show that the ranges obtained overlap for three of the considered regions, this is NR (1.72-7.68), SR (1.39-23.08) and ER (3.56-19.54). As a criterion for differentiating the only group whose phenotype was different correspond to samples collected in the CR (40.09- 1094.50), see Figure 6.

Given the above, the Waller relationship is not useful for defining geographical origin of the samples under study. In contrast, when considering the relationship THC vs THC/CBN, this could have a greater discriminatory power as the criterion of origin of the samples under study. Through this relationship we obtain four distinct groups being the most appropriate criteria and discriminate source for the region of origin of the analyzed samples, this relationship is shown in Figure 7.

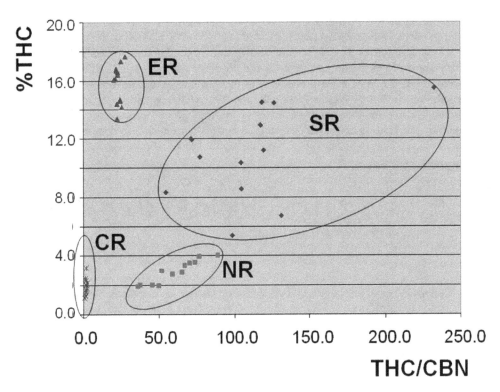

Fig. 7. THC vs THC/CBN diagram for the four analyzed regions.

By the way the ratio THC vs THC / CBN is obtain this could be an appropriate tool to discriminate between the studied samples. SR was the most dispersed group, CR was the less dispersed group, these samples had antagonistic behavior. NR and ER had antagonistic behavior too, these were the medium dispersed groups. Through this relationship we obtain four sample distinct groups, it could be the most appropriate criteria and discriminate source of the region of origin of the samples analyzed.

5. Conclusions

The present study analyzed the cannabinoids content in samples of *Cannabis sativa* L. cultivated illicitly in Colombia. The physicochemical conditions were optimized for the extraction and quantification of cannabidiol (CBD), Δ9-tetrahidrocannabinol (THC) and cannabinol (CBN) starting from a vegetable sample using gas chromatography with flame ionization detector GC-FID, validating the respective analytic method.

The THC average content (% dry basis) in the East region (ER) and South region (SR) samples showed very high concentrations of THC, 15.74% and 10.98%, respectively. The higher content of THC in vegetable samples of cannabis from the ER and SR could be indicative of the employment of improved varieties. This fact generates a great concern about the potential effects produced by the drug in consumers.

Additionally, variation in the contents of CBD, THC and CBN in the thirteen samples collected in each region implies that the value of the Waller phenotype for each group of samples was not found around a value but becomes an interval for each source. The numerical value of this phenotype, by itself, did not serve as a criterion for differentiation on the geographical origin of the samples, due to overlapping ranges for three of the regions concerned, this is North region (NR), SR and ER. The most appropriate criteria and discriminate source for the region of origin of the four groups of samples analyzed was plotting the ratio THC vs THC / CBN, as this allows to distinguish four distinct groups.

6. Acknowledgment

Grateful acknowledgements are expressed to the Dirección de Investigación sede Bogotá (DIB) of the Universidad Nacional de Colombia for their financial support (DIB project No. 8003170) and the Departamento Administrativo de Seguridad (DAS) and the Fiscalía General de la Nación for their technique support.

7. References

Baker, P.; Bagon, K. & Gough T. (1980). Variation in the THC content illicitly imported Cannabis products. *Bulletin on Narcotics*, Vol. 32, pp. 31-40.

Davis, T. & Farmilo, C. (1963). Identification and origin determinations of Cannabis by gas and paper chromatography. *Analytical Chemistry*, Vol. 35, pp. 751–755.

De Backer, B.; Debrus, B.; Lebrun, P.; Theunis, L.; Dubois, N.; Decock, L.; Verstraete, A.; Hubert, P. & Charlier, C. (2009) Innovative development and validation of an HPLC/DAD method for the qualitative and quantitative determination of major cannabinoids in cannabis plant material. *Journal of Chromatography B*, Vol. 877 pp. 4115-4124.

De Meijer, E. ; Bagatta, M.; Carboni, A.; Crucitti, P.; Moliterni, V. & Mandolino, G. (2003). The inheritance of chemical phenotype in Cannabis sativa L. *Genetics*, Vol. 163, pp. 335-346.

Dirección Nacional de Estupefacientes (2002). La Lucha de Colombia contra las Drogas Ilícitas. Acciones y Resultados 2001. Ed.: Ministerio del Interior y de Justicia. Bogotá-Colombia.

Dirección Nacional de Estupefacientes (2007). Observatorio de Drogas de Colombia. Acciones y Resultados. Resultados Operativos 2006. Ed.: Ministerio del Interior y de Justicia. Bogotá-Colombia.

ElSohly, M. & Slade, D. (2005). Chemical constituents of marijuana: The complex mixture of natural cannabinoids. *Life Sciences*, Vol. 78, pp. 539 – 548.

Fernandez-Alonso, J. ; Galindo, A. & Idrobo, J. (2007). Las plantas como Evidencia Legal: Desarrollo de la Botánica Forense en Colombia. *Rev. Acad. Colomb. Cienc.*, Vol. 31, pp. 181-198.

Florian, N.M.; Parada, F. & Garzón, W.F. (2009). Study of cannabinoids content in marihuana samples (Cannabis sativa L.) cultivated in several regions of Colombia. *Vitae*, Vol. 16, pp. 237-244.

Gambaro, V.; Dell'Acqua, L.; Farè, F.; Froldi, R.; Saligari, E. & Tassoni, G. (2002). Determination of primary active constituents in Cannabis preparations by high-resolution gas chromatography/flame ionization detection and high-perfomance liquid chromatography/UV detection. *Analytica Chimica Acta*, Vol. 468, pp. 245-254.

Hemphill, J.; Turner, J. & Mahlberg, P. (1980). Cannabinoid content of individual plant organs from different geographical strains of Cannabis sativa L. *Journal of Natural Products*, Vol. 43, pp. 112-122.

Lewis, R.; Ward, S.; Johnson, R. & Thorburn, D. (2005). Distribution of the principal cannabinoids within bars of compressed Cannabis resin. *Analytica Chimica Acta*, Vol. 538, pp. 399–405.

Licata, M.; Verri, P. & Beduschi, G. (2005). δ9 THC content in illicit cannabis products over the period 1997 – 2004 (first four months). *Ann 1s Super Sanitù*, Vol. 41, pp. 483 -485.

Mechtler, K.; Bailer, J. & de Hueber, K. (2004). Variations of Δ9-THC content in single plants of hemp varieties. *Industrial Crops and Products*, Vol. 19, pp. 19-24.

Milman, G.; Barnes, A.J.; Lowe, R.H. & Huestis, M.A. (2010). Simultaneous quantification of cannabinoids and metabolites in oral fluid by two-dimensional gas chromatography mass spectrometry. *Journal of Chromatography A*, Vol. 1217, pp. 1513–1521.

Montero, D. 2008. Colombia de nuevo país exportador. In: El espectador, Reportaje : (15, jun., 2008): 19, c. 1.

Müller, R.K.; Grosse, J.; Thieme, D.; Lang, R.; Teske, J. & Trauer, H. (1999). Introduction to the application of capillary gas chromatography of performance-enhancing drugs in doping control. *Journal of Chromatography A*, Vol. 843 pp. 275–285.

Pellegrini, M.; Marchei, E.; Pacifi, R. & Pichini, S. (2005). A rapid simple procedure for the determination of cannabinoids in hemp food products by gas chromatography-mass spectrometry. *Journal of Pharmaceutical and Biomedical Analysis*, Vol. 36, pp. 939–946.

Perrotin-Brunel, H.; Kroon, M.C.; van Roosmalen, M.J.E.; van Spronsen, J.; Peters, C.J. & Witkamp, G.J. (2010). Solubility of non-psychoactive cannabinoids in supercritical carbon dioxide and comparison with psychoactive cannabinoids. *J. of Supercritical Fluids*, Vol. 55, pp. 603–608.

Tipparat, P.; Natakankitkul, S.; Chamnivikaipong, P. & Chutiwat, S. (2011). Characteristics of cannabinoids composition of Cannabis plants grown in Northern Thailand and its forensic application. *Forensic Science International*. doi:10.1016/j.forsciint.2011.05.006

Turner, C.; Elsohly, M. & Boeren, E. (1980). Constituents of Canabis sativa L. XVII: A review of natural constituents. *Journal of Natural Products*, Vol. 43, pp. 169-234.

Turner, C.; Elsohly, M.; Cheng, P. & Lewis, G. (1979). Constituents of Cannabis sativa L. XIV. Intrinsic problems in classifying Cannabis based on a single cannabinoid analysis. *Journal of Natural Products,* Vol. 42, pp. 317-319.

9

Chromatographic Methodologies for Analysis of Cocaine and Its Metabolites in Biological Matrices

Maria João Valente[1], Félix Carvalho[1], Maria de Lourdes Bastos[1],
Márcia Carvalho[1,2] and Paula Guedes de Pinho[1]
[1]REQUIMTE, Laboratory of Toxicology, Department of Biological Sciences,
Faculty of Pharmacy, University of Porto,
[2]CEBIMED, Faculty of Health Sciences,
University Fernando Pessoa, Porto
Portugal

1. Introduction

Cocaine is the main active alkaloid extracted from the leaves of the coca plant, *Erythroxylum coca*. It is a widely abused psychotropic drug, for its immediate neurological effects, including euphoria, reduced fatigue and increased mental acuity and sexual desire (Devlin & Henry, 2008; Goldstein et al., 2009; Small et al., 2009). However, cocaine abuse is usually followed by many pathophysiological consequences, namely central and peripheral neurochemical changes that result in hypertension-related morbidity and mortality, including myocardial infarction and cerebrovascular accidents, as well as liver and kidney toxicity, tissue ischemia and adverse psychotic effects such as paranoia and hallucinations (Devlin & Henry, 2008; Glauser & Queen, 2007; Heard et al., 2008; Karch, 2005; Lombard et al., 1988; Ndikum-Moffor et al., 1998; Tang et al., 2009; White & Lambe, 2003).

According to a recent report on drug abuse, and despite a visible decrease of production and consumption in the last few years, in 2008 cocaine abuse still affected up to 0.5% of the adult population (15-64 years old) worldwide. Cocaine remains the second most problematic drug in the world, after opiates (UNODC, 2011).

In Europe, cocaine ranks second in most abused illicit drugs, after cannabis. It revealed a mean prevalence of 1.3% of the adult population by the same year, with national prevalence reaching over 6% of the young adult population (15-34 years old) (EMCDDA, 2010).

In this chapter we will point out the clinical and forensic relevance of measuring cocaine and its metabolites in different biological matrices, and provide a bibliographic review on techniques for sample preparation and existing chromatographic methodologies for cocaine analysis.

2. Toxicokinetics

Chemically, cocaine may exist in two forms: a hydrochloride salt or a free-base rock ("crack"). "Crack" melts at 98 °C and volatizes above 90 °C, but is not very soluble in water, making it possible to smoke but not to inject. In contrast, the salt is easily soluble in water but has a high melting point (195 °C) and decomposes when smoked, being suitable for intravenous (i.v.) injection, nasal insufflation (or snorting) or ingestion (Cook, 1991; Favrod-Coune & Broers, 2010).

Subsequently to absorption, cocaine is easily diffused in the blood into most body organs including heart, brain, liver, kidneys and adrenal glands (Favrod-Coune & Broers, 2010; Fowler et al., 1989; Volkow et al., 1992). However, its bioavailability depends on the route of administration, being of about 90% if injected or smoked, 25 to 94% when snorted, depending on the dose, and only up to 30% after ingestion (Cook, 1991; Leikin & Paloucek, 2008).

The onset of action, the intensity and the duration of the effects experienced by consumers are also affected by the type of consumption. For the smoked form, the onset occurs almost immediately, and the intensity of the neurological effects is nearly two-fold higher than for the other means of abuse (Favrod-Coune & Broers, 2010; Freye & Levy, 2009), most likely the reasons why "crack" is the most consumed form of cocaine.

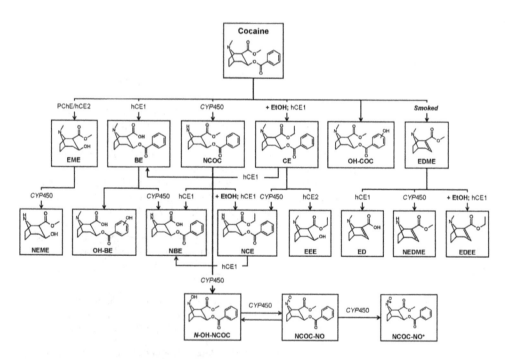

BE - benzoylecgonine; CE - cocaethylene; CYP450 - cytocrome P450; ED - ecgonidine; EDEE - ecgonidine ethyl ester; EDME - ecgonidine methyl ester; EEE - ecgonine ethyl ester; EME - ecgonine methyl ester; EtOH - ethanol; hCE1 - human carboxylesterase type 1; hCE2 - human carboxylesterase type 2; NBE - norbenzoylecgonine; NCE - norcocaethylene; NCOC - norcocaine; NCOC-NO - norcocaine nitroxide; NCOC-NO+ - norcocaine nitrosonium; NEDME - norecgonidine methyl ester; NEME - norecgonine methyl ester; N-OH-NCOC - N-hydroxynorcocaine; OH-BE - hydroxybenzoylecgonine; OH-COC - hydroxycocaine; PChE - pseudocholinesterase.

Fig. 1. Cocaine metabolic pathways.

The i.v. injection takes a few seconds (15 - 30 s) to onset the first effects, following the snorted form with over a minute, and finally the oral form, the most unusual one among addicts, which takes over 20 minutes to produce effects (Freye & Levy, 2009; Heard et al., 2008; Jeffcoat et al., 1989).

The psychotropic effect usually lasts 2 to 3 hours after cocaine ingestion, approximately 1 hour when snorted, and less than 30 minutes for the injected and smoked forms (Favrod-Coune & Broers, 2010; Jeffcoat et al., 1989).

Figure 1 represents cocaine metabolic profile, which strongly depends on both form of consumption and administration route.

Following administration, cocaine is primarily metabolized into two major metabolites, benzoylecgonine (BE) and ecgonine methyl ester (EME), and two minor metabolites, norcocaine (NCOC) and m- and p-hydroxycocaine (OH-COC) (Goldstein et al., 2009; Maurer et al., 2006; Zhang & Foltz, 1990).

BE is mainly produced in the liver, through human carboxylesterase type 1 (hCE1), whereas EME may be formed in the liver by hCE2, and in the plasma via a pseudocholinesterase (PChE), namely butyrylcholinesterase (Goldstein et al., 2009). Both free metabolites are excreted in urine, together representing up to 95% of the excretion products (Kanel et al., 1990).

NCOC results from hepatic N-demethylation of the drug through the cytocrome P450 ($CYP450$) system, in particular $CYP3A4$ in human liver, and represents no more than 5% of the administered dose (Goldstein et al., 2009; Kloss et al., 1983; LeDuc et al., 1993). The same enzyme mediates further oxidations, yielding the secondary metabolites N-hydroxynorcocaine (N-OH-NCOC), norcocaine nitroxide (NCOC-NO) and norcocaine nitrosonium (NCOC-NO+), which are described as responsible for cocaine-induced hepatotoxicity (Kovacic, 2005; Ndikum-Moffor et al., 1998; Pellinen et al., 1994; Thompson et al., 1979).

Regarding OH-COC, despite being produced at very low levels (less than 12% that of NCOC in hepatic microssomes), the isomer p-OH-COC was proven to be pharmacologically active in mice (Watanabe et al., 1993).

Polydrug abuse is a common pattern among cocaine users. In fact, by 2009, over 40% of them simultaneously consumed ethanol (UNODC, 2011). From this combination results the formation of the biologically active metabolite cocaethylene (CE), transesterification product via hCE1 between cocaine and alcohol (Harris et al., 2003; Hearn et al., 1991; Laizure et al., 2003).

Like cocaine, CE can undergo further N-demethylation via $CYP450$ or hCE2-mediated hydrolysis, yielding two unique ethanol-related cocaine metabolites, norcocaethylene (NCE) and ecgonine ethyl ester (EEE), respectively (Boyer & Petersen, 1990; Dean et al., 1992; Wu et al., 1992). NCE may also be a NCOC transesterification product in the concurrent use with alcohol (Maurer et al., 2006).

Besides cocaine, both EME and BE can undergo a N-demethylation as well, producing norecgonine methyl ester (NEME) and norbenzoylecgonine (NBE). This last metabolite can also be formed by hydrolysis of NCOC or NCE (Maurer et al., 2006).

During "crack" smoking, ecgonidine methyl ester (EDME) is formed in large quantities as a thermal breakdown product of cocaine (Jacob et al., 1990; Kintz et al., 1997). EDME may be metabolized by identical pathways as for cocaine: it can be oxidized into norecgonidine methyl ester (NEDME) via $CYP450$, or hydrolyzed through hCE1 into ecgonidine (ED) or ecgonidine ethyl ester (EDEE) in the presence of ethanol. This last one may be analyzed as a specific biomarker of the concomitant use of "crack" and ethanol (Fandino et al., 2002).

3. Clinical and forensic relevance of cocaine analysis

Over decades, cocaine abuse has reached epidemic proportions, and health complications related to cocaine use continue to be a major social burden worldwide.

According to the World Drug Report 2011 (UNODC, 2011), drug of abuse-related deaths are estimated between 104,000-263,000 per year, and they include fatal overdoses (over 50% of all deaths), accidents, suicides, deaths from infectious diseases transmitted through the use of contaminated needles, including hepatitis C and HIV, or complications due to chronic use, namely organ failure and myocardial infarction (Kloner et al., 1992; Shanti & Lucas, 2003; UNODC, 2011).

In Europe, cocaine-related deaths represent 21% of all deaths related to illicit drug abuse, with a report of approximately 1,000 deaths per year (EMCDDA, 2010; UNODC, 2011).

Of note, the reported mean purity of traded cocaine rounding 50% by 2009 and a lowering trend along the years, as well as the common mixture with several active adulterants like painkillers, may complicate the scenario of cocaine intoxications (EMCDDA, 2010; UNODC, 2011). In addition, since the polydrug use includes approximately 62% of cocaine users, drug combination often results in complex clinical patterns which are difficult to discriminate and treat (UNODC, 2011).

Thus, a thorough methodology for detection and quantification of cocaine, alongside with other drugs, may be crucial for an accurate evaluation of cocaine intoxication cases and contribute for a positive outcome.

For human performance forensic toxicology purposes, also defined as behavioral toxicology, cocaine is frequently tested in urine samples and swabs of oral fluid from drivers and applicants for driving licenses with a history of drug use (Brookoff et al., 1994; Gjerde et al., 2008; Montagna et al., 2000; Samyn et al., 2002; Tagliaro et al., 2000; Wylie et al., 2005).

Cocaine detection is also a common procedure in the context of workplace drug testing, more often in pre-employment and post-accidental screening, but also in random screenings, usually in urine samples (George, 2005; Verstraete & Pierce, 2001; Zwerling et al., 1990).

Another area of forensic toxicology is *postmortem* forensic toxicology, which involves in suspected drug-related deaths. These may include suspected drug intoxication cases (overdoses or accidental), suicides, homicides, motor vehicle accidents, arson fire fatalities and apparent deaths due to natural causes. In these cases, cocaine and its metabolites may be analyzed in several specimens including blood, vitreous humor, bile, urine, stomach contents or organ tissues (Bertol et al., 2008; Darke & Duflou, 2008; Dias et al., 2008; Garlow et al., 2007; Graham & Hanzlick, 2008; Simonsen et al., 2011).

4. Determination of cocaine and its metabolites in biological specimens

The development of a procedure for the quantitative analysis of a biological matrix includes several steps, from sampling, to sample preparation, chromatographic analysis and finally analysis of the results (figure 2).

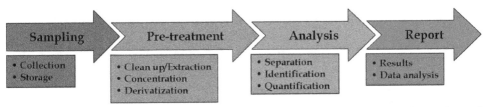

Fig. 2. Schematic representation of the steps included in the overall procedure for analysis of exogenous compounds in a biological specimen.

One of the main concerns regarding biological sampling for cocaine determination involves its instability in many matrices. At room temperature, cocaine can be quickly hydrolyzed into BE, and it is even more susceptible in cholinesterase-containing samples, including plasma and whole blood, in which the parent drug easily degrades into EME (Garrett & Seyda, 1983; Isenschmid et al., 1989).

The stability issue is not as significant in urine specimens as it is for plasma or blood. While in blood stability appears to be dependent on cocaine initial concentration, in urine it depends mainly on pH (Baselt, 1983). It was shown that cocaine concentration in urine may fall down to 37% when stored at -20 °C, for a 12-month period of time (Dugan et al., 1994), but by acidifying the samples to a pH of 5.0, cocaine and BE levels in the frozen urine samples may be stable for at least 110 days (Hippenstiel & Gerson, 1994). In these samples, the use of preservatives, such as sodium fluoride, appears to have only minor effects on the specimen stability (Baselt, 1983). In blood and plasma without preservation, most cocaine is hydrolyzed into EME. This may be prevented with the addition of a PChE inhibitor (Isenschmid et al., 1989).

Urine specimens are the most commonly used for general drug screening (Leyton et al., 2011; Marchei et al., 2008; Zwerling et al., 1990). However, for cocaine detection, there are some limitations, including limited window of detection, occurrence of false-negatives as a consequence of very low cocaine concentrations in samples, specific requirements for storage, possibility of sample dilution *in vivo* by excessive fluid ingestion, requirement of collection under observation to avoid adulteration or sample exchange, or even absence of urine specimens in *postmortem* cases (Cone et al., 1998; Cone et al., 2003; Musshoff et al., 2006; Polla et al., 2009).

The analysis of oral fluid swabs, sweat patches and hair samples has become a viable substitute to urinalysis, specifically in the context of behavioral toxicology and workplace drug testing (Samyn et al., 2002; Toennes et al., 2005; Verstraete, 2005).

The main advantages of oral fluid include not only the non-invasiveness of the collection, but also the higher concentration of the parent drug found in saliva when compared to blood and urine. For the same individual, cocaine concentration in oral fluid is approximately three-fold of that found in plasma and over five-fold in urine (Cone et al., 1994a; Moolchan et al., 2000; Samyn et al., 2002; Schramm et al., 1993). In addition, cocaine elimination half life is lower in saliva, which makes oral fluid analysis suitable for determination of very recent use (Dolan et al., 2004; Jufer et al., 2000). Moreover, saliva can provide an unequivocal screen result within minutes and has demonstrated a good correlation with impairment symptoms of drivers under the influence of drugs, reasons that make saliva the preferred matrix for roadside analysis (Kidwell et al., 1998; Verstraete, 2005).

However, oral fluid use has some limitations, such as the limited volume of specimen when compared to urine sampling, especially considering that recent use often results in the production of little amounts of saliva or even none at all, and the variability of salivary pH (Cognard et al., 2006; Kidwell et al., 1998; Verstraete, 2005).

Similarly to oral fluid, sweat is occasionally chosen for on-site testing (Samyn et al., 2002; Samyn & van Haeren, 2000). The sweat samples may be collected as skin swabs or through patches similar to bandages attached to the skin (Kacinko et al., 2005; Kidwell et al., 2003). These patches can be worn comfortably for several days (usually one week). This allows an accumulation of cocaine in the patch over the days, which is very useful, for example, for monitoring patients on drug-abuse treatment or epidemiologic surveys on cocaine-use in a given population (Burns & Baselt, 1995; Chawarski et al., 2007; Kidwell et al., 1997; Preston et al., 1999).

The main limitations of sweat analysis include the lower amount secreted at a given time in comparison to saliva, the great variability of results between doses and individuals, the variation of drug disposition between sites of collection and collection devices, and the occurrence of false positives from prior skin contamination or external patch contamination. Part of the drug may as well be reabsorbed into the skin or degraded in the patch (Burns & Baselt, 1995; Donovan et al., 2011; Huestis et al., 1999; Kidwell et al., 1998; Kidwell et al., 2003).

An early controlled study demonstrated that cocaine is detected in sweat samples up to 48 hours after administration (Cone et al., 1994b), but subsequent works suggested a window of detection as long as one week (Burns & Baselt, 1995; Kintz, 1996). Nonetheless, cocaine concentration in sweat is an indicator of a relatively recent use (Chawarski et al., 2007; Kidwell et al., 1997).

For past drug abuse, hair samples present the wider window of detection, allowing a higher rate of positive results than urine (Dolan et al., 2004; Kline et al., 1997; Scheidweiler et al., 2005). A study on hair cocaine and BE incorporation showed that a single 25-35 mg intravenous cocaine dose may be detected in hair for up to 6 months (Henderson et al., 1996).

A segmental hair analysis, meaning a determination of cocaine content in the length of the hair shaft, provides useful information about the individual history of drug abuse and may be used to estimate time of exposure back to a few months (Scheidweiler et al., 2005; Strano-Rossi et al., 1995). This characteristic makes hair analysis a suitable alternative matrix for long-term studies such as monitoring relapses during treatment programs or follow-up of treatment outcomes (Moeller et al., 1993; Simpson et al., 2002; Strano-Rossi et al., 1995; Wish et al., 1997).

Hair samples are not easily adultered and collection procedure does not violate the individual privacy. The hair fibers are preferentially obtained from the posterior vertex area of the scalp and as close as possible to the skin. Due to its stability, there are no specific criteria for transportation or storing, although it is recommendable to wrap the samples in aluminum foil to avoid contamination and store at room temperature.

A general critical step of hair analysis is the interpretation of the results. At this point, a few issues must be taken into account. One potential problem inherent to hair cocaine interpretation concerns the racial bias. Some studies have demonstrated that ethnicity must be considered, since the incorporation of cocaine into the human hair seems to be more extent in non-Caucasian than in Caucasian subjects, possibly due to pigmentation differences (Henderson et al., 1998; Joseph et al., 1996; Joseph et al., 1997; Reid et al., 1996). Hair cosmetic treatments, like bleaching or dying, can also interfere with the analytical results as they may affect the drug stability, leading to a partial or total loss of hair cocaine contents (Skender et al., 2002; Wennig, 2000).

Hair cocaine may reflect not only chronic cocaine abuse, but also environmental contamination. This last one includes passive contamination, for example cocaine from dust or sprays deposited on the hair surface, and passive ingestion, which may be related to passive "crack" smoking or unknowingly oral ingestion, by contact with persons who have consumed cocaine or with contaminated objects (Mieczkowski, 1997).

Several studies have demonstrated that the inclusion of an efficient washing step prior to hair analysis, typically with an organic solvent such as dichloromethane, will effectively eliminate the environmental drug contamination component (Kintz, 1998; Koren et al., 1992; Schaffer et al., 2002; Skender et al., 2002). However, Kidwell & Blank (1996) showed that heavy hair cocaine contamination cannot be completely eliminated with any of the washing solutions tested (from water and methanol, ionic or non-ionic solutions, to dimethylformamide). Romano et al. (2001) also demonstrated that even a rather small amount of cocaine (10 mg) applied to the hair persists despite using decontamination procedures.

In order to distinguish systemic exposure from environmental contamination, Koren et al. (1992) suggested the determination of the major metabolite BE in hair samples, which allegedly is detected only as a result of cocaine abuse and not contamination, whereas Cone et al. (1991) identified NCOC and CE more suitable to classify hair cocaine as a reflection of drug abuse.

Postmortem cocaine determination and interpretation can involve additional problems. As defined by Mckinney et al. (1995), "the interpretation of *postmortem* cocaine concentrations is made in an attempt to estimate drug concentrations present at the time of death and thus infer not only drug presence but also drug toxicity".

For instance, when the *postmortem* interval is excessively prolonged, or when the autopsy or laboratory analysis takes too long to be processed, cocaine can be completely hydrolyzed, chemically or enzymatically. Moreover, *postmortem* cocaine redistribution and release from tissues is a reality and has to be taken into account (Drummer, 2004; Yarema & Becker, 2005).

Several studies have demonstrated the lack of predictability of *postmortem* redistribution rates of cocaine and its metabolites over time. Also, *postmortem* blood and urine cocaine and its metabolites levels do not reflect the *antemortem* or *perimortem* values, and thus should not be used to establish cause of death (Karch et al., 1998; McKinney et al., 1995; Stephens et al., 2004; Yarema & Becker, 2005).

In alternative, samples from gastric contents and vitreous humor, nails, either fingernails or toenails, bone, and tissues such as brain, lung, liver and muscle may be analyzed to determine *postmortem* drug levels (Garside et al., 1998; McGrath & Jenkins, 2009; Stephens et al., 2004; Yarema & Becker, 2005).

Due to its isolation in the eye cavity, vitreous humor seems to be less susceptible to *postmortem* redistribution and putrefaction than other biological fluids. Despite the small amount of sample that can be collected, this specimen can be useful when the body undergoes massive bleeding or burning, or when it is in a state of prolonged decomposition (De Martinis & Martin, 2002).

4.1 Sample preparation

Due to the short half-life of cocaine in most biological specimens and its extensive metabolism, it is important to include into the analysis cocaine metabolites as well, increasing thus the detection window for drug abuse.

In order to obtain "clean" samples for analysis and increase the chromatographic sensibility towards specific drugs and their metabolites, most biological matrices require pre-treatment and concentration steps prior to chromatographic analysis. This is accomplished by extraction procedures that include mainly liquid-liquid extraction (LLE), solid-phase extraction (SPE) and more recently solid-phase microextraction (SPME).

4.1.1 Extraction procedures

The variation of acid-base properties among cocaine and its metabolites, as displayed in table 1, may challenge the selection of the most efficient extraction procedure.

The LLE consists on the separation of analytes based on their solubilities, with extraction occurring between two liquid immiscible phases (one aqueous and one organic) by adding adequate solvents.

Analyte	Acid-base properties
Cocaine	weak base; pKa = 8.6
Benzoylecgonine	amphoteric; pKa = 2.2, 11.2
Ecgonine methyl ester	weak base; pka > 8.0
Norcocaine	weak base; pka = 8.0
Hidroxycocaine	weak base
Cocaethylene	weak base; pka > 8.0
Ecgonidine methyl ester	weak base
Hydroxybenzoylecgonine	amphoteric
Benzoylnorecgonine	amphoteric
Norcocaethylene	weak base
Ecgonine ethyl ester	weak base
Ecgonidine	amphoteric

Table 1. Acid-base properties of cocaine and its metabolites.

Through LLE, the weak base analytes, such as cocaine, NCOC and EME, are the most easily extracted from biological matrices. On the other hand, isolation of amphoteric compounds, including BE, is more complex and requires a careful choice of the appropriate solvent and regulation of the pH.

Wallace et al. (1976) described a method for cocaine and BE determination in urine samples of patients who undergone surgery with cocaine anaesthesia. After extraction into a chloroform-ethanol solution (80/20%), the organic phase was evaporated to dryness at 55 °C, under a stream of filtered air. Recovered extracts were analyzed by gas chromatography (GC) coupled to a flame ionization detector (FID), and using this LLE method it was attained a recovery of 93 and 65% for cocaine and BE, respectively, and a limit of detection (LOD) of <0.1 and 0.2 µg/mL.

This relatively low recovery of amphoteric species may be a great limitation of LLE. However, it can be useful when the aim is to quantify the parent drug in a matrix where one or more metabolites are known to be present in large amounts. An example of this application is the determination of cocaine levels in urine samples, in which BE is the major analyte present.

With this purpose, Garside et al. (1997) reported a single-step LLE method using petroleum ether as the only solvent for quantification of cocaine in urine through GC coupled to mass spectrometry (MS) detection. The method has a considerably low cost and since only cocaine and other non-polar metabolites were isolated, it was not necessary to use the time-consuming and expensive derivatization step. However, a mean recovery of only 48.8% for cocaine was achieved.

A following study by Farina et al. (2002) using as solvent an ethyl ether-isopropanol mixture led to a 74.4% recovery of cocaine from urine samples, as measured by a GC method with nitrogen-phosphorous detector (NPD).

LLE was also efficiently applied to cocaine and its metabolites determination in other matrices including hair (Kintz & Mangin, 1995), nails (Engelhart & Jenkins, 2002), serum (Williams et al., 1996), plasma (Dawling et al., 1990), whole blood (Gunnar et al., 2004) and organ tissues (Hime et al., 1991).

Nonetheless, there are obvious limitations inherent to the LLE, including the use of large amounts of possibly hazardous solvents and the low recovery as a result of poor separation of the organic and aqueous phases or even formation of emulsions (Ferrera et al., 2004; Franke & de Zeeuw, 1998; Ulrich, 2000).

The SPE technique has been efficiently used to extract cocaine and its metabolites from several biological matrices, including whole blood, plasma, urine, saliva, hair and sweat, with recoveries over 80% for all analytes (Badawi et al., 2009; Bjork et al., 2010; Brunet et al., 2008; Cordero & Paterson, 2007; Lin et al., 2001; Ohshima & Takayasu, 1999). Despite the advantages, SPE still requires organic solvents, though in lower quantities compared to LLE, and the columns' price can increase the costs of the extraction procedure. When comparing extraction efficiencies of LLE and SPE applied to the same samples, for the same purpose, it is generally observed that both recovery and quality of chromatograms are superior for the SPE technique (Clauwaert et al., 1997).

For both liquid chromatography (LC) and GC, SPE appears to be the preferred extraction method through which all cocaine analytes may be isolated from a single sample with very reasonable recovery rates.

SPE allows the extraction of compounds dissolved in a liquid matrix by adsorption of the analytes in a solid porous phase. The compounds are then separated based on their affinity to the stationary phase. Therefore, the selection of the appropriate SPE column type depends on the analytes chemical and physical properties. For cocaine analysis the most usual phases used include strong cation-exchange phases, non polar C8 or C18 and mixed-mode phases that combine the other two, allowing the extraction of both polar and non polar cocaine analytes in the same column.

In a recent study, Jagerdeo and Abdel-Rahim (2009) compared the specificity and extraction efficiency of different SPE columns for cocaine and its metabolites from urine samples. They showed that a non polar C8 sorbent efficiently extracted the parent drug and CE, but no EME and only a trace amount of BE. On the other hand, both divinylbenzene copolymers ENV+ (for aliphatic and aromatic polar analytes) and Oasis MCX (strong cation-exchange phase) enabled the extraction of all analytes, with improved signal to noise ratio but with a lower extraction rate than C8. The mixed-mode phase showed the best results, with better recoveries, cleaner chromatograms and great mass accuracy.

In the last few decades, a solvent-free extraction method, the solid-phase microextraction (SPME), first designed for isolation of volatile chlorinated organic chemicals in water (Arthur & Pawliszyn, 1990), has been applied to the analysis of biological samples.

SPME can be used both in laboratory context and on-site, and it consists of a syringe-like device with a fused silica fiber coated with a polymeric stationary phase, like polyacrylate or polydimethylsiloxane, which adsorbs the analytes by direct immersion on liquid samples or by head-space (HS) extraction. The fiber is then placed in the injection port of a chromatography equipment and the analytes are recovered through desorption at elevated temperatures (Manini & Andreoli, 2002).

For cocaine analysis, SPME allows the detection of cocaine analytes at parts per billion (ppb) levels (ng/mL) in variable specimens such as urine, plasma, sweat, saliva and hair (Alvarez et al., 2007; Bermejo et al., 2006; Follador et al., 2004; Yonamine & Saviano, 2006; Yonamine et al., 2003).

Besides the low LOD values, SPME is considered easy to automate and involves little equipment. It can be used for the extraction of either liquid or solid matrices and it can be performed on very small samples (Ulrich, 2000). However, there are several disadvantages inherent to SPME technique, namely the possibility of carry-over from one sample to next one, the cost and fragility of the fiber, and the prolonged equilibration time prior to extraction (Ferrera et al., 2004).

Table 2 presents the main advantages and limitations of each extraction method.

The choice of extraction method will depend on the matrix to be analyzed, the analytes to detect, and the budget and material existent in the laboratory.

Method	Advantages	Limitations
LLE	Inexpensive May be good for solid matrices	Large amounts of organic solvents Difficult separation Poor and variable recoveries Emulsion formation Not appropriate for extraction of chemically different analytes
SPE	Fast Easy to automate Good for extraction of chemically different analytes Clean extracts Handles small samples	Use of organic solvents May require derivatization Expensive
SPME	Solvent-free Easy to automate Little equipment required May be used on-site Adequate for solid matrices High sensitive Small volume samples	Expensive Fragile polymer coating Prolonged extraction Requires procedure optimization (extraction time and temperature) Possible carry-over between samples Low recoveries

LLE – liquid-liquid extraction; SPE – solid-phase extraction; SPME – solid-phase microextraction.

Table 2. Advantages and limitations of the extraction procedures.

4.1.2 Derivatization procedures

Derivatization is a reaction by which a compound is chemically modified through reaction with a so called derivatizing agent, with a specific functional group. The reaction product is a compound (or derivative) with new properties that include different volatility, solubility, aggregation state or reactivity. It may be performed for several reasons, such as increasing compatibility with the chromatographic equipment (e.g. by decreasing polarity and increasing volatility), improve separation and resolution efficiency and attain lower detection limits (Wang et al., 2006).

Derivatization can also be useful when isotopically labeled analogs of the analytes are chosen as internal standard (IS). In these cases, it is required that the analytes and the IS generate sufficiently separated peaks, and that derivatization of the analytes allows the elimination of the phenomenon of "cross-contribution", i.e. "contribution of the analyte and the IS to the intensities of ions designated for the IS and the analyte" (Chang et al., 2001).

For GC analysis, the derivatizing agents include silyl, acyl or alkyl groups that will substitute the proton from a terminal -N-H, -S-H and/or a -O-H polar group, producing non-polar and more volatile derivatives (Segura et al., 1998; Wang et al., 2006).

The ability of the analytes to form silyl or acyl derivatives depends on their functional group. While the TMS derivatives have large affinity towards hydroxyl and carboxyl groups and much lower towards amines, the acylating agents promptly targets highly polar groups including amines and both alcohols and phenols (Segura et al., 1998; Soderholm et al., 2010).

The overall derivatization technique is described in figure 3.

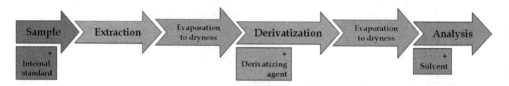

Fig. 3. Schematic representation of the steps included in the overall derivatization procedure of the analytes.

After adding the IS to the sample, extraction procedure is carried out. Subsequently, the solvent is evaporated to dryness with a gentle nitrogen stream, optionally with heating. The derivatizing agent is added and the derivatization of the analytes is performed by heating the sample. For silylating agents, the procedure ends at this time point and the sample is ready to analyze, right after cooling to room temperature. When performing acylation or alkylation, the samples have to be evaporated once more to eliminate the excess of agent, and the residue is further recovered by a solvent for posterior chromatographic analysis.

Cocaine and CE are not prone to derivatization. On the other hand, all N-demethylated metabolites, such as NCOC, NBE and NCE, can produce derivatives from the –N-H substitution, while BE, EME and the metabolites OH-BE and OH-COC may undergo a hydroxyl substitution.

Table 3 summarizes some studies on determination of cocaine and its metabolites in biological samples using either acylation, alkylation or silylation as derivatization methods for GC analysis.

The most usual agents for silylation are the trimethylsilyl (TMS) derivates, which confer to the new compounds high volatility and stability. Several TMS derivates with different chemical and physical characteristics have been produced and commercialized so far, but the TMS-amides N-methyl-N-trimethylsilyltrifluoracetamide (MSTFA) and N,O-bis(trimethylsilyl)trifluoroacetamide (BSTFA) are still the most commonly used, generally and particularly for cocaine analysis in biological specimen (Brunet et al., 2008; Romolo et al., 2003; Segura et al., 1998).

In addition, many studies employ a catalyst like trimethylchlorosilane (TMCS) with BSTFA (Brunet et al., 2008; Brunetto et al., 2010; Cone et al., 1994a; Kintz & Mangin, 1995; Romolo et al., 2003), or less commonly t-butyldimethylchlorosilane (TBDMCS) with the silylating agent N-methyl-N-t-butyldimethylsilyltrifluoroacetamide (MTBSTFA) (Lowe et al., 2006), to improve the silylating potential of the derivatizing agents (Segura et al., 1998).

Acylation is frequently applied to cocaine and its metabolites determination as well, however the acidic by-products generated in this reaction requires the elimination of the excess of derivatizing agent prior to analysis, while silylating agents can be directly injected into the GC equipment for analysis. The shorter time of preparation and the less amount of solvents required for analysis are the main advantages of silylation over acylation (Segura et al., 1998).

Sample	Analytes	Derivatizing agents	LOD	Reference
Blood Urine	Cocaine, BE	PFPA/HFIP	20 ng/mL	(Aderjan et al., 1993)
Plasma Saliva Urine	Cocaine, BE, EME Other metabolites	BSTFA/TMCS	1 ng/mL 3-6 ng/mL	(Cone et al., 1994a)
Hair	Cocaine, BE	HFBA/HFIP	0.03 ng/mg	(Jurado et al., 1995)
Hair	Cocaine, BE, EME, CE	BSTFA/TMCS	0.05 – 0.8 ng/mg	(Kintz & Mangin, 1995)
Hair	Cocaine, BE	BSTFA/TMCS	0.05 – 0.2 ng/mg	(Romolo et al., 2003)
Blood Urine Muscle tissue	Cocaine, BE, EME, CE Other metabolites	PFPA/ PFPOH	2 ng/mL 2-640 ng/mL	(Cardona et al., 2006)
Brain tissue	Cocaine, BE, EME, CE, EDME, EEE	MTBSTFA/TBDMCS	50 ng/g	(Lowe et al., 2006)
Urine	Cocaine, EME	PFPA/PFPOH	12.5 - 50 ng/mL	(Saito et al., 2007)
Sweat	Cocaine, BE, EME, EDME	BSTFA/TMCS	2.5 ng/patch	(Brunet et al., 2008)
Urine	Cocaine, BE	BSTFA/TMCS	3 - 10 ng/mL	(Brunetto et al., 2010)
Adipose tissue	Cocaine, BE, EME, CE	BSTFA	5 - 20 ng/g	(Colucci et al., 2010)
Hair	Cocaine, CE	MSTFA	0.08 – 0.09 ng/mg	(Merola et al., 2010)

BE - benzoylecgonine; BSTFA - N,O-bis(trimethylsilyl)trifluoroacetamide; CE - cocaethylene;
EDME - ecgonidine methyl ester; EEE - ecgonine ethyl ester; EME - ecgonine methyl ester;
HFBA - heptafluorobutyric anhydride; HFIP - 1,1,1,3,3,3-hexafluoroisopropanol; MSTFA - N-methyl-N-trimethylsilyltrifluoroacetamide; MTBSTFA - N-methyl-N-t-butyldimethylsilyltrifluoroacetamide;
PFPA - pentafluoropropionic anhydride; PFPOH - 2,2,3,3,3-pentafluoro-1-propanol;
TBDMCS - t-butyldimethylchlorosilane; TMCS - trimethylchlorosilane.

Table 3. Studies on determination of cocaine and its metabolites in biological samples by gas chromatography using derivatization procedures.

For acylation of cocaine analytes, the haloalkylacyl derivates, particularly the fluorinated ones like pentafluoropropionic anhydride (PFPA) and heptafluorobutyric anhydride (HFBA), are widely applied to several biological matrices, like blood and urine (Aderjan et al., 1993; Cardona et al., 2006; Saito et al., 2007), and hair and tissues (Cardona et al., 2006; Jurado et al., 1995).

Due to the weak reaction of acyl derivatizing agents with carboxyl groups, most of the studies on cocaine determination using a GC method combines to the acylating agent an alkylating one, such as 1,1,1,3,3,3-hexafluoroisopropanol (HFIP) and 2,2,3,3,3-pentafluoro-1-propanol (PFPOH) (see table 3), which easily displace the reactive proton of carboxyl groups, increasing thus the efficiency of the derivatization of all hydroxyl, carboxyl and amine functional groups (Cardona et al., 2006).

4.2 Chromatographic analysis

Over the years, several chromatographic methods have been developed to determine cocaine analytes in biological samples. For screening, one of the easiest and less expensive methods is the thin-layer chromatography (TLC), presenting a good alternative to immunoassays. Gas chromatography (GC) and liquid chromatography (LC) are more appropriate for confirmation and quantification.

4.2.1 Thin-layer chromatography

Since early before the 1980's, TLC has been systematically used for urine drug screening. Through this method, cocaine, the major urinary metabolite BE, and the transesterification product CE, can be detected. However, the method presents low sensitivity, with LOD values over 1 µg/mL, even when methylating BE back into the parent drug is performed (Bailey, 1994; Budd et al., 1980; Wolff et al., 1990).

The simplicity of the method, the rapid analysis time, and the ability to detect not only the parent drug but also metabolites and other interfering drugs made TLC very useful for forensic purposes. However, due to its proven low sensitivity and lack of specificity, as the conventional methodology may not distinguish cocaine from other compounds usually present in biological samples (for example, nicotine and caffeine), TLC is not as much applied to drug screening as immunoassays are (Janicka et al., 2010; Yonamine & Sampaio, 2006).

More recently, an improved and computerized TLC technique denominated high-performance thin-layer chromatography (HPTLC) was developed. HPTLC presents better resolution, allowing the separation of cocaine and its metabolites from interferences, and is more sensitive, reaching LOD values down to 50-550 ng/mL. In addition, the association to an advanced densitometer and a detector, such as the ultraviolet (UV) detector, makes HPTLC suitable for quantitative analysis in cases of high cocaine doses, as for example in cocaine overdoses (Antonilli et al., 2001; Yonamine & Sampaio, 2006).

4.2.2 High-perfomance liquid chromatography

For many decades, LC has been widely applied to the separation of organic compounds. The separation through LC is based on the analytes distribution between a liquid mobile phase and a stationary phase. Nowadays, the LC is usually equipped with pumps that apply relatively elevated pressures to force the mobile phase through the very small packing particles that forms the stationary phase, being referred as high-performance liquid chromatography (HPLC). Table 4 summarizes some studies on cocaine and its metabolites determination in biological material through LC or HPLC, with variable detection equipment.

Sample	Analytes	Column	Mobile phase	Detector	LOD	Reference
Urine	Cocaine, BE	C18	(A) NH_4HCO_3 (B) CH_3OH 5-90-5% B in A	MS/MS	1 – 1.4 ng/mL	(Berg et al., 2009)
Serum	Cocaine, BE	PFPP	(A) 0.1% HCOOH + $HCOONH_4$ 1mM (B) CH_3CN + 0.1% HCOOH + $HCOONH_4$ 1mM	MS/MS	0.1 -0.4 ng/mL	(Bouzas et al., 2009)
Urine	Cocaine, BE, CE	C18	(A) CH_3COONH_4 in $H_2O:CH_3OH:CH_3CN$ (8:1:1) (B) CH_3COONH_4 in $H2O:CH_3OH:CH_3CN$ (1:2:2) 100-47.2% A in B	DAD	20 ng/mL	(Clauwaert et al., 1996)
Plasma	Cocaine, BE, CE	C8	(A) CH_3CN (B) PO_4^{3-} buffer 10-50-10% A in B	DAD	10 ng/mL	(Fernandez et al., 2006)
Blood Urine	Cocaine, BE, EME, CE, NCOC	C18	(A) 5% CH_3CN + 0.05% HCOOH (B) 100% CH_3CN + 0.05% HCOOH 95-60% B in A	MS/MS	0.001 – 0.003 mg/kg	(Johansen & Bhatia, 2007)
Hair	Cocaine, BE	C18	(A) 0.01% HCOOH (B) CH_3OH (C) CH_3CN 10% B + 70-30-70% A + 20-60-20% C	MS/MS	1 – 10 pg/mg	(Lopez et al., 2010)
Hair	Cocaine, BE	C18	$CH_3OH:CH_3CN:KH_2PO_4$ buffer (10:15:75) + 0.25% $N(CH_2CH_3)_3$	FD	1 ng/mL	(Mercolini et al., 2008)
Saliva	Cocaine, BE	Phenyl	(A) CH_3OH + $HCOONH_4$ 10 mM (B) H_2O + $HCOONH_4$ 10 mM 6-41.2% A in B	MS/MS	0.22 - 0.29 ng/mL	(Mortier et al., 2002)
Urine	Cocaine, EME	PFPP	CH_3CN + $HCOONH_4$ 5mM + HCOOH	MS/MS	1.6 – 2.8 pg on column	(Needham et al., 2000)
Plasma	Cocaine, BE, EME, NCOC	C18	$HOC(COOH)(CH_2COOH)_2$ 0.05 M:Na_2HPO_3 (4:1) + 18% CH_3CN + 0.3% $N(CH_2CH_3)_3$	UV	35 – 90 ng/mL	(Virag et al., 1996)
Serum	Cocaine, CE	Cyanopro pyl	CH_3CN: PO_4^{3-} buffer (38:62)	UV	25 ng/mL	(Williams et al., 1996)

BE - benzoylecgonine; CE - cocaethylene; DAD - diode array detector; EME - ecgonine methyl ester; FD - fluorescence detector; MS - mass spectrometry; NCOC - norcocaine; PFPP - pentafluorophenylpropyl; UV - ultraviolet.

Table 4. Studies on cocaine and its metabolites determination in biological samples by liquid chromatography.

The separation of cocaine analytes is usually performed in reversed-phase columns, such as C8 and C18. However, other stationary phases may be used, depending on the properties of the analytes in study. For example, Needham et al. (2000), after observing the unsuccessful retention in a C18 column of EME, a very polar cocaine metabolite, demonstrated that a pentafluorophenylpropyl (PFPP) bonded silica column increased the retention and improved the peak shape of both metabolite and parent drug.

Among the detection equipment used with chromatographic methods, two of the most popular for LC cocaine analysis include the UV detectors and the fluorescence detector (FD), due to their low cost and easy automation (Janicka et al., 2010).

The weak UV absorption of polar cocaine metabolites diminishes the usefulness of an UV detector. Nonetheless, some studies have shown acceptable results using an UV or a diode array detector (DAD), but with visible lack of sensitivity when compared to other detectors (see table 4).

Mass spectrometry (MS) greatly improved the detection and identification of analytes after chromatographic elution, providing identification based on mass-spectral data. More common than the simple MS detection, many LC methods use tandem MS (or MS/MS, or MS^2) in which multiple steps of MS selection enable a more accurate identification.

The elevated sensitivity of MS allows detection of compounds at concentrations below ppb levels, as found for several biological specimens (see table 4).

4.2.3 Gas chromatography

GC is a widely used methodology for drug abuse analysis. In this chromatographic technique the mobile phase is a carrier gas, typically an unreactive gas like nitrogen, hydrogen or helium. The sample is carried through a liquid or a polymeric stationary phase bounded to a solid support inside a column. This column is located inside an oven that controls the temperature of the mobile phase, and the analytes in the sample are separated based on polarity and vapor pressure differences.

Either liquid or gaseous (extracted through HS-SPME) samples may be analyzed by GC, however, only volatile compounds can be detected. Thus, while cocaine and its non polar metabolite CE are easily determined in biological samples extracts without prior preparation techniques (Cognard et al., 2005; Hime et al., 1991), most of the other cocaine analytes requires previous derivatization.

Like LC techniques, GC presents high selectivity and low detection levels. Table 5 presents some analytical studies on cocaine and its metabolites by several GC techniques in different biological matrices.

Before the development of the MS detector, cocaine analysis in biological samples by GC methods used essentially a nitrogen-phosphorus detector (NPD). This detector is moderately priced and provides a quite sensitive analysis, with cocaine LOD values below 100 ng/mL.

Among the studies using GC-NPD, the use of extraction methods slightly improves the sensitivity of the chromatographic method towards cocaine analytes. Urine samples extracted by SPME shows a somewhat lower cocaine LOD than urine samples treated by LLE (12 vs. 15 ng/mL), while the analysis of non extracted blood samples presents lower sensitivity than the other two (LOD = 20 ng/mL) (Farina et al., 2002; Hime et al., 1991; Kumazawa et al., 1995).

Sample	Analytes	Extraction/ Derivatization	Detector	LOD	Reference
Plasma	Cocaine, CE	SPME/-	MS	11 – 19 ng/mL	(Alvarez et al., 2007)
Hair	Cocaine, BE	SPE/MSTFA+TMCS	MS	15 – 20 pg/mg	(Barroso et al., 2008)
Blood Urine Muscle tissue	Cocaine, BE, EME, NCOC, CE, EDME, NBE, NCE, OH-BE, EEE	SPE/PFPA+ PFPOH	MS	2-640 ng/mL	(Cardona et al., 2006)
Hair	Cocaine, EME, CE, EDME	SPE/-	MS/MS	5 – 50 pg/mg	(Cognard et al., 2005)
Saliva	Cocaine, EME, CE, EDME	SPE/-	MS/MS	0.1 – 0.5 ng/mL	(Cognard et al., 2006)
Urine	Cocaine	LLE/-	NPD	15 ng/mL	(Farina et al., 2002)
Blood	Cocaine, CE	-/-	NPD	20 ng/mL	(Hime et al., 1991)
Placenta	Cocaine, BE, CE	SPE/MSTFA	MS	0.2 – 0.7 ng/mL	(Joya et al., 2010)
Urine	Cocaine	SPME/-	NPD	12 ng/mL	(Kumazawa et al., 1995)
Saliva Urine	Cocaine, BE, EME	LLE/PFPA + PFPOH	MS	2 ng/mL	(Strano-Rossi et al., 2008)
HPTECs	Cocaine, BE, EME, NCOC	SPE/MSTFA	MS	0.4 – 20.9 ng/mL	(Valente et al., 2010)
Nails	Cocaine, BE, NCOC	SPE/PFPA + PFPOH	MS	3 – 3.5 ng/mg	(Valente-Campos et al., 2006)
Skin biopsy	Cocaine, BE, EME, NCOC, CE, EDME, NCE, EEE	SPE/MTBSTFA+TBDMCS	MS	1.25 – 5 ng/biopsy	(Yang et al., 2006)

BE - benzoylecgonine; CE - cocaethylene; EDME - ecgonidine methyl ester; EEE - ecgonine ethyl ester; EME - ecgonine methyl ester; HPTECs - human proximal tubular epithelial cells; LLE - liquid-liquid extraction; MS - mass spectrometry; MSTFA - N-methyl-N-trimethylsilyltrifluoracetamide; MTBSTFA - N-methyl-N-t-butyldimethylsilyltrifluoroacetamide; NBE - norbenzoylecgonine; NCE - norcocaethylene; NCOC - norcocaine; NPD - nitrogen-phosphorus detector; OH-BE - hydroxybenzoylecgonine; PFPA - pentafluoropropionic anhydride; PFPOH - 2,2,3,3,3-pentafluoro-1-propanol; SPE - solid-phase extraction; SPME - solid-phase microextraction; TBDMCS - t-butyldimethylchlorosilane; TMCS - trimethylchlorosilane.

Table 5. Gas chromatography studies measuring cocaine and its metabolites in biological samples.

MS development for GC analysis greatly improved the detection of cocaine analytes. In fact, when comparing equal specimens analyzed by GC-NPD and GC-MS, LOD values for the parent compound using the second method may be ten-fold lower than those seen with the conventional NPD. Taking the example of blood samples once more, while Hime et al. (1991) described a cocaine LOD of 20 ng/mL by GC-NPD analysis, Cardona et al. (2006) obtained a cocaine LOD of 2 ng/mL using a GC-MS equipment with prior SPE extraction and combined acylation/methylation of the analytes.

GC-MS is more often applied to the analysis of less conventional biological matrices than LC. These include nails and even biopsy material (Joya et al., 2010; Valente-Campos et al., 2006; Yang et al., 2006). Moreover, GC-MS allows the determination of metabolites of specific consumption patterns, like EDME and EEE for "crack" abuse, and CE and NCE for concomitant use with alcohol, and other secondary minor metabolites, such as NBE and OH-BE (Cardona et al., 2006; Yang et al., 2006).

Furthermore, and as described for LC techniques, the detection of analytes eluted through GC can be performed by tandem MS as well. MS/MS improvement over MS is visible when comparing equal samples analyzed by both methods. For instance, saliva samples analyzed by GC-MS showed a cocaine LOD of 2 ng/mL, whereas GC-MS/MS was sensible for cocaine concentrations below ppb levels (0.1 ng/mL) (Cognard et al., 2006; Strano-Rossi et al., 2008).

In our laboratory, we have recently developed and validated a GC method for detection and quantification of cocaine and its metabolites in primary cultured human proximal tubular epithelial cells (HPTECs) (Valente et al., 2010). As far as we know, this was the first chromatographic technique described for the analysis of cocaine analytes in a cellular matrix.

This *in vitro* cellular model, which was previously optimized and characterized by our group (Valente et al., 2011a) as well, was used to evaluate the specificity and sensitivity of a GC-MS method for the quantification of cocaine, its major metabolites BE and EME, and the minor metabolite NCOC, particularly known for its cytotoxic effects on the liver.

Samples of confluent cells cultured at physiological conditions (supplemented medium, at 37 °C and a humidified environment with 95% O_2 and 5% CO_2) were used as matrix for analysis in which standard solutions of cocaine and its metabolites were prepared. Extraction was then performed through strong cation-exchange phase SPE columns (OASIS MCX), allowing the pre-concentration of the cocaine analytes in the samples. The compounds were then submitted to derivatization with MSTFA, which generated well resolved chromatographic peaks for all the analytes in study.

The method was proven to be accurate, linear for a wide range of concentrations (0 - 100 µg/mL) and specific for cocaine analytes. It provided very low LOD values for all cocaine analytes (0.4 – 20.9 ng/mL).

This validated GC-MS technique was further successfully applied to a toxicokinetics study on renal cocaine metabolism, in which we were able to demonstrate that, unlike what happens in the liver, cocaine is metabolized in the kidney into EME and NCOC in lesser extent, but not into BE (Valente et al., 2011b).

This study demonstrated the usefulness of GC, and particularly GC-MS, not only for the determination of drugs of abuse in biological samples, for either clinical or forensic purposes, but also for physiological evaluations and development of toxicological models.

4.3 Data analysis

After a complete chromatographic separation, a chromatogram is obtained as the example shown in figure 4a. The identification of each peak in the chromatogram can be attained through comparison of the retention times of the compounds in the sample and standard compounds analyzed at the same chromatographic conditions. Another way is the comparison of the mass spectrum of the analyte, provided by a MS detector, with the existing mass spectra in a database.

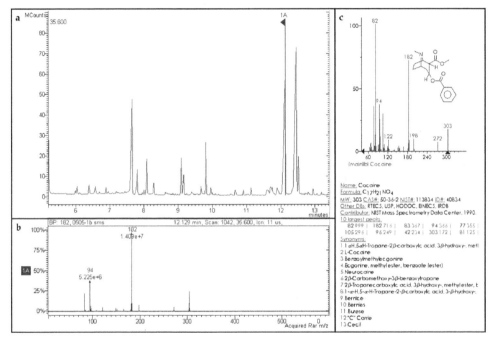

(a) full scan chromatogram, (b) mass spectrum of indicated peak and (c) cocaine identification through a mass spectrum database (National Institute of Standards and Technology, NIST 05 database).

Fig. 4. Analysis of a biological matrix containing cocaine and its metabolites, through gas chromatography with detection by mass spectrometry.

Figure 4 represents the identification of the cocaine peak in a biological sample eluted in a GC-MS equipment. In figure 4a is pointed out a peak (1A) and the respective mass spectrum in figure 4b, indicating the relative abundance of each mass-to-charge ratio (m/z) in that peak. The m/z profile of the selected peak is then compared to those existing in the database, and the compounds with approximated spectrum are presented in a decreasing order of similarity. In this case, cocaine m/z profile is given as the most resembling to the 1A peak (figure 4c).

Independently of the detector used, the quantification of an analyte requires the use of calibration curves obtained from standard solutions of the compounds in study analyzed at the same chromatographic conditions of the samples, and preferably prepared in an equal matrix to eliminate matrix effects.

To avoid miscalculations resulting from errors inherent to steps prior to analysis, for example injection of variable sample volumes in the chromatographic equipment, it is recommendable to use an IS. The IS is added to each sample at the same time point, its concentration should not alter with further preparation procedures, and the IS chromatographic peak cannot interfere or elute at the same time of any analyte of the sample.

Using an appropriate IS, for both samples and SS, the determination of the compounds takes into account the area of the IS chromatographic peak, and the calibration curves are presented as [standard solution area/IS area = f(concentration of the standard solution)]. Finally, the concentration of each analyte will be extrapolated using the ratio [analyte area/IS area].

5. References

Aderjan, R. E., Schmitt, G., Wu, M., & Meyer, C. (1993). Determination of cocaine and benzoylecgonine by derivatization with iodomethane-D3 or PFPA/HFIP in human blood and urine using GC/MS (EI or PCI mode). *J Anal Toxicol*, 17(1), 51-55, 0146-4760.

Alvarez, I., Bermejo, A. M., Tabernero, M. J., Fernandez, P., & Lopez, P. (2007). Determination of cocaine and cocaethylene in plasma by solid-phase microextraction and gas chromatography-mass spectrometry. *J Chromatogr B Analyt Technol Biomed Life Sci*, 845(1), 90-94, 1570-0232.

Antonilli, L., Suriano, C., Grassi, M. C., & Nencini, P. (2001). Analysis of cocaethylene, benzoylecgonine and cocaine in human urine by high-performance thin-layer chromatography with ultraviolet detection: a comparison with high-performance liquid chromatography. *J Chromatogr B Biomed Sci Appl*, 751(1), 19-27, 1387-2273.

Arthur, C. L., & Pawliszyn, J. (1990). Solid phase microextraction with thermal desorption using fused silica optical fibers. *Anal Chem*, 62(19), 2145-2148 0003-2700.

Badawi, N., Simonsen, K. W., Steentoft, A., Bernhoft, I. M., & Linnet, K. (2009). Simultaneous screening and quantification of 29 drugs of abuse in oral fluid by solid-phase extraction and ultraperformance LC-MS/MS. *Clin Chem*, 55(11), 2004-2018, 1530-8561.

Bailey, D. N. (1994). Thin-layer chromatographic detection of cocaethylene in human urine. *Am J Clin Pathol*, 101(3), 342-345, 0002-9173.

Barroso, M., Dias, M., Vieira, D. N., Queiroz, J. A., & Lopez-Rivadulla, M. (2008). Development and validation of an analytical method for the simultaneous determination of cocaine and its main metabolite, benzoylecgonine, in human hair by gas chromatography/mass spectrometry. *Rapid Commun Mass Spectrom*, 22(20), 3320-3326, 0951-4198.

Baselt, R. C. (1983). Stability of cocaine in biological fluids. *J Chromatogr*, 268(3), 502-505, 0021-9673.

Berg, T., Lundanes, E., Christophersen, A. S., & Strand, D. H. (2009). Determination of opiates and cocaine in urine by high pH mobile phase reversed phase UPLC-MS/MS. *J Chromatogr B Analyt Technol Biomed Life Sci*, 877(4), 421-432, 1570-0232.

Bermejo, A. M., Lopez, P., Alvarez, I., Tabernero, M. J., & Fernandez, P. (2006). Solid-phase microextraction for the determination of cocaine and cocaethylene in human hair by gas chromatography-mass spectrometry. *Forensic Sci Int*, 156(1), 2-8, 0379-0738.

Bertol, E., Trignano, C., Di Milia, M. G., Di Padua, M., & Mari, F. (2008). Cocaine-related deaths: an enigma still under investigation. *Forensic Sci Int*, 176(2-3), 121-123, 1872-6283.

Bjork, M. K., Nielsen, M. K., Markussen, L. O., Klinke, H. B., & Linnet, K. (2010). Determination of 19 drugs of abuse and metabolites in whole blood by high-performance liquid chromatography-tandem mass spectrometry. *Anal Bioanal Chem*, 396(7), 2393-2401, 1618-2650.

Bouzas, N. F., Dresen, S., Munz, B., & Weinmann, W. (2009). Determination of basic drugs of abuse in human serum by online extraction and LC-MS/MS. *Anal Bioanal Chem*, 395(8), 2499-2507, 1618-2650 (Electronic).

Boyer, C. S., & Petersen, D. R. (1990). Potentiation of cocaine-mediated hepatotoxicity by acute and chronic ethanol. *Alcohol Clin Exp Res*, 14(1), 28-31.

Brookoff, D., Cook, C. S., Williams, C., & Mann, C. S. (1994). Testing reckless drivers for cocaine and marijuana. *N Engl J Med*, 331(8), 518-522, 0028-4793.

Brunet, B. R., Barnes, A. J., Scheidweiler, K. B., Mura, P., & Huestis, M. A. (2008). Development and validation of a solid-phase extraction gas chromatography-mass spectrometry method for the simultaneous quantification of methadone, heroin, cocaine and metabolites in sweat. *Anal Bioanal Chem*, 392(1-2), 115-127, 1618-2650.

Brunetto, M. d. R., Delgado, Y., Clavijo, S., Contreras, Y., Torres, D., Ayala, C., Gallignani, M., Forteza, R., & Cerda Martin, V. (2010). Analysis of cocaine and benzoylecgonine in urine by using multisyringe flow injection analysis-gas chromatography-mass spectrometry system. *J Sep Sci*, 33(12), 1779-1786, 1615-9314.

Budd, R. D., Mathis, D. F., & Yang, F. C. (1980). TLC analysis of urine for benzoylecgonine and norpropoxyphene. *Clin Toxicol*, 16(1), 1-5, 0009-9309.

Burns, M., & Baselt, R. C. (1995). Monitoring drug use with a sweat patch: an experiment with cocaine. *J Anal Toxicol*, 19(1), 41-48, 0146-4760.

Cardona, P. S., Chaturvedi, A. K., Soper, J. W., & Canfield, D. V. (2006). Simultaneous analyses of cocaine, cocaethylene, and their possible metabolic and pyrolytic products. *Forensic Sci Int*, 157(1), 46-56, 0379-0738.

Chang, W. T., Lin, D. L., & Liu, R. H. (2001). Isotopic analogs as internal standards for quantitative analyses by GC/MS--evaluation of cross-contribution to ions designated for the analyte and the isotopic internal standard. *Forensic Sci Int*, 121(3), 174-182, 0379-0738.

Chawarski, M. C., Fiellin, D. A., O'Connor, P. G., Bernard, M., & Schottenfeld, R. S. (2007). Utility of sweat patch testing for drug use monitoring in outpatient treatment for opiate dependence. *J Subst Abuse Treat*, 33(4), 411-415, 0740-5472.

Clauwaert, K. M., Van Bocxlaer, J. F., Lambert, W. E., & De Leenheer, A. P. (1996). Analysis of cocaine, benzoylecgonine, and cocaethylene in urine by HPLC with diode array detection. *Anal Chem*, 68(17), 3021-3028, 0003-2700.

Clauwaert, K. M., Van Bocxlaer, J. F., Lambert, W. E., & De Leenheer, A. P. (1997). Liquid chromatographic determination of cocaine, benzoylecgonine, and cocaethylene in whole blood and serum samples with diode-array detection. *J Chromatogr Sci*, 35(7), 321-328, 0021-9665.

Cognard, E., Bouchonnet, S., & Staub, C. (2006). Validation of a gas chromatography--ion trap tandem mass spectrometry for simultaneous analyse of cocaine and its metabolites in saliva. *J Pharm Biomed Anal*, 41(3), 925-934, 0731-7085.

Cognard, E., Rudaz, S., Bouchonnet, S., & Staub, C. (2005). Analysis of cocaine and three of its metabolites in hair by gas chromatography-mass spectrometry using ion-trap detection for CI/MS/MS. *J Chromatogr B Analyt Technol Biomed Life Sci*, 826(1-2), 17-25, 1570-0232.

Colucci, A. P., Aventaggiato, L., Centrone, M., & Gagliano-Candela, R. (2010). Validation of an extraction and gas chromatography-mass spectrometry quantification method for cocaine, methadone, and morphine in postmortem adipose tissue. *J Anal Toxicol*, 34(6), 342-346, 1945-2403.

Cone, E. J., Hillsgrove, M., & Darwin, W. D. (1994a). Simultaneous measurement of cocaine, cocaethylene, their metabolites, and "crack" pyrolysis products by gas chromatography-mass spectrometry. *Clin Chem*, 40(7 Pt 1), 1299-1305, 0009-9147.

Cone, E. J., Hillsgrove, M. J., Jenkins, A. J., Keenan, R. M., & Darwin, W. D. (1994b). Sweat testing for heroin, cocaine, and metabolites. *J Anal Toxicol*, 18(6), 298-305, 0146-4760.

Cone, E. J., Lange, R., & Darwin, W. D. (1998). In vivo adulteration: excess fluid ingestion causes false-negative marijuana and cocaine urine test results. *J Anal Toxicol*, 22(6), 460-473, 0146-4760.

Cone, E. J., Sampson-Cone, A. H., Darwin, W. D., Huestis, M. A., & Oyler, J. M. (2003). Urine testing for cocaine abuse: metabolic and excretion patterns following different routes of administration and methods for detection of false-negative results. *J Anal Toxicol*, 27(7), 386-401, 0146-4760.

Cone, E. J., Yousefnejad, D., Darwin, W. D., & Maguire, T. (1991). Testing human hair for drugs of abuse. II. Identification of unique cocaine metabolites in hair of drug abusers and evaluation of decontamination procedures. *J Anal Toxicol*, 15(5), 250-255, 0146-4760.

Cook, C. E. (1991). Pyrolytic characteristics, pharmacokinetics, and bioavailability of smoked heroin, cocaine, phencyclidine, and methamphetamine. *NIDA Res Monogr*, 115, 6-23.

Cordero, R., & Paterson, S. (2007). Simultaneous quantification of opiates, amphetamines, cocaine and metabolites and diazepam and metabolite in a single hair sample using GC-MS. *J Chromatogr B Analyt Technol Biomed Life Sci*, 850(1-2), 423-431, 1570-0232.

Darke, S., & Duflou, J. (2008). Toxicology and circumstances of death of homicide victims in New South Wales, Australia 1996-2005. *J Forensic Sci*, 53(2), 447-451, 0022-1198.

Dawling, S., Essex, E. G., Ward, N., & Widdop, B. (1990). Gas chromatographic measurement of cocaine in serum, plasma and whole blood. *Ann Clin Biochem*, 27 (Pt 5), 478-481, 0004-5632.

De Martinis, B. S., & Martin, C. C. (2002). Automated headspace solid-phase microextraction and capillary gas chromatography analysis of ethanol in postmortem specimens. *Forensic Sci Int*, 128(3), 115-119, 0379-0738.

Dean, R. A., Harper, E. T., Dumaual, N., Stoeckel, D. A., & Bosron, W. F. (1992). Effects of ethanol on cocaine metabolism: formation of cocaethylene and norcocaethylene. *Toxicol Appl Pharmacol*, 117(1), 1-8, 0041-008X.

Devlin, R. J., & Henry, J. A. (2008). Clinical review: Major consequences of illicit drug consumption. *Crit Care*, 12(1), 202.

Dias, A. C., Ribeiro, M., Dunn, J., Sesso, R., & Laranjeira, R. (2008). Follow-up study of crack cocaine users: situation of the patients after 2, 5, and 12 years. *Subst Abus*, 29(3), 71-79, 0889-7077.

Dolan, K., Rouen, D., & Kimber, J. (2004). An overview of the use of urine, hair, sweat and saliva to detect drug use. *Drug Alcohol Rev*, 23(2), 213-217, 0959-5236.

Donovan, D. M., Bigelow, G. E., Brigham, G. S., Carroll, K. M., Cohen, A. J., Gardin, J. G., Hamilton, J. A., Huestis, M. A., Hughes, J. R., Lindblad, R., Marlatt, G. A., Preston, K. L., Selzer, J. A., Somoza, E. C., Wakim, P. G., & Wells, E. A. (2011). Primary outcome indices in illicit drug dependence treatment research: systematic approach to selection and measurement of drug use end-points in clinical trials. *Addiction*, 1360-0443.

Drummer, O. H. (2004). Postmortem toxicology of drugs of abuse. *Forensic Sci Int*, 142(2-3), 101-113, 0379-0738.

Dugan, S., Bogema, S., Schwartz, R. W., & Lappas, N. T. (1994). Stability of drugs of abuse in urine samples stored at -20 degrees C. *J Anal Toxicol*, 18(7), 391-396, 0146-4760.

EMCDDA. (2010). The EMCDDA annual report 2010: the state of the drugs problem in Europe. *Euro Surveill*, 15(46), 61-70, 1560-7917.

Engelhart, D. A., & Jenkins, A. J. (2002). Detection of cocaine analytes and opiates in nails from postmortem cases. *J Anal Toxicol*, 26(7), 489-492, 0146-4760.

Fandino, A. S., Toennes, S. W., & Kauert, G. F. (2002). Studies on hydrolytic and oxidative metabolic pathways of anhydroecgonine methyl ester (methylecgonidine) using microsomal preparations from rat organs. *Chem Res Toxicol*, 15(12), 1543-1548, 0893-228X.

Farina, M., Yonamine, M., & Silva, O. A. (2002). One-step liquid-liquid extraction of cocaine from urine samples for gas chromatographic analysis. *Forensic Sci Int*, 127(3), 204-207, 0379-0738.

Favrod-Coune, T., & Broers, B. (2010). The Health Effect of Psychostimulants: A Literature Review. *Pharmaceuticals*, 3, 2333-2361.

Fernandez, P., Morales, L., Vazquez, C., Bermejo, A. M., & Tabernero, M. J. (2006). HPLC-DAD determination of opioids, cocaine and their metabolites in plasma. *Forensic Sci Int*, 161(1), 31-35, 0379-0738.

Ferrera, Z. S., Sans, C. P., Santana, C. M., & Rodriguez, J. J. (2004). The use of micellar systems in the extraction and pre-concentration of organic pollutants in environmental samples. *TrAC*, 23(7), 469-479.

Follador, M. J., Yonamine, M., de Moraes Moreau, R. L., & Silva, O. A. (2004). Detection of cocaine and cocaethylene in sweat by solid-phase microextraction and gas chromatography/mass spectrometry. *J Chromatogr B Analyt Technol Biomed Life Sci*, 811(1), 37-40, 1570-0232.

Fowler, J. S., Volkow, N. D., Wolf, A. P., Dewey, S. L., Schlyer, D. J., Macgregor, R. R., Hitzemann, R., Logan, J., Bendriem, B., Gatley, S. J., & et al. (1989). Mapping cocaine binding sites in human and baboon brain in vivo. *Synapse*, 4(4), 371-377, 0887-4476.

Franke, J. P., & de Zeeuw, R. A. (1998). Solid-phase extraction procedures in systematic toxicological analysis. *J Chromatogr B Biomed Sci Appl*, 713(1), 51-59, 1387-2273.

Freye, E., & Levy, J. V. (2009). *Pharmacology and Abuse of Cocaine, Amphetamines, Ecstasy and Related Designer Drugs*. Springer, 978-90-481-2447-3, New York.

Garlow, S. J., Purselle, D. C., & Heninger, M. (2007). Cocaine and alcohol use preceding suicide in African American and white adolescents. *J Psychiatr Res,* 41(6), 530-536, 0022-3956.

Garrett, E. R., & Seyda, K. (1983). Prediction of stability in pharmaceutical preparations XX: stability evaluation and bioanalysis of cocaine and benzoylecgonine by high-performance liquid chromatography. *J Pharm Sci,* 72(3), 258-271, 0022-3549.

Garside, D., Goldberger, B. A., Preston, K. L., & Cone, E. J. (1997). Rapid liquid-liquid extraction of cocaine from urine for gas chromatographic-mass spectrometric analysis. *J Chromatogr B Biomed Sci Appl,* 692(1), 61-65, 1387-2273.

Garside, D., Ropero-Miller, J. D., Goldberger, B. A., Hamilton, W. F., & Maples, W. R. (1998). Identification of cocaine analytes in fingernail and toenail specimens. *J Forensic Sci,* 43(5), 974-979, 0022-1198.

George, S. (2005). A snapshot of workplace drug testing in the UK. *Occup Med (Lond),* 55(1), 69-71, 0962-7480.

Gjerde, H., Normann, P. T., Pettersen, B. S., Assum, T., Aldrin, M., Johansen, U., Kristoffersen, L., Oiestad, E. L., Christophersen, A. S., & Morland, J. (2008). Prevalence of alcohol and drugs among Norwegian motor vehicle drivers: a roadside survey. *Accid Anal Prev,* 40(5), 1765-1772, 1879-2057.

Glauser, J., & Queen, J. R. (2007). An overview of non-cardiac cocaine toxicity. *J Emerg Med,* 32(2), 181-186.

Goldstein, R. A., DesLauriers, C., Burda, A., & Johnson-Arbor, K. (2009). Cocaine: history, social implications, and toxicity: a review. *Semin Diagn Pathol,* 26(1), 10-17.

Graham, J. K., & Hanzlick, R. (2008). Accidental drug deaths in Fulton County, Georgia, 2002: characteristics, case management and certification issues. *Am J Forensic Med Pathol,* 29(3), 224-230, 1533-404X.

Gunnar, T., Mykkanen, S., Ariniemi, K., & Lillsunde, P. (2004). Validated semiquantitative/quantitative screening of 51 drugs in whole blood as silylated derivatives by gas chromatography-selected ion monitoring mass spectrometry and gas chromatography electron capture detection. *J Chromatogr B Analyt Technol Biomed Life Sci,* 806(2), 205-219, 1570-0232.

Harris, D. S., Everhart, E. T., Mendelson, J., & Jones, R. T. (2003). The pharmacology of cocaethylene in humans following cocaine and ethanol administration. *Drug Alcohol Depend,* 72(2), 169-182.

Heard, K., Palmer, R., & Zahniser, N. R. (2008). Mechanisms of acute cocaine toxicity. *Open Pharmacol J,* 2(9), 70-78.

Hearn, W. L., Rose, S., Wagner, J., Ciarleglio, A., & Mash, D. C. (1991). Cocaethylene is more potent than cocaine in mediating lethality. *Pharmacol Biochem Behav,* 39(2), 531-533, 0091-3057.

Henderson, G. L., Harkey, M. R., Zhou, C., Jones, R. T., & Jacob, P., 3rd. (1996). Incorporation of isotopically labeled cocaine and metabolites into human hair: 1. dose-response relationships. *J Anal Toxicol,* 20(1), 1-12, 0146-4760.

Henderson, G. L., Harkey, M. R., Zhou, C., Jones, R. T., & Jacob, P., 3rd. (1998). Incorporation of isotopically labeled cocaine into human hair: race as a factor. *J Anal Toxicol,* 22(2), 156-165, 0146-4760.

Hime, G. W., Hearn, W. L., Rose, S., & Cofino, J. (1991). Analysis of cocaine and cocaethylene in blood and tissues by GC-NPD and GC-ion trap mass spectrometry. *J Anal Toxicol*, 15(5), 241-245, 0146-4760.

Hippenstiel, M. J., & Gerson, B. (1994). Optimization of storage conditions for cocaine and benzoylecgonine in urine: a review. *J Anal Toxicol*, 18(2), 104-109, 0146-4760.

Huestis, M. A., Oyler, J. M., Cone, E. J., Wstadik, A. T., Schoendorfer, D., & Joseph, R. E., Jr. (1999). Sweat testing for cocaine, codeine and metabolites by gas chromatography-mass spectrometry. *J Chromatogr B Biomed Sci Appl*, 733(1-2), 247-264, 1387-2273.

Isenschmid, D. S., Levine, B. S., & Caplan, Y. H. (1989). A comprehensive study of the stability of cocaine and its metabolites. *J Anal Toxicol*, 13(5), 250-256, 0146-4760.

Jacob, P., 3rd, Lewis, E. R., Elias-Baker, B. A., & Jones, R. T. (1990). A pyrolysis product, anhydroecgonine methyl ester (methylecgonidine), is in the urine of cocaine smokers. *J Anal Toxicol*, 14(6), 353-357, 0146-4760.

Jagerdeo, E., & Abdel-Rehim, M. (2009). Screening of cocaine and its metabolites in human urine samples by direct analysis in real-time source coupled to time-of-flight mass spectrometry after online preconcentration utilizing microextraction by packed sorbent. *J Am Soc Mass Spectrom*, 20(5), 891-899, 1879-1123.

Janicka, M., Kot-Wasik, A., & Namiesnik, J. (2010). Analytical procedures for determination of cocaine and its metabolites in biological samples. *TrAC*, 29(3), 209-224.

Jeffcoat, A. R., Perez-Reyes, M., Hill, J. M., Sadler, B. M., & Cook, C. E. (1989). Cocaine disposition in humans after intravenous injection, nasal insufflation (snorting), or smoking. *Drug Metab Dispos*, 17(2), 153-159.

Johansen, S. S., & Bhatia, H. M. (2007). Quantitative analysis of cocaine and its metabolites in whole blood and urine by high-performance liquid chromatography coupled with tandem mass spectrometry. *J Chromatogr B Analyt Technol Biomed Life Sci*, 852(1-2), 338-344, 1570-0232.

Joseph, R. E., Jr., Su, T. P., & Cone, E. J. (1996). In vitro binding studies of drugs to hair: influence of melanin and lipids on cocaine binding to Caucasoid and Africoid hair. *J Anal Toxicol*, 20(6), 338-344, 0146-4760.

Joseph, R. E., Jr., Tsai, W. J., Tsao, L. I., Su, T. P., & Cone, E. J. (1997). In vitro characterization of cocaine binding sites in human hair. *J Pharmacol Exp Ther*, 282(3), 1228-1241, 0022-3565.

Joya, X., Pujadas, M., Falcon, M., Civit, E., Garcia-Algar, O., Vall, O., Pichini, S., Luna, A., & de la Torre, R. (2010). Gas chromatography-mass spectrometry assay for the simultaneous quantification of drugs of abuse in human placenta at 12th week of gestation. *Forensic Sci Int*, 196(1-3), 38-42, 1872-6283.

Jufer, R. A., Wstadik, A., Walsh, S. L., Levine, B. S., & Cone, E. J. (2000). Elimination of cocaine and metabolites in plasma, saliva, and urine following repeated oral administration to human volunteers. *J Anal Toxicol*, 24(7), 467-477, 0146-4760.

Jurado, C., Gimenez, M. P., Menendez, M., & Repetto, M. (1995). Simultaneous quantification of opiates, cocaine and cannabinoids in hair. *Forensic Sci Int*, 70(1-3), 165-174, 0379-0738.

Kacinko, S. L., Barnes, A. J., Schwilke, E. W., Cone, E. J., Moolchan, E. T., & Huestis, M. A. (2005). Disposition of cocaine and its metabolites in human sweat after controlled cocaine administration. *Clin Chem*, 51(11), 2085-2094, 0009-9147.

Kanel, G. C., Cassidy, W., Shuster, L., & Reynolds, T. B. (1990). Cocaine-induced liver cell injury: comparison of morphological features in man and in experimental models. *Hepatology*, 11(4), 646-651, 0270-9139.

Karch, S. B. (2005). Cocaine cardiovascular toxicity. *South Med J*, 98(8), 794-799.

Karch, S. B., Stephens, B., & Ho, C. H. (1998). Relating cocaine blood concentrations to toxicity--an autopsy study of 99 cases. *J Forensic Sci*, 43(1), 41-45, 0022-1198.

Kidwell, D. A., Blanco, M. A., & Smith, F. P. (1997). Cocaine detection in a university population by hair analysis and skin swab testing. *Forensic Sci Int*, 84(1-3), 75-86, 0379-0738.

Kidwell, D. A., & Blank, D. L. (1996). Environmental Exposure - The Stumbling Block Of Hair Testing. In P. Kintz (Ed.), *Drug Testing in Hair* (pp. 17-68). CRC Press, 0-8493-9112-6, Boca Raton.

Kidwell, D. A., Holland, J. C., & Athanaselis, S. (1998). Testing for drugs of abuse in saliva and sweat. *J Chromatogr B Biomed Sci Appl*, 713(1), 111-135, 1387-2273.

Kidwell, D. A., Kidwell, J. D., Shinohara, F., Harper, C., Roarty, K., Bernadt, K., McCaulley, R. A., & Smith, F. P. (2003). Comparison of daily urine, sweat, and skin swabs among cocaine users. *Forensic Sci Int*, 133(1-2), 63-78, 0379-0738.

Kintz, P. (1996). Drug testing in addicts: a comparison between urine, sweat, and hair. *Ther Drug Monit*, 18(4), 450-455, 0163-4356.

Kintz, P. (1998). Hair testing and doping control in sport. *Toxicol Lett*, 102-103, 109-113, 0378-4274.

Kintz, P., & Mangin, P. (1995). Simultaneous determination of opiates, cocaine and major metabolites of cocaine in human hair by gas chromotography/mass spectrometry (GC/MS). *Forensic Sci Int*, 73(2), 93-100, 0379-0738.

Kintz, P., Sengler, C., Cirimele, V., & Mangin, P. (1997). Evidence of crack use by anhydroecgonine methylester identification. *Hum Exp Toxicol*, 16(2), 123-127, 0960-3271.

Kline, J., Ng, S. K., Schittini, M., Levin, B., & Susser, M. (1997). Cocaine use during pregnancy: sensitive detection by hair assay. *Am J Public Health*, 87(3), 352-358, 0090-0036.

Kloner, R. A., Hale, S., Alker, K., & Rezkalla, S. (1992). The effects of acute and chronic cocaine use on the heart. *Circulation*, 85(2), 407-419, 0009-7322.

Kloss, M. W., Rosen, G. M., & Rauckman, E. J. (1983). N-demethylation of cocaine to norcocaine. Evidence for participation by cytochrome P-450 and FAD-containing monooxygenase. *Mol Pharmacol*, 23(2), 482-485, 0026-895X.

Koren, G., Klein, J., Forman, R., & Graham, K. (1992). Hair analysis of cocaine: differentiation between systemic exposure and external contamination. *J Clin Pharmacol*, 32(7), 671-675, 0091-2700.

Kovacic, P. (2005). Role of oxidative metabolites of cocaine in toxicity and addiction: oxidative stress and electron transfer. *Med Hypotheses*, 64(2), 350-356, 0306-9877.

Kumazawa, J., Watanabe, K., Sato, T., Seno, H., Ishii, A., & Suzuki, O. (1995). Detection of cocaine in human urine by solid-phase microextraction and capillary gas chromatography with nitrogen-phosphorus detection. *Jpn. J. Forensic Toxicol.*, 13(3), 207-210, 0915-9606.

Laizure, S. C., Mandrell, T., Gades, N. M., & Parker, R. B. (2003). Cocaethylene metabolism and interaction with cocaine and ethanol: role of carboxylesterases. *Drug Metab Dispos,* 31(1), 16-20, 0090-9556.

LeDuc, B. W., Sinclair, P. R., Shuster, L., Sinclair, J. F., Evans, J. E., & Greenblatt, D. J. (1993). Norcocaine and N-hydroxynorcocaine formation in human liver microsomes: role of cytochrome P-450 3A4. *Pharmacology,* 46(5), 294-300, 0031-7012.

Leikin, J. B., & Paloucek, F. P. (2008). Cocaine. In J. B. Leikin & F. P. Paloucek (Eds.), *Poisoning and Toxicology Handbook* (4th ed., pp. 84-85; 212-215). CRC Press, 1420044796, Boca Raton.

Leyton, V., Sinagawa, D. M., Oliveira, K. C., Schmitz, W., Andreuccetti, G., De Martinis, B. S., Yonamine, M., & Munoz, D. R. (2011). Amphetamine, cocaine and cannabinoids use among truck drivers on the roads in the State of Sao Paulo, Brazil. *Forensic Sci Int,* 1872-6283.

Lin, S. N., Moody, D. E., Bigelow, G. E., & Foltz, R. L. (2001). A validated liquid chromatography-atmospheric pressure chemical ionization-tandem mass spectrometry method for quantitation of cocaine and benzoylecgonine in human plasma. *J Anal Toxicol,* 25(7), 497-503, 0146-4760.

Lombard, J., Wong, B., & Young, J. H. (1988). Acute renal failure due to rhabdomyolysis associated with cocaine toxicity. *West J Med,* 148(4), 466-468, 0093-0415.

Lopez, P., Martello, S., Bermejo, A. M., De Vincenzi, E., Tabernero, M. J., & Chiarotti, M. (2010). Validation of ELISA screening and LC-MS/MS confirmation methods for cocaine in hair after simple extraction. *Anal Bioanal Chem,* 397(4), 1539-1548, 1618-2650.

Lowe, R. H., Barnes, A. J., Lehrmann, E., Freed, W. J., Kleinman, J. E., Hyde, T. M., Herman, M. M., & Huestis, M. A. (2006). A validated positive chemical ionization GC/MS method for the identification and quantification of amphetamine, opiates, cocaine, and metabolites in human postmortem brain. *J Mass Spectrom,* 41(2), 175-184, 1076-5174.

Manini, P., & Andreoli, R. (2002). Application of solid-phase microextraction–gas chromatography–mass spectrometry in quantitative bioanalysis. In W. M. A. Niessen (Ed.), *Current practice of gas-chromatography-mass spectrometry* (Vol. 229-246). Marcel Dekker, Inc., 0-8247-0473-8, Basel, Switzerland.

Marchei, E., Colone, P., Nastasi, G. G., Calabro, C., Pellegrini, M., Pacifici, R., Zuccaro, P., & Pichini, S. (2008). On-site screening and GC-MS analysis of cocaine and heroin metabolites in body-packers urine. *J Pharm Biomed Anal,* 48(2), 383-387, 0731-7085.

Maurer, H. H., Sauer, C., & Theobald, D. S. (2006). Toxicokinetics of drugs of abuse: current knowledge of the isoenzymes involved in the human metabolism of tetrahydrocannabinol, cocaine, heroin, morphine, and codeine. *Ther Drug Monit,* 28(3), 447-453, 0163-4356.

McGrath, K. K., & Jenkins, A. J. (2009). Detection of drugs of forensic importance in postmortem bone. *Am J Forensic Med Pathol,* 30(1), 40-44, 1533-404X.

McKinney, P. E., Phillips, S., Gomez, H. F., Brent, J., MacIntyre, M., & Watson, W. A. (1995). Vitreous humor cocaine and metabolite concentrations: do postmortem specimens reflect blood levels at the time of death? *J Forensic Sci,* 40(1), 102-107, 0022-1198.

Mercolini, L., Mandrioli, R., Saladini, B., Conti, M., Baccini, C., & Raggi, M. A. (2008). Quantitative analysis of cocaine in human hair by HPLC with fluorescence detection. *J Pharm Biomed Anal*, 48(2), 456-461, 0731-7085.

Merola, G., Gentili, S., Tagliaro, F., & Macchia, T. (2010). Determination of different recreational drugs in hair by HS-SPME and GC/MS. *Anal Bioanal Chem*, 397(7), 2987-2995, 1618-2650.

Mieczkowski, T. (1997). Distinguishing passive contamination from active cocaine consumption: assessing the occupational exposure of narcotics officers to cocaine. *Forensic Sci Int*, 84(1-3), 87-111, 0379-0738.

Moeller, M. R., Fey, P., & Wennig, R. (1993). Simultaneous determination of drugs of abuse (opiates, cocaine and amphetamine) in human hair by GC/MS and its application to a methadone treatment program. *Forensic Sci Int*, 63(1-3), 185-206, 0379-0738.

Montagna, M., Stramesi, C., Vignali, C., Groppi, A., & Polettini, A. (2000). Simultaneous hair testing for opiates, cocaine, and metabolites by GC-MS: a survey of applicants for driving licenses with a history of drug use. *Forensic Sci Int*, 107(1-3), 157-167, 0379-0738.

Moolchan, E. T., Cone, E. J., Wstadik, A., Huestis, M. A., & Preston, K. L. (2000). Cocaine and metabolite elimination patterns in chronic cocaine users during cessation: plasma and saliva analysis. *J Anal Toxicol*, 24(7), 458-466, 0146-4760.

Mortier, K. A., Maudens, K. E., Lambert, W. E., Clauwaert, K. M., Van Bocxlaer, J. F., Deforce, D. L., Van Peteghem, C. H., & De Leenheer, A. P. (2002). Simultaneous, quantitative determination of opiates, amphetamines, cocaine and benzoylecgonine in oral fluid by liquid chromatography quadrupole-time-of-flight mass spectrometry. *J Chromatogr B Analyt Technol Biomed Life Sci*, 779(2), 321-330, 1570-0232.

Musshoff, F., Driever, F., Lachenmeier, K., Lachenmeier, D. W., Banger, M., & Madea, B. (2006). Results of hair analyses for drugs of abuse and comparison with self-reports and urine tests. *Forensic Sci Int*, 156(2-3), 118-123, 0379-0738.

Ndikum-Moffor, F. M., Schoeb, T. R., & Roberts, S. M. (1998). Liver toxicity from norcocaine nitroxide, an N-oxidative metabolite of cocaine. *J Pharmacol Exp Ther*, 284(1), 413-419.

Needham, S. R., Jeanville, P. M., Brown, P. R., & Estape, E. S. (2000). Performance of a pentafluorophenylpropyl stationary phase for the electrospray ionization high-performance liquid chromatography-mass spectrometry-mass spectrometry assay of cocaine and its metabolite ecgonine methyl ester in human urine. *J Chromatogr B Biomed Sci Appl*, 748(1), 77-87, 1387-2273.

Ohshima, T., & Takayasu, T. (1999). Simultaneous determination of local anesthetics including ester-type anesthetics in human plasma and urine by gas chromatography-mass spectrometry with solid-phase extraction. *J Chromatogr B Biomed Sci Appl*, 726(1-2), 185-194, 1387-2273.

Pellinen, P., Honkakoski, P., Stenback, F., Niemitz, M., Alhava, E., Pelkonen, O., Lang, M. A., & Pasanen, M. (1994). Cocaine N-demethylation and the metabolism-related hepatotoxicity can be prevented by cytochrome P450 3A inhibitors. *Eur J Pharmacol*, 270(1), 35-43.

Polla, M., Stramesi, C., Pichini, S., Palmi, I., Vignali, C., & Dall'Olio, G. (2009). Hair testing is superior to urine to disclose cocaine consumption in driver's licence regranting. *Forensic Sci Int*, 189(1-3), e41-43, 1872-6283.

Preston, K. L., Huestis, M. A., Wong, C. J., Umbricht, A., Goldberger, B. A., & Cone, E. J. (1999). Monitoring cocaine use in substance-abuse-treatment patients by sweat and urine testing. *J Anal Toxicol*, 23(5), 313-322, 0146-4760.

Reid, R. W., O'Connor, F. L., Deakin, A. G., Ivery, D. M., & Crayton, J. W. (1996). Cocaine and metabolites in human graying hair: pigmentary relationship. *J Toxicol Clin Toxicol*, 34(6), 685-690, 0731-3810.

Romano, G., Barbera, N., & Lombardo, I. (2001). Hair testing for drugs of abuse: evaluation of external cocaine contamination and risk of false positives. *Forensic Sci Int*, 123(2-3), 119-129, 0379-0738.

Romolo, F. S., Rotolo, M. C., Palmi, I., Pacifici, R., & Lopez, A. (2003). Optimized conditions for simultaneous determination of opiates, cocaine and benzoylecgonine in hair samples by GC--MS. *Forensic Sci Int*, 138(1-3), 17-26, 0379-0738.

Saito, T., Mase, H., Takeichi, S., & Inokuchi, S. (2007). Rapid simultaneous determination of ephedrines, amphetamines, cocaine, cocaine metabolites, and opiates in human urine by GC-MS. *J Pharm Biomed Anal*, 43(1), 358-363, 0731-7085.

Samyn, N., De Boeck, G., & Verstraete, A. G. (2002). The use of oral fluid and sweat wipes for the detection of drugs of abuse in drivers. *J Forensic Sci*, 47(6), 1380-1387, 0022-1198.

Samyn, N., & van Haeren, C. (2000). On-site testing of saliva and sweat with Drugwipe and determination of concentrations of drugs of abuse in saliva, plasma and urine of suspected users. *Int J Legal Med*, 113(3), 150-154, 0937-9827.

Schaffer, M. I., Wang, W. L., & Irving, J. (2002). An evaluation of two wash procedures for the differentiation of external contamination versus ingestion in the analysis of human hair samples for cocaine. *J Anal Toxicol*, 26(7), 485-488, 0146-4760.

Scheidweiler, K. B., Cone, E. J., Moolchan, E. T., & Huestis, M. A. (2005). Dose-related distribution of codeine, cocaine, and metabolites into human hair following controlled oral codeine and subcutaneous cocaine administration. *J Pharmacol Exp Ther*, 313(2), 909-915, 0022-3565.

Schramm, W., Craig, P. A., Smith, R. H., & Berger, G. E. (1993). Cocaine and benzoylecgonine in saliva, serum, and urine. *Clin Chem*, 39(3), 481-487, 0009-9147.

Segura, J., Ventura, R., & Jurado, C. (1998). Derivatization procedures for gas chromatographic-mass spectrometric determination of xenobiotics in biological samples, with special attention to drugs of abuse and doping agents. *J Chromatogr B Biomed Sci Appl*, 713(1), 61-90, 1387-2273.

Shanti, C. M., & Lucas, C. E. (2003). Cocaine and the critical care challenge. *Crit Care Med*, 31(6), 1851-1859, 0090-3493.

Simonsen, K. W., Normann, P. T., Ceder, G., Vuori, E., Thordardottir, S., Thelander, G., Hansen, A. C., Teige, B., & Rollmann, D. (2011). Fatal poisoning in drug addicts in the Nordic countries in 2007. *Forensic Sci Int*, 207(1-3), 170-176, 1872-6283.

Simpson, D. D., Joe, G. W., & Broome, K. M. (2002). A national 5-year follow-up of treatment outcomes for cocaine dependence. *Arch Gen Psychiatry*, 59(6), 538-544, 0003-990X.

Skender, L., Karacic, V., Brcic, I., & Bagaric, A. (2002). Quantitative determination of amphetamines, cocaine, and opiates in human hair by gas chromatography/mass spectrometry. *Forensic Sci Int*, 125(2-3), 120-126, 0379-0738.

Small, A. C., Kampman, K. M., Plebani, J., De Jesus Quinn, M., Peoples, L., & Lynch, K. G. (2009). Tolerance and sensitization to the effects of cocaine use in humans: a retrospective study of long-term cocaine users in Philadelphia. *Subst Use Misuse*, 44(13), 1888-1898.

Soderholm, S. L., Damm, M., & Kappe, C. O. (2010). Microwave-assisted derivatization procedures for gas chromatography/mass spectrometry analysis. *Mol Divers*, 14(4), 869-888, 1573-501X.

Stephens, B. G., Jentzen, J. M., Karch, S., Mash, D. C., & Wetli, C. V. (2004). Criteria for the interpretation of cocaine levels in human biological samples and their relation to the cause of death. *Am J Forensic Med Pathol*, 25(1), 1-10, 0195-7910.

Strano-Rossi, S., Bermejo-Barrera, A., & Chiarotti, M. (1995). Segmental hair analysis for cocaine and heroin abuse determination. *Forensic Sci Int*, 70(1-3), 211-216, 0379-0738.

Strano-Rossi, S., Colamonici, C., & Botre, F. (2008). Parallel analysis of stimulants in saliva and urine by gas chromatography/mass spectrometry: perspectives for "in competition" anti-doping analysis. *Anal Chim Acta*, 606(2), 217-222, 1873-4324.

Tagliaro, F., Valentini, R., Manetto, G., Crivellente, F., Carli, G., & Marigo, M. (2000). Hair analysis by using radioimmunoassay, high-performance liquid chromatography and capillary electrophoresis to investigate chronic exposure to heroin, cocaine and/or ecstasy in applicants for driving licences. *Forensic Sci Int*, 107(1-3), 121-128, 0379-0738.

Tang, Y. L., Kranzler, H. R., Gelernter, J., Farrer, L. A., Pearson, D., & Cubells, J. F. (2009). Transient cocaine-associated behavioral symptoms rated with a new instrument, the scale for assessment of positive symptoms for cocaine-induced psychosis (SAPS-CIP). *Am J Addict*, 18(5), 339-345, 1521-0391.

Thompson, M. L., Shuster, L., & Shaw, K. (1979). Cocaine-induced hepatic necrosis in mice-- the role of cocaine metabolism. *Biochem Pharmacol*, 28(15), 2389-2395.

Toennes, S. W., Kauert, G. F., Steinmeyer, S., & Moeller, M. R. (2005). Driving under the influence of drugs -- evaluation of analytical data of drugs in oral fluid, serum and urine, and correlation with impairment symptoms. *Forensic Sci Int*, 152(2-3), 149-155, 0379-0738.

Ulrich, S. (2000). Solid-phase microextraction in biomedical analysis. *J Chromatogr A*, 902(1), 167-194, 0021-9673.

UNODC. (2011). UNODC World Drug Report 2011. *United Nations Publication*, Sales No. E.10.XI.10, 978-92-1-148262-1.

Valente-Campos, S., Yonamine, M., de Moraes Moreau, R. L., & Silva, O. A. (2006). Validation of a method to detect cocaine and its metabolites in nails by gas chromatography-mass spectrometry. *Forensic Sci Int*, 159(2-3), 218-222, 0379-0738.

Valente, M. J., Carvalho, F., Bastos, M. L., Carvalho, M., & de Pinho, P. G. (2010). Development and validation of a gas chromatography/ion trap-mass spectrometry method for simultaneous quantification of cocaine and its metabolites benzoylecgonine and norcocaine: application to the study of cocaine metabolism in human primary cultured renal cells. *J Chromatogr B Analyt Technol Biomed Life Sci*, 878(30), 3083-3088, 1873-376X.

Valente, M. J., Henrique, R., Costa, V. L., Jeronimo, C., Carvalho, F., Bastos, M. L., de Pinho, P. G., & Carvalho, M. (2011a). A rapid and simple procedure for the establishment of human normal and cancer renal primary cell cultures from surgical specimens. *PLoS One*, 6(5), e19337, 1932-6203.

Valente, M. J., Henrique, R., Vilas-Boas, V., Silva, R., Bastos, M. L., Carvalho, F., Guedes de Pinho, P., & Carvalho, M. (2011b). Cocaine-induced kidney toxicity: an in vitro study using primary cultured human proximal tubular epithelial cells. *Arch Toxicol*, 1432-0738.

Verstraete, A. G. (2005). Oral fluid testing for driving under the influence of drugs: history, recent progress and remaining challenges. *Forensic Sci Int*, 150(2-3), 143-150, 0379-0738.

Verstraete, A. G., & Pierce, A. (2001). Workplace drug testing in Europe. *Forensic Sci Int*, 121(1-2), 2-6, 0379-0738.

Virag, L., Mets, B., & Jamdar, S. (1996). Determination of cocaine, norcocaine, benzoylecgonine and ecgonine methyl ester in rat plasma by high-performance liquid chromatography with ultraviolet detection. *J Chromatogr B Biomed Appl*, 681(2), 263-269, 1572-6495.

Volkow, N. D., Fowler, J. S., Wolf, A. P., Wang, G. J., Logan, J., MacGregor, R., Dewey, S. L., Schlyer, D., & Hitzemann, R. (1992). Distribution and kinetics of carbon-11-cocaine in the human body measured with PET. *J Nucl Med*, 33(4), 521-525.

Wallace, J. E., Hamilton, H. E., King, D. E., Bason, D. J., Schwertner, H. A., & Harris, S. C. (1976). Gas-liquid chromatographic determination of cocaine and benzoylecgonine in urine. *Anal Chem*, 48(1), 34-38, 0003-2700.

Wang, S.-M., Wu, M.-Y., Liu, R. H., Lewis, R. J., & Canfield, D. V. (2006). Evaluation of isotopically labeled internal standards and methods of derivatization for quantitative determination of cocaine and related compounds. *Forensic Toxicol*, 24(1), 23-35, 1860-8965.

Watanabe, K., Hida, Y., Matsunaga, T., Yamamoto, I., & Yoshimura, H. (1993). Formation of p-hydroxycocaine from cocaine by hepatic microsomes of animals and its pharmacological effects in mice. *Biol Pharm Bull*, 16(10), 1041-1043, 0918-6158.

Wennig, R. (2000). Potential problems with the interpretation of hair analysis results. *Forensic Sci Int*, 107(1-3), 5-12, 0379-0738.

White, S. M., & Lambe, C. J. (2003). The pathophysiology of cocaine abuse. *J Clin Forensic Med*, 10(1), 27-39, 1353-1131.

Williams, C. L., Laizure, S. C., Parker, R. B., & Lima, J. J. (1996). Quantitation of cocaine and cocaethylene in canine serum by high-performance liquid chromatography. *J Chromatogr B Biomed Appl*, 681(2), 271-276, 1572-6495.

Wish, E. D., Hoffman, J. A., & Nemes, S. (1997). The validity of self-reports of drug use at treatment admission and at followup: comparisons with urinalysis and hair assays. *NIDA Res Monogr*, 167, 200-226, 1046-9516.

Wolff, K., Sanderson, M. J., & Hay, A. W. (1990). A rapid horizontal TLC method for detecting drugs of abuse. *Ann Clin Biochem*, 27 (Pt 5), 482-488, 0004-5632.

Wu, A. H., Onigbinde, T. A., Johnson, K. G., & Wimbish, G. H. (1992). Alcohol-specific cocaine metabolites in serum and urine of hospitalized patients. *J Anal Toxicol*, 16(2), 132-136, 0146-4760.

Wylie, F. M., Torrance, H., Seymour, A., Buttress, S., & Oliver, J. S. (2005). Drugs in oral fluid Part II. Investigation of drugs in drivers. *Forensic Sci Int*, 150(2-3), 199-204, 0379-0738.

Yang, W., Barnes, A. J., Ripple, M. G., Fowler, D. R., Cone, E. J., Moolchan, E. T., Chung, H., & Huestis, M. A. (2006). Simultaneous quantification of methamphetamine, cocaine, codeine, and metabolites in skin by positive chemical ionization gas chromatography-mass spectrometry. *J Chromatogr B Analyt Technol Biomed Life Sci*, 833(2), 210-218, 1570-0232.

Yarema, M. C., & Becker, C. E. (2005). Key concepts in postmortem drug redistribution. *Clin Toxicol (Phila)*, 43(4), 235-241, 1556-3650.

Yonamine, M., & Sampaio, M. C. (2006). A high-performance thin-layer chromatographic technique to screen cocaine in urine samples. *Leg Med (Tokyo)*, 8(3), 184-187, 1344-6223.

Yonamine, M., & Saviano, A. M. (2006). Determination of cocaine and cocaethylene in urine by solid-phase microextraction and gas chromatography-mass spectrometry. *Biomed Chromatogr*, 20(10), 1071-1075, 0269-3879.

Yonamine, M., Tawil, N., Moreau, R. L., & Silva, O. A. (2003). Solid-phase micro-extraction-gas chromatography-mass spectrometry and headspace-gas chromatography of tetrahydrocannabinol, amphetamine, methamphetamine, cocaine and ethanol in saliva samples. *J Chromatogr B Analyt Technol Biomed Life Sci*, 789(1), 73-78, 1570-0232.

Zhang, J. Y., & Foltz, R. L. (1990). Cocaine metabolism in man: identification of four previously unreported cocaine metabolites in human urine. *J Anal Toxicol*, 14(4), 201-205, 0146-4760.

Zwerling, C., Ryan, J., & Orav, E. J. (1990). The efficacy of preemployment drug screening for marijuana and cocaine in predicting employment outcome. *JAMA*, 264(20), 2639-2643, 0098-7484.

Part 3

Essential Oils

10

Development and Validation of Analytical Methodology and Evaluation of the Impact of Culture Conditions and Collection Associated with the Seasonality in the Production of Essential Oil of *Plectranthus amboinicus* (Lour) Spreng

Fabíola B. Carneiro[1,*], Irinaldo D. Júnior[2],
Pablo Q. Lopes[1] and Rui O. Macêdo[2]

[1]*Depto. de Ciências Farmacêuticas, Universidade Federal de Pernambuco - UFPE,*
- Cidade Universitária, Recife – PE,
[2]*Laboratórios Unificados de Desenvolvimento e Ensaios de Medicamentos – LUDEM –*
Universidade Federal da Paraíba - UFPB,João Pessoa – PB,
Brazil

1. Introduction

The species of Plectranthus (Lamiaceae) are used in folk medicine in many parts of the world (Hedge, 1992). The genus occurs in four continents: Africa, America, Oceania and Asia and phytochemical studies reported that abietane diterpenes and triterpenoids are the most common metabolites in the genus (Albuquerque, 2000). The β-caryophyllene, a constituent of essential oil of Plectranthus amboinicus, according to Haslam (1996), can be used in traditional medicine as a remedy for the treatment of various organic illnesses. The caryophyllene showed the following properties: antiedêmico (Shimizu, 1990), fagorrepelente (Keeler et al., 1991), antiinflammatory (Shimizu, 1990), antitumor (Zheng et al., 1992), bactericidal (Kang et al., 1992), insetífugo (Jacobson et al., 1990) and spasmolytic (Duke, 1992). Some of these activities have been given to its oxide derivative (Shimizu, 1990, Zheng et al., 1992). The registration of herbal medicines (Anvisa, 2004) establishes quality control of raw material is a prerequisite for obtaining essential herbal with reproducibility of action. The development of an analytical method, the adaptation method known or the implementation of the evaluation process involves estimating its efficiency in the laboratory routine. Particular method is considered valid if its characteristics are in accordance with the prerequisites established (Brito et al, 2003). Validation is essential to determine whether methods developed are completely suitable for their intended purposes, in order to obtain reliable results that can be satisfactorily interpreted. In this way, it allows the knowledge of the limitations and reliability of the measurements made in the analysis. The main objective

---

* Corresponding Author

of the validation of analytical methodologies is to ensure that the method is accurate, specific, reproducible and durable within the range specified for the substance under examination, thus ensuring its credibility in routine use (USP 1994). One of the difficulties to develop herbal medicines that have these characteristics of reproducibility, batch to batch, the therapeutic action is the variability in the content of major constituents, due to the effects of seasonality in the cultivation of plants. Temporal and spatial variations in the total content as well as the relative proportions of secondary metabolites in plants occur at different levels (seasonal and daily, intraplantar, inter-and intraspecific) and, despite the existence of a genetic control, the expression may undergo changes resulting from interaction of biochemical processes, physiological, ecological and evolutionary (Lindroth et al. 1987; Hartmann, 1996). In fact, secondary metabolites represent an interface between chemistry and the environment surrounding the plants (Kutchan, 2001). The constituents of essential oil are biosynthesized mainly in glandular trichomes of leaves and floral cups (Lawrence, 1992) and depend, in addition to genetic factors, also of physiological and environmental factors (Davis et al., 2004). Whereas the amount of constituents present in plants vary considerably depending on external factors, including temperature, irrigation, solar irradiation, soil nutrients, time of collection, plant age, among others, it is necessary a detailed study of these features aiming at the quality of the raw plant, ensuring product quality and clinical effectiveness of herbal medicine. Our objectives are the development and validation of analytical methodology for the determination of β-caryophyllene in essential oil extracted from Plectranthus amboinicus (Lour) Spreng and the relationship between some variables that may be related to variations in the amount of essential oil of the species Plectranthus amboinicus (Lour.) Spreng., Lamiaceae, grown. We evaluated the influences of the following variables: incidence solar, irrigation, time of collection, stages of plant development and different types of fertilization. For this was monitored quantitatively by gas chromatography, the essential oil β-caryophyllene.

2. Materials and methods

2.1 Development and validation of analytical method

The chemical used reference (β-caryophyllene) was obtained from the manufacturer Sigma-Aldrich (USA). The fresh plant Plectranthus amboinicus (Lour) Spreng was collected in an experimental plot of the Laboratory Rabelo. The quantitative and qualitative studies of essential oils obtained from Plectranthus amboinicus (Lour) Spreng were performed on two types of detection techniques

2.1.1 Identification of volatile constituents of Plectranthus amboinicus (Lour) Spreng

The analysis of identification of volatiles were performed by gas chromatography coupled to a mass spectrometer. Was performed on a Shimadzu GC/MS-QP5050A system equipped with a capillary column DB-5 5 phenyl methylpolysiloxane (30 mx 0.25 mm, 0.10 micron). The carrier gas used was helium with a flow of 1.6 ml / min, split 1:200, injector temperature 260 ° C, initial column temperature equal to 60 ° C heated at a rate of 10 ° C / min up to 280 ° C. The injection volume was 1.0 µl. Time analysis 30 minutes. The mass spectrometer was operated in SCAN mode with scan range of 50-400 with an electron impact (70 eV) and the detector temperature equal to 280 ° C.

2.1.2 Quantification of β-caryophyllene

Quantification studies were performed on a Shimadzu Gas Chromatograph (model GC-17A) equipped with a capillary column DB-1 dimethylpolysiloxane (30 mx 0.25 mm, 0.25 micron) and flame ionization detector (FID) was used for analysis of samples. The carrier gas was N2 with a flow of 1.3 ml / min, split 1:5, injector temperature 260 ° C, detector temperature 280 ° C, initial column temperature equal to 60 ° C heated at a rate of 8 ° C / min up to 280 ° C for 10 minutes remaining at this temperature. The injection volume was 1.0 μl. Analysis time of 30 minutes.

2.1.3 Preparation of standard solution

The standard solution of 0.180 g / ml was prepared by diluting 0.20 ml of the chemical reference (β-caryophyllene) in 20 ml of hexane and an aliquot of 0.5 ml of this solution was diluted with 25 ml of hexane.

2.1.4 Preparation of sample solution

The essential oils obtained from fresh plant Plectranthus amboinicus (Lour) Spreng were analyzed in order to observe the presence of markers (β-caryophyllene) and possible interferences.

The plants were collected at 7 am, 12h and 16h and submitted separately to hydrodistillation in Clevenger apparatus, according to the parameters of the Brazilian Pharmacopoeia Fourth Edition, using approximately 60 g of plant material. The extraction time was determined by quantifying the hexane fractions obtained by clavenger at intervals of 1, 2, 3, 4, 5, 6 and 7 hours and analyzed by gas chromatography with flame ionization detector (GC / FID) for so after 7 hours of analysis was no longer observed the presence of volatile components, and thus set the time of 7 hours for the extraction of essential oils.

2.1.5 Procedure for validation of analytical methodology

The analytical validation protocol applied in the study was based on the recommendations of Resolution 899 RE, the National Agency of Sanitary Surveillance for raw materials, establishing the following parameters: The specificity of the method was demonstrated by injecting the standard solution, sample solution and the solvent (hexane). The injections were performed to demonstrate the absence of interference that may promote the elution peak of the chemical reference (β-caryophyllene). The linearity of the curve area versus concentration of β-caryophyllene (standard) was obtained in the following concentration range 0.050 mg / mL to 0.450 mg / mL. The linear curve was evaluated by linear regression analysis by the least squares of the midpoints of three (3) calibration curves authentic. Studies of repeatability and intermediate precision were performed using six replicas of dilutions of the standard solution. The repeatability was evaluated on the same day, for each sample, and intermediate precision was performed on three consecutive days, also for each sample. The studies followed the proposed method and the data were expressed as relative standard deviation (rsd). The repeatability and intermediate precision of the standard solution were analyzed at concentrations of β-caryophyllene equivalent to 0.180 mg / ml, respectively. The accuracy of the method was performed through studies of recovery of β-caryophyllene from the parent plant through the addition of a known amount of standard β-caryophyllene to extract amboinicus Plectranthus (Lour) Spreng (Brazil, 2003) being expressed from the results found (mean concentration experimental) divided by value added (theoretical concentration) multiplied by 100, and the average recovery obtained from the analysis in triplicate. The limits of detection and quantification were obtained with data from the calibration curve of the chemical reference and mathematically calculated using the equations below:

Equation 1. Calculation for determining the limits of detection (LOD) and quantification (LQ).

$$LD = \frac{(3,3 \cdot S)}{I}$$

$$LQ = \frac{(10 \cdot S)}{I}$$

where:

S = standard deviation of the intercept of the three standard curves;
I = average slope of the three curves.

Were performed on the strength variations related to different temperatures and flows evaluating the impact of these changes in the areas obtained. The studies were performed with dilutions of standard solution with concentration level of 0.180 mg/ml. Analyses were prepared in triplicate and the data evaluated using the relative standard deviation (RSD).

2.2 Study of seasonality

2.2.1 Plant material

The plant material [amboinicus Plectranthus (Lour.) Spreng. Lamiaceae] was grown in experimental plots on the premises of the Laboratory Rabelo, in the municipality of Cabedelo that is inserted into the drive and geoenvironmental coastal tableland has an average altitude from 50 to 100 m. In general, the soils are deep and low natural fertility. The climate is wet tropical with dry summer. The rainy season begins in the fall starting in February and ending in October, as rainfall amounts shown in Table 1. A sample of the plant material was identified, herborized and incorporated into the collection of the Agricultural Research Institute of Pernambuco. Proof: MB Costa e Silva, excicata No. 024/2006.

MONTHS	*PRECIPITATION (mm)	**TEMPERATURE (°C)
December/2006	42,2	29,8
January/2007	31,7	30,0
February/2007	162,1	30,0
March/2007	194,0	29,9
April/2007	262,7	29,5
May/2007	224,7	28,8
June/2007	616,9	27,4
July/2007	127,8	27,3
August/2007	203,1	27,0
September/2007	201,2	28,0

* Source: EMATER. ** Source: Solar energy laboratory UFPB.

Table 1. Precipitation of rain (rainfall) and temperature.

2.2.2 Local cultivation and preparation of the substrate: Experimental groups

P. amboinicus were grown in experimental plots during the months of December 2006 to September 2007. The substrate was organic fertilizer (manure), a mineral fertilizer [NPK (nitrogen, phosphorus and potassium)] and two mineral fertilizer (NPK with limestone) under different treatments. In each experimental plot eight seedlings were grown, distributed in groups of three, according to Table 2.

	Sun	*Shadow	Organic fertilizer	NPK	NPK c/ limestone	Daily irrigation (DI)	**Alternating irrigation (AI)
Bed 1	X		X			X	
Bed 2	X		X				X
Bed 3		X	X			X	
Bed 4		X	X				X
Bed 6	X			X			X
Bed 7	X				X		X

* Shadow = Covered with shade 50% ** Irrigation alternating = Three times a week.

Table 2. Experimental groups.

2.2.3 Collection of plant material

It was regarded as the start date of the experiment 12 December 2006, samples were collected in two vegetative cycles DAP (days after planting or plant age) according to the dates of Table 3. In each treatment were collected aerial parts of six plants to obtain a homogeneous sample, at 7, 12 and 16

	DAYS AFTER PLANTING (dap)			
	60	90	120	150
Vegetative cycle 1	12 a 23/02/07	12 a 23/03/07	16 a 2 7/04/07	14 a 25/05/07
Vegetative cycle 2	11 a 22/06/07	16 a 27/07/07	13 a 24/08/07	10 a 21/09/07

Table 3. Colection period.

2.2.4 Essential oil extraction and determination of extraction time

The collected material was submitted separately to hydrodistillation in Clevenger apparatus, according to the parameters of the Brazilian Pharmacopoeia Fourth Edition, using approximately 60 g of plant material. The extraction time was determined by quantifying the hexane fractions obtained by Clevenger at intervals of 1 h and analyzed by gas chromatography with flame ionization detector (GC / FID) for both after 7 h of analysis was no longer observed the presence of volatile components, and thus set the time of 7 h for extraction of essential oils.

2.2.5 Analysis of essential oil

2.2.5.1 Identification of essential oils

The analyses of identification of volatiles were performed by gas chromatography coupled to a mass spectrometer. Was performed on a Shimadzu GC/MS-QP5050A system equipped with a capillary column DB-1 dimethylpolysiloxane (30 mx 0.25 mm, 0.25 micron). The carrier gas used was helium with a flow of 1.6 mL/min, split 1:200, injector temperature 260 ° C, and initial column temperature equal to 60 ° C heated at a rate of 10 ° C/min up to 280 ° C. The injection volume was 1.0 µl. The mass spectrometer was operated in SCAN mode with scan range of 50 to 400 with an electron impact (70 eV) and the detector temperature equal to 280 ° C.

2.2.5.2 Quantification of β-caryophyllene

Quantification studies were performed on a Shimadzu Gas Chromatograph (model GC-17A) equipped with a capillary column DB-1 dimethylpolysiloxane (30 m x 0.25 mm, 0.25 micron) and flame ionization detector (FID) was used for analysis of samples. The carrier gas was N2 with a flow of 1.3 mL/min, split 1:5, injector temperature 260 ° C, detector temperature 280 ° C, initial column temperature equal to 60 ° C heated at a rate of 8 ° C / min remaining up to 280 ° C for 10 min at this temperature. The injection volume was 1.0 mL.

3. Results and e discussion

3.1 Validation

Validation testing

The solutions of the samples and blank (placebo) were analyzed by the proposed method. No interference of other constituents was observed in time corresponding to the peak marker β-caryophyllene. The data area and retention time of peak marker showed no co-eluting peaks, indicating that the method is specific. The identification of peaks was based on comparison of retention time of standard and sample, and comparing their mass spectra obtained under the same conditions and with the mass spectrum of the instrument library (Wiley, 6th Edition for Class-5000 1999, 229,119 spectra). After optimization of analytical conditions, the parameters selectivity, linearity, precision, accuracy, limits of detection and quantification and robustness were evaluated. Linearity was tested with a diluted solution in hexane, concentration of 0.050 mg / mL to 0.450 mg / mL of β-caryophyllene, performing three replicates authentic as described in Table 4. The results shown graphically in Figure 1 showed linearity in the track, where the correlation coefficient should not be less than 0.99.

Conc. de β-caryophyllene (µg/ml)	Avarege* ± dp**	dpr(%)***
0,45	201926,6 ± 2736,0	1,35
0,23	98561,0 ±1169,5	1,19
0,18	71092,6 ±1415,3	1,99
0,09	41221,3 ±797,1	1,93
0,05	18300,6 ±1812,6	9,90

*Average, ** Coefficient of variation, ***Relative standard deviation

Table 4. Linearity of the method from a dilute solution in hexane their respective areas, medium and three replicates authentic.

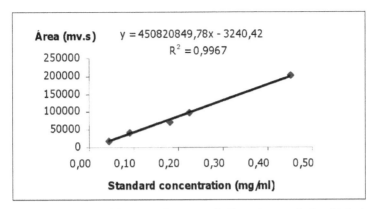

Fig. 1. Equation of the linear calibration of β-caryophyllene (y = a + bx), which shows the relative peak area (Y) and the standard concentration (X).

The repeatability (intra-run precision) was evaluated from injections in six times as Brazil (2003) where the concentration used was 0.180 mg / ml, representing the midpoint of the range specified by the procedure (0.050 mg / mL to 0.450 mg / mL), whose results are shown in Table 5, and expressed as mean (Xm), relative standard deviation (RSD).

Repeatability	Área of the peaks of the sampls	Concentration (μg/ml)
1	70803	0,1803
2	70768	0,1802
3	71438	0,1819
4	71260	0,1814
5	70577	0,1897
6	70093	0,1785
Média	70823,2	0,1804
dpr (%)*	0,68	0,67

*Relative standard deviation

Table 5. Test of repeatability (precision intra-run) with the average values of peak areas.

Tests of intermediate precision (inter-run precision) were performed on different days and the values shown in Table 6.

Parâmetros	Conc. of β-caryophyllene in essential oils of Plectranthus amboinicus (μ g / ml)		
	Day 03	Day 02	Day 03
Médias*	0,1759	0,1751	0,1737
Dpr**	0,65	1,19	1,31

*N=3 **Relative standard deviation

Table 6. Assessment of intermediate precision expressed on different days at concentrations of β-caryophyllene.

Intermediate precision expresses the effects of variations between different conditions. The value of standard deviation below 5% indicates that the method has an acceptable level of accuracy.

The recovery of the matrix plant through the addition of a known amount of standard β-caryophyllene to extract amboinicus Plectranthus (Lour) Spreng. The mean recovery performed in triplicate with the results shown in Table 7 indicates that the method has an acceptable level of accuracy.

Samples*	Concentration		Recovery (%)
	Added (mg/ml)	Recovered (mg/ml)	
1	0,35	0,32±0,00031	91,42±4,5
2	0,18	0,17±0,00007	94,44±4,12
3	0,06	0,05±0,00002	83,33±4,00

Table 7. Results of the analysis accuracy.

By measuring the linearity, the data obtained by area, we calculated the limit of detection and quantification using the equations:

$$LD = \frac{(3,3 \cdot S)}{I}$$

$$LQ = \frac{(10 \cdot S)}{I}$$

where:
S = standard deviation of the intercept of the three standard curves;
I = average slope of the three curves.

Having obtained the values 0.00004 mg/ml and the detection limit of 0.00012 mg/ml for the limit of quantification. These results demonstrate that the proposed method is sensitive enough to detect and quantify the sample. Studies of robustness were evaluated by the recovery results obtained after the change of the analytical parameters showed that it is possible to consider the robust method for these parameters because the data showed an accuracy of the method in the range 90 to 100%, according to the table 8.

Parameters	Retention time (min.) ± dp*	Average content (%) ± dp*	dpr (%)**
Temperature			
detector	11,6±0,20	99,32 ± 777,35	1,13
injector	11,5±0,12	98,88 ± 882,85	1,28
flow through the column	11,7±0,05	100,27 ± 429,82	0,62

* Standard Deviation ** Relative standard deviation.

Table 8. Results of the robustness.

3.1.1 Performance equipment

3.1.1.1 System Suitability

Validation using the software Class -5000 (SHIMADZU, Japan) were calculated area and retention time according to the USP-24. The standard deviation (SD) of replicate injections of standard solution was calculated for the peak areas of the marker. The system suitability data are shown in Table 9 and indicate an acceptable level with a standard deviation less than 2%.

Injections	Area	Retention time
1	78807	9,786
2	75724	9,404
3	77134	9,579
4	76887	9,548
5	78583	9,759
Media	77427	9,615
dpr (%)*	1,65	1,64

** Standard deviation

Table 9. Suitability of the chromatographic system for the β-caryophyllene.

3.2 Seasonality

3.2.1 Vegetative cycle 1

In the period from February to May, corresponding to a growth cycle according to Figure 2, one can deduce that the highest concentration of β-caryophyllene as a function of

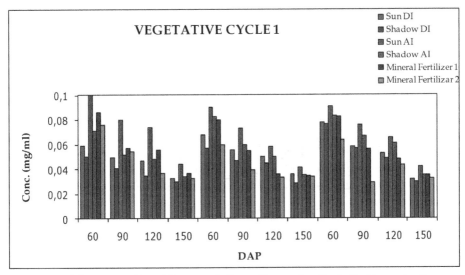

Fig. 2. Study of the seasonality of β-caryophyllene (Vegetative Cycle 1). Concentration (mg / mL) x time of collection X DAP (days after planting).

experimental conditions was provided with irrigation alternating sun followed by shade conditions with irrigation alternating sun with daily irrigation, a mineral fertilizer, mineral fertilizer with irrigation and shade 2 daily, respectively. Relating the best experimental condition (alternating sun with irrigation) in terms of DAP shows that the highest concentration of β-caryophyllene occurred at 60 DAP followed by a descending manner for 90, 120 and 150 DAP, respectively. This fact is also observed in other conditions. In relation to the different collection times (7, 12 and 16 h), the data show that the schedule with the highest concentration of β-caryophyllene related to experimental condition was the best of 7h with a concentration of (0.1005 mg/mL) 60 DAP (0.0805 mg/mL) 90 DAP (0.0742 mg/mL) and 120 DAP (0.0444 mg/mL) 150 DAP, followed by 16 and 12 p.m., respectively.

3.2.2 Vegetative cycle 2

In the period from June to September corresponding to the second growing season, according to Figure 3, shows that the highest concentration of β-caryophyllene as a function of experimental conditions was provided a mineral fertilizer [nitrogen, phosphorus and potassium (NPK)] followed by two conditions mineral fertilizer (NPK with lime), alternating with irrigation sun, sun with daily irrigation, shade and shadow alternated with irrigation with daily irrigation respectively. Relating the best experimental condition (a mineral fertilizer) as a function of days after planting (DAP) shows that the highest

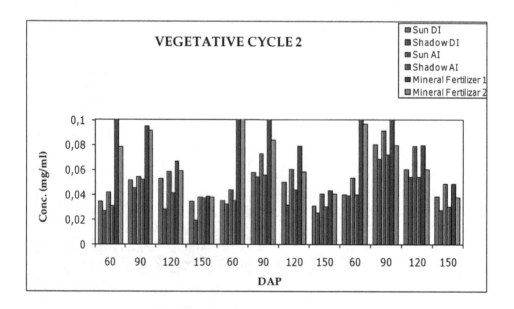

Fig. 3. Graph of the seasonal study of β-caryophyllene (Vegetative Cycle 2). Concentration (mg / mL) x time of collection X DAP (days after planting).

concentration of β-caryophyllene occurred at 60 DAP followed by a descending manner for 90, 120 and 150 DAP, respectively. This fact is also observed in mineral fertilizer condition 2. In other conditions can be observed that the highest concentration of β-caryophyllene occurred at 90 DAP followed in decreasing order by 120, 60 and 150 DAP, respectively. In relation to the different collection times (7, 12 and 16 h), the data show that the schedule with the highest concentration of β-caryophyllene related to better the condition of the experiment was 16 h at concentrations (0.1175 mg/mL) 60 DAP (0.1133 mg/mL) 90 DAP (0.0799 mg/mL) and 120 DAP (0.0493 mg/mL) 150 DAP, followed by the time 12 and 7 p.m., respectively.

Figure 4 lists the concentration of β-caryophyllene condition of the sun in different irrigation alternating cycles of vegetation to rainfall in terms of DAP, the data confirm that the production of essential oil yield has increased in the months of low rainfall and low income in months of high precipitation. It is observed that a higher concentration of rainfall in June (vegetative cycle 2/60DAP), with an index of 616.9 mm, recorded the lowest yield of essential oil (0.0541 mg/mL), confirming that the production of essential oil is reduced in the presence of excess water.

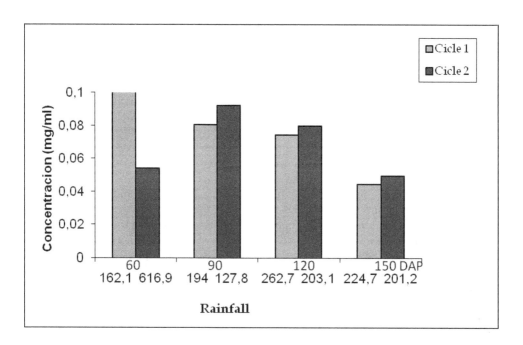

Fig. 4. Concentration (mg / mL) x rainfall in terms of DAP (days after planting).

According to Salisbury & Ross (1991) and Bazaaz et al. (1987), critical physiological factors, such as photosynthesis, stomatal behavior, mobilization of reserves, leaf expansion and growth can be altered by water stress and consequently lead to changes in secondary metabolism.

The effects of rain on the vegetation should be considered in relation to the annual rate, its distribution throughout the year, its effect on humidity and its effect together with the ability to absorb water from the soil (Evans, 1996). Examples of the influence of rainfall on the production of secondary metabolites are the positive correlation of some components of the essential oil of Santolina rosmarinifolia (Pala-Paul et al., 2001) and the negative correlation between the production of saponins, as in Phytolacca dodecandra lemmatoxina with precipitation levels (Ndamba et al., 1994). The effect of drought on the concentration of metabolites is sometimes dependent on the degree of water stress and the period in which, while short-term effects seem to lead to increased production, while the long term the opposite effect is observed (Waterman & Mole, 1989; Horner, 1990, Mattson & Haack, 1987; Waterman & Mole, 1994, Medina et al., 1986).

The age and plant development, as well as different plant organs, are also of considerable importance and may influence not only the total amount of metabolites produced, but also the relative proportions of the mixture (Bowers & Stamp, 1993; Hendriks, 1997 , Evans, 1996, Jenks et al. 1996; Kasperbauer & Wilkinson, 1972). It is also known that younger tissues generally have higher rates of biosynthetic metabolites (Hartmann, 1996), such as essential oils (Hall & Langenheim, 1986, Gershenzon et al., 1989), sesquiterpene lactones (Spring & Bienert, 1987) phenolic acids (Koeppe et al., 1970), alkaloids (Hoft et al., 1998), flavonoids and stilbenes (Slimestad, 1998). Duriyaprapan et al. (1986) and Tuomi et al. (1991) also argue that the concentration of secondary metabolites used for defense plant concentration tends to reverse the growth rates, and then there is deviation of the primary metabolism of compounds (sugars, proteins, lipids) for the production of secondary metabolites, such as terpenoids. Martins & Santos (1995) mentioned that, according to the active ingredient of the plant, there are times when the concentration of these principles is greater. In the mornings it is recommended to crop plants with essential oils and alkaloids, and in the afternoon, plant glucosides. It was noted, for example, a variation of more than 80% in the concentration of eugenol in essential oil of basil (Ocimum gratissimum), which reaches a maximum around noon time that is responsible for 98% of essential oil in contrast to a concentration of 11% around 17h (Silva et al., 1999). Coniina levels are higher in Conium maculatum when collections are made in the morning to dusk (Suwal & Fairbairn, 1961). The total content of taxanes in Taxus media was lower in the morning, increasing during the day and peaking in the late evening (ElSohly et al., 1997). Regarding the second growing season is observed that these results strengthen the hypothesis that between June and September, with milder temperatures, soils with a mineral fertilizer (NPK) and two mineral fertilizer (NPK with limestone) provide better absorption of nutrients than organic fertilizers. According to Evans (1996) to track changes that occur yearly, monthly and daily temperature is a factor that exerts the greatest influence on plant development, thus affecting the production of secondary metabolites. In general, the formation of volatile oils appears to increase at higher temperatures, although very hot day lead to an excessive loss of these metabolites.

Koeppe et al. (1970) demonstrated in tobacco leaves (Nicotiana tabacum), an increase of four to five times the content of scopolamine, chlorogenic acid and its isomers (antioxidant compounds) after submission to low temperatures. It should be noted that the application of organic fertilizer is intended to improve the physical properties (density, aeration and drainage, water retention), chemical (nutrient supply, remediation of toxic substances, pH index) and physicochemical (adsorption of nutrients, cation exchange capacity) of soil. The organic fertilizer involved in water retention and regulation of soil temperature (Manlio, 2006). The organic manure should not be considered the main source of nitrogen, phosphorus and potassium from the soil-plant system, although it contains these macronutrients, but not enough to meet the needs of the plant (Roberto et al., 2007). Mineral fertilizers are suppliers of plant nutrients, although they do not provide improvements to the physical properties of soil, only by improving the supply of chemical nutrients. Another important factor related to increased yield of essential oil in mineral fertilizer is related to loss of nitrogen by volatilization as ammonia, being lower in the second growing season probably due to lower temperatures (Table 2). According to Marschner (1995) and Malavolta et al. (1989), deficiency of nitrogen in the plant is involved in the reduction of growth as a result of the functions that the nutrient plays in the plant. Typically, nitrogen is the nutrient required by most cultures, since the molecular structure acts as the amino acids, proteins, enzymes, pigments and byproducts. Correa (1994) states that the completion of organic manure with mineral fertilizer is able to guarantee an optimal concentration of active ingredients in the plant. The first mineral fertilizer obtained higher rates than the two mineral fertilizers, because, according to Oliveira et al. (2005) the amount of essential oil produced is negatively influenced by liming (lime application). The highest yields of oil were obtained in mineral and mixed treatments (organic fertilizer with NPK) without liming. Among the treatments recommended fertilization without liming mixed by combining high yield of essential oil.

4. Conclusion

In this study we can conclude that this chromatographic method developed and validated, was sensitive, accurate, reproducible, robust and linear, can be used, therefore, to evaluate the concentration of β-caryophyllene in extracts obtained from plant Plectranthus amboinicus (Lour) Spreng. The settings allow you to evaluate the relative content of β-caryophyllene in the essential oil from aerial parts of Plectranthus amboinicus (Lour.) Spreng., Lamiaceae, it is observed that in the months of lowest rainfall has been a higher oil yield essential, and the months of heaviest rainfall showed lower yields. The assessment of the amount of essential oil in different collection times showed that in the months of lowest rainfall the best time of collection was at 7 pm where the plant had a greater retention of oil, followed by 16 and 12 h, respectively. In the rainy season, the best time of collection was at 16 h where the plant had a greater retention of oil, followed by 12 and 7h, respectively.

The results presented here strengthen the hypothesis that the organic fertilizer in the summer had the best performance in relation to essential oil yield because of its characteristic term regulatory and water retention. At high temperatures the mineral fertilizer is less efficient compared to lower temperatures.

5. References

Albuquerque RL 2000. Contribuição ao estudo químico de plantas medicinais do Brasil: *Plectranthus barbatus* Andr. E *Plectranthus amboinicus* (Lour) Spreng. Fortaleza, 166 p. Dissertação de Mestrado, Universidade Federal do Ceará

Anvisa 2004. Resolução da Diretoria Colegiada da Anvisa n° 48, 16 de março

Bazaaz F, Chiariello N, Coley P, Pitelka L 1978. Herbal medicine in Jordan with special emphasis on commonly used herbs. *Bioscience 37*: 58

Bowers MD, Stamp NE 1993. Commonly used herbal medicines in the United States: a review. *Ecology 74*: 54-57

Brito, N.M.; Amarante Júnior, P.; Polose, L.; Ribeiro, M.L. Validação De Métodos analíticos: Estratégia e discussão. R. Ecotoxicol. E Meio Ambiente, Curitiba, V.13, p. 129-143, Jan./Dez. 2003

Corrêa JC, Sharma RD 1994. Produtividade do algodoeiro herbáceo em plantio direto no Cerrado com rotação de culturas. *Pesq Agropec Bras 39*: 41-46

Duke JA 1992 Handbook of Biologically Active Phytochemicals and their Activities, *CRC Press: Boca Raton*

Duriyaprapan S, Kessler RC, Foster C, Norlock FE, Calkins DR 1986. The effect of temperature on growth, oil yeld and oil quality of Japanese mint. *Ann Bot-London 58*: 729-736

ElSohly HN, Croom EM, Kopycki WJ, Joshi AS, McChesney JD 1997. The medicinal plants in our time. *Phytochem Anal 8*: 124-127

Evans WC 1996. Trease and Evans' Pharmacognosy. *14. ed. London:* WB Saunders Company.

Fairbairn JW, Suwal PN 1961. Bioactive natural products. *Phytochemistry 1*: 38-41

Freitas MS, Souza KCB, Resende UP 2004. Produção e qualidade de óleos essenciais de *Mentha arvensis* em resposta à inoculação de fungos micorrízicos arbusculares. *Pesq. Agropecu Bras 39*: 887-894

Gershenzon J, Maffei M, Croteau R 1989. Some aspects of toxic contaminants in herbal medicines. *Plant Physiol 89*: 1351-1352

Hall GD, Langenheim JH 1986. Bioactive natural products. *Biochem Syst Ecol 14*: 61-64

Hartmann T 1996. Global harmonization of herbal health claims. *Ent Exp Appl 80*: 177-179

Haslam E 1996. Natural polyphenols (vegetable tannins) as drugs: possible modes of action. *J Nat Prod 59*: 205-215

Hedge IC 1992. Country borage (*Coleus amboinicus* Lour.): a potent flavoring and medicinal plant. *Adv Lab Sci 6*: 7-17

Hendriks H, Anderson-Wildeboer Y, Engels G, Bos R, Woerdenbag H J 1997. Malay ethnomedico botany in Machang. *Planta Med 63*: 356 369

Höft M, Verpoorte R, Beck E 1998. Screening of medicinal plants for induction of somatic segregation activity. *Planta Med 64*: 148-152

Horner JD 1990. Kannada District in Karnataka, India; plants in treatment of skin diseases. *Biochem Syst Ecol 18*: 211-217

Jacobson M 1990. Glossary of Plant-Derived Insect Deterrents, *CRC Press: Boca Raton*

Jenks MA, Tuttle HA, Feldmann KA 1996. The vegetable material medica of western India. *Phytochemistry 42*: 29-33

Kang R, Helms R, Stout MJ, Jaber H, Nakatsu T 1992. Vietnamese culinary herbs in the United States. *J Agric Food Chem 40*: 2328-2332

Keeler RF, Tu AT 1991. Toxicological of Plant and Fungal Compounds; Handbook of Natural Toxins; *Marcel Dekker: Nova York,c* p. 665

Koeppe DE, Rohrbaugh LM, Rice EL, Wender SH 1970. Pharmacological screening of plant decoctions commonly used in Cuban folk medicine. *Plant Physiol 23*: 258-261

Kutchan TM 2001. Herbal mixtures in the traditional medicine of Eastern Cuba. *Plant Physiol 125*: 58-62

Lawrence BM 1992. Chemical components of Labiatae oils and their exploitation. In: Harley RM, Reynolds T (eds.). *Advances in Labiatae science.* Kew: Royal Botanic Gardens, p. 399-436

Lindroth RL, Hsia MTS, Scriber JM 1987. Tropical Plants. *Biochem Syst Ecol 15*: 681-682

Malavolta E, Vitti, GC, Oliveira SA 1989. Avaliação do estado nutricional das plantas: princípios e aplicações. *Piracicaba,* POTAFOS, 201p

Manlio SF 2006. Nutrição Mineral de Plantas. Viçosa, MG: *Sociedade Brasileira de Ciência do Solo.* p. 89-114

Marschner, H. 1995. Mineral Nutrition of Higher Plants. *Academic Press, London,* UK

Martins ER, Santos RHS 1995. Plantas medicinais: uma alternativa terapêutica de baixo custo. Viçosa: *UFV Imprensa Universitária*

Mattson WJ, Haack RA 1987. Amazonian uses of some plants growing in India. *Bioscience 37*: 110-112

Medina E, Olivares E, Diaz M 1986. Herbal mixtures in the traditional medicine of Eastern Cuba. *Oecologia 70*: 441-443

Ndamba J, Lemmich E, Mølgaard P 1994. African traditional medicine. A dictionary of plant use and applications. *Phytochemistry 35*: 95-98

Oliveira TK, Carvalho GJ, Moraes RNS 2005 Plantas de cobertura e seus efeitos sobre o feijoeiro em plantio direto. *Pesq Agropec Bras 37*: 1079-1087

Palá-Paúl J, Pérez-Alonso MJ, Velasco-Negueruela A, Palá-Paúl R, Sanz J, Conejero F 2001. Plants as source of drugs. *Biochem Syst Ecol 29*: 663-665

Roberto CS, Francisco EAB, Câmara GMS, 2007. Estado nutricional e produção do capim-pé-de-galinha e da soja cultivada em sucessão em sistema antecipado de adubação. *Bragantia 66*: 259-266

Salisbury FB, Ross CW 1991. Inhibition of the growth of cariogenic bacteria *in vitro* by plant flavanones. *Plant Physiol 2*: 71-75

Shimizu M 1990. Quantity estimation of some contaminants in commonly used medicinal plants. *Chem Pharm Bull 38*:2283-2287

Silva MGV, Craveiro AA, Matos FJA, Machado MIL, Alencar JW 1999. Efeito antioxidante de extrato fluido de *Plectranthus amboinicus. Fitoterapia 70*: 32-34

Slimestad R 1998. Assessing African medicinal plants for efficacy and safety: agricultural and storage practices. *Biochem Syst Ecol 26*: 225-229

United States Pharmacopeia 23ª Ed. Rockville: United States. Pharmacopeial convention Inc, 1994. 2391p

11

Chemical Composition, Antioxidant and Antimicrobial Activities of Essential Oil of *Warionia saharae* from Oases of Morocco

K. Sellam[1], M. Ramchoun[2], C. Alem[2], Farid Khallouki[2]
B. El Moualij[2] and L. El Rhaffari[1]

[1]*Environment and Health Research Team (EHRT),*
Faculty of the Sciences and Technology, Er-Rachidia Morocco,
[2]*Reseach Team in Natural Product Biochemistry,*
Faculty of the Sciences and Technology, Er-Rachidia Morocco,
Morocco

1. Introduction

The *Warionia saharae* which belongs to the important composite's family is an endemic species of North Africa, characterized by a discerning odour (Bonnet and Maurry, 1889). *Warionia saharae* was reported for the first time in the Oranais Sahara (Beni ounif in Algeria) by Dr.Warrion as a shrub of 1 to 3 m of height. The thick trunk, is covered of a gray peel, structural of very wavy terminal leaf bouquets, and of capitulate of yellow flowers (Photos 1 and 2). The picking of stems leafed of this bush, clear a very heady and spicy odour; the latex that flows out of injuries of the peel, glue to hands in a very tenacious way (Lebrun, 1979).

In Morocco, *Warionia saharae* is growing wild in various regions (Benabid and Fennane, 1994). The habitat is between schistose rocks (Watillon et al., 1988). *Warionia saharae* is known in Morocco by the Berber vernacular names of "âfessas" and "Tazart n-îfiss". In the Moroccan traditional medicine, the leaves of the plant are used to treat inflammatory diseases, such as rheumatoid arthritis, and for gastrointestinal disorders (Bellakhdar, 1997), inflammation of the womb, colds, Jaundice and cardiac pains (El Rhaffari et Zaid, 2002).

This work aims to study and characterise a new bioactive natural products from medicinal plants of Moroccan oases. Information concerning in vitro antioxidant, antimicrobial activities of the essential oil from the *Warionia saharae* has not been reported earlier.

2. Materials and methods

2.1 Plant material

The aerial part of *Warionia saharae* was collected in Er-rachidia (Morocco), during the flowering period (April/June, 2009). A duplicate specimen is held at the FST Er-rachidia. The dried plant material is stored in the laboratory at room temperature (25°C) and in the shade before the extraction.

Photo 1. *Warionia saharae* (Plant) (Taken y El Rhaffari L.)

Photo 2. *Warionia saharae* (Flower) (Taken y El Rhaffari L.)

Chemical Composition, Antioxidant and Antimicrobial Activities of Essential Oil of Warionia saharae from
Oases of Morocco

231

2.2 Steam distillation apparatus and procedure

The extraction of essential oil of the aerial part of *Warionia saharae* was conducted by steam distillation in a Clevenger apparatus. The essential oil obtained was dried under anhydrous sodium sulfate and stored at 4°C in the dark before analysis.

2.3 Essential oil analysis

Analyses of volatile compounds were run on a Trace GC Ultra gas chromatography equipped with a VB-5 capillary column (Methylpolysiloxane with 5% phenyl; 25 m x 0.2 mm; film thickness 0.2μm) was directly coupled to the mass spectrometer (Polaris Q. MS). Helium (1 mL/min) was used as carrier gas. The program was 2 min isothermal at 40 °C, then the temperature increased by 4 °C/min to 180 °C and 2 min isothermal at 180 °C. The injection port temperature was 250 °C and that of the detector was 280 °C. Ionization of sample components was performed in the EI mode (70 eV).

Kovat's retention indices were calculated using cochromatographed standards hydrocarbons. The individual compounds were identified by MS and their identity was confirmed by comparing their retention indices relatives to C8–C32 n-alkanes and by comparing their mass spectra and retention times with those of authentic samples or with data already available in the NIST library and literature (Adams, 1989 et 2001).

2.4 Antimicrobial activity

The antimicrobial activity was evaluated by paper disc diffusion and dilution methods against three selected Gram-positive and Gram-negative species: *Staphylococcus aureus* ATCC 29213, *Escherichia coli* ATCC 35218, *Pseudomonas aeruginosa* ATCC 27853, and against the phytopathogenic fungi Penicillium sp. and *Candida albicans*.

The qualitative antimicrobial essay of the volatile fraction of W. saharae was carried out by the disc diffusion method (Anonymous, 1995).

The minimal inhibitory concentration (MIC) and minimal bactericidal concentration (MBC) of tested volatile fractions were determined using the Mueller Hinton broth (MHB) dilution method(Anonymous, 1995).

All samples were tested in triplicate. The MIC was defined as the lowest concentration preventing visible growth (May et al., 2000; Burt, 2004).

2.5 Antioxidant activity

The antioxidant activity was assessed by 1,1-diphenyl-2-picrylhydrazyl (DPPH), ferric reducing ability (FRAP) and β-carotene bleaching method systems. Data collected for each assay was an average of three experiments.

The method is based on the radical scavenging activity of the antioxidant (Mensor et al., 2001; Tepe et al., 2004).

The FRAP assay was performed as described by Benzie and Nilsson (Benzie and Strain, 1996; Nilsson et al., 2005).

The β -carotene method was carried out according to Shahidi technique (Shahidi et al., 2001).

2.6 Statistical analysis

For antioxidant assays the results were expressed as value±error using statistical analysis formulas referring to the value of $p < 0.05$ of confidence interval for the Student's t-test law.

3. Results and discussion

3.1 Chemical composition of the essential oil

The essential oil was extracted by the hydrodistillation of the dried leaves of *Warionia saharae* from Errachidia region (Morocco), were analyzed by GC-MS.

The essential oil yields, calculated on a dry weight basis, was about 1.1%(w/w). The identified combinations in essential oil, retention time (RT) and quantitative percentage of the compounds are presented in table 1. A total of 32 compounds, amounting 92.5% of the oils, were identified.

Peak number	RT[b]	Compound[a]	Composition (%)
1	6.63	α-Thujene	0.9
2	6.75	α-Pinene	0.6
3	10.19	Camphene	0.99
4	10.68	Sabinene	2.22
5	11.79	α-Phellandrene	1.50
6	12.24	α-Carene	0.60
7	12.54	p-Cymene	3.77
8	12.77	1,8-Cineole	6.12
9	13.80	3-Carene	0.77
10	14.88	α-Terpinene	0.41
11	15.49	Linalool	16.79
12	18.12	Terpinen-4-ol	3.40
13	18.60	α-Fenchol	2.59
14	18.86	Ethyl 8-fluorooctanoate	0.55
15	20.84	Geraniol	1.14
16	22.56	Carvacrol	1.34
17	23.92	Ethyl 3-phenylpropionate	0.85
18	23.99	Ocimenyl acetate	0.81
19	28.51	Tridecan-1-ol	0.55
20	28.73	Longipinane	2.05
21	29.45	Allo-Aromadendrene	0.44
22	30.20	α-Farnesene	0.51
23	30.75	Trans nerolidol	17.95
24	31.19	Caryophyllene oxide	1.91
25	33.24	β-Eudesmol	23.74
Total			92.5

[a] Compounds listed in order of elution from a DB-5 column.
[b] Retention time (as minutes).

Table 1. Composition of essential oil of *Warionia saharae* (%).

The major components of *Warionia saharae* essential oils were β-Eudesmol (23.74%), Trans-Nerolidol (17.95%), Linalool (16.79%), 1,8 cineole (6.12%).

The high content of oxygenated identified compounds might explain the characteristic and fragrant odor of the oil (Figure 1 and Table 1).

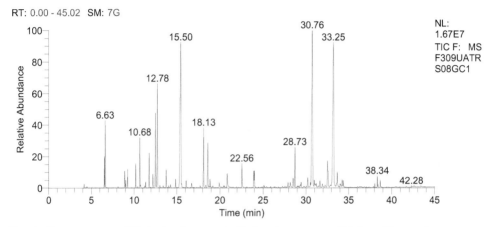

Fig. 1. Chromatogram *Warionia saharae* volatile fractions run on VB-5 capillary column.

3.2 Antimicrobial activity

Antimicrobial activities of *Warionia saharae* essential oil were evaluated by a paper disc diffusion method against bacterial strains including Gram+ and Gram-, and fungal strains (Table 2).

microorganism	source	Inhibition (mm)[a]					MIC (µg/ml)		
		Essential oil (EO)			STA[b](µg/disk)		EO	Amp	Nys
		5µg /disc	10µg /disc	15µg /disc	A10	N10			
E. coli	ATCC25922	12	13	16	16	Nd	20	25	Nd
P. aeruginosa	ATCC27853	16	20	22	18	Nd	10	5	Nd
S. aureus	ATCC25923	10	14	15	14	Nd	5	10	Nd
Candida albicans	BTCE	18	18	20	Nd	22	2.5	Nd	5
Penicillium sp.	BTCE	14	16	20	Nd	18	5	Nd	10

[a] Diameter of the zone of inhibition (mm) including disk diameter of 6mm – average of three experiments.
[b] standards antibiotic; A10 : Ampicillin (Amp), N10 : Nystatin (Nys).
MIC, minimum inhibitory concentration; values given as mg/ml for the essential oils and as µg/ml for antibiotics.
BTCE, Blood Transfusion Center Errachidia (Morocco).
Nd, not determined.

Table 2. Antimicrobial activity of *Warionia saharae* essential oil.

The comparison between in vitro bacteriostatic activity of essential oil of *Warionia saharae*, and the inhibition zone formed by standard antibiotic disc (Ampicillin (A10) and Nystatin (N10)) showed that the essential oils were more active against the microorganisms tested. These results suggest that the volatile oil of *Warionia saharae* would probably be a good therapeutic agent against these bacteria and fungus.

The bacteriostatic properties of the oil are suspected to be associated with the high Eudesmol and Nerolidol content, which has been tested previously and was found to have a significant antibiotic activity (Brehm-Stecher and Johnson, 2003; Lattaoui and Tantaoui-Elaraki, 1994).

3.3 Antioxidant activity

The study of the antioxidant activities showed that the essential oil have a higher antioxidant capacity relative to the antioxidant of reference BHT and BHA (94.2 ± 0.4%).

In contrast to β-carotene/linoleic acid system, essential oil showed a moderate activity in this system (65.2 ± 0.4%).

The ferric reducing ability of plasma (FRAP) assay was used for assessing "antioxidant power" of *Warionia saharae* essential oil. The antioxidant power of the leaves oil samples were compared with BHA and BHT as reference antioxidants.

The results reported here can be considered as the first information on the antimicrobial and antioxidant properties of *Warionia saharae*. Further studies are needed to evaluate the in vivo potential of these oils in animal models.

4. Conclusion

This work interested *Warionia saharae* from Er-rachidia region (south eastern Morocco). It showed that the essential oil yield is about 1.1% (w/w) in dried leaves.

The essential oils is analysed by GC-MS. Thirty two compounds were identified, they are amounting 92.5% of the oils. The major components of essential oils were ß-Eudesmol (23.74%), Trans-Nerolidol (17.95%), Linalool (16.79%), 1,8 cineole (6.12%), p-Cymene (3.77%) and terpinen 4-ol (3.40%).

Warionia oil exhibited a significant antimicrobial activity against *Staphylococcus aureus, Candida albicans and Bacillus cereus.* In each case the activity of essential oil was higher than those of the standard antibiotic.

The essential oil showed a relatively high radical scavenging ability and antioxidant activity determined by 1,1-diphenyl-1-picrylhydrazyl (DPPH) assay, ferric reducing (FRAP) assay and ß -carotene bleaching test.

The GC and GC-MS analysis was useful in order to identify the components partially involved in antimicrobial and antioxidant activities of the essential oils obtained from *Warionia saharae* leaves.

Therefore, *Warionia saharae* essential oil could be a source of pharmaceutical materials required for the preparation of new therapeutic and antimicrobial agents. This is the object of our future investigations.

5. References

Adams, R.P., 1989. Identification of Essential Oil by Ion Trap Mass Spectroscopy. Academic Press, San Diego.

Adams, R.P., 2001. Identification of Essential Oil Components by Gas Chromatography/ QUADrupole Mass Spectroscopy. Allured, Illinois.

Anonymous: National Committee for Clinical Laboratory, 1995. Standards methods for dilution antimicrobial susceptibility tests for bacteria that grow aerobically. Wayne Pa., vol. 15, p. 15 (approved standards no. M7-A3).

Bellakhdar, J. (1997). La pharmacopée marocaine traditionnelle, Médecine arabe ancienne et savoirs populaires, Ibis press, p. 208.

Benabid, A. and Fennane, M. (1994). Connaissance sur la végétation du Maroc: phytogéographie, phytosociologie et séries de végétations. Lazaroa 14: 21-97.

Benzie, I.F., Strain, J.J., 1996. The ferric reducing ability of plasma (FRAP) as ameasure of "Antioxidant Power": The FRAP Assay. Anal. Biochem. 239, 70–76.

Bonnet and Maurry, P. (1889). Etude sur le *Warionia saharae*. Benth & coss. Assoc. Fr. Avanc. Sc. Congrès de Paris.

Brehm-Stecher B F and Johnson E A. (2003). Sensitization of Staphylococcus aureus and Escherichia coli to antibiotics by the sesquiterpenoids nerolidol, farnesol, bisabolol and apritone. Antimicrobial Agents Chemotherapy 47(10): 3357–3360.

Burt S. (2004). Essential oils: their antibacterial properties and potential applications in foods-a review. Int J Food Microbiol. 94:223–53.

El Rhaffari L. et Zaid A. 2002 Pratique de la phytothérapie dans le sud-est du Maroc (Tafilalet). Un savoir empirique pour une pharmacopée rénovée. Actes du 4'ème congrès Européen d'Ethnopharmacologie: origine des pharmacopées traditionnelles et élaboration des pharmacopées savantes, Metz, 11-13 mai 2000. Publiés par le CRD Montpellier, 2002, pp 295-304.

Hicheri F, Ben Jannet H, Cheriaa J, Jegham S, & Mighri Z. (2003) Antibacterial activities of a few prepared derivatives of oleanolic acid and of other natural triterpenic compounds. C R Chim .6:473–83.

Lattaoui N, Tantaoui-Elaraki A. (1994). Individual and combined antibacterial activity of the main components of three thyme essential oils. Revista Italiana Eppos 13: 13-19.

Lebrun, J.P. (1979). Eléments pour un Atlas des plantes vasculaires de l'Afrique sèche, 2: 11-12. J.E.M.V.P.T., Bot 6.

May J, Chan CH, King A, Williams L, & French GL. (200). Time-kill studies of tea tree oils on clinical isolates. J Antimicrob Chemother.45:639–43.

Mensor, L.L., Menezes, F.S., Leitão, G.G., Reis, A.S., Dos Santos, T.C., Coube, C.S., & Leitao, S.G., 2001. Screening of Brazilian plant extracts for antioxidant activity by the use of DPPH free radical method. Phytotherapy Res. 15, 127–130.

Nilsson, J., Pillai, D., Önning, G., Persson, C., Nilsson, A., & Akesson, B., 2005. Comparison of the 2,2'-azinobis-3-ethylbenzothiazoline-6-sulphonic acid (ABTS) and ferric reducing antioxidant power (FRAP) methods to asses the total antioxidant capacity in extracts of fruit and vegetables. Mol. Nutrit. Food Res. 49, 239–246.

Shahidi F, Chavan UD, Naczk M & Amarowicz R (2001). Nutrient distribution and phenolic antioxidants in air-classified fractions of beach pea (*Lathyrus maritimus* L.). J. Agric. Food Chem. 49: 926-933.

Tepe, B., Daferera, D., Sokmen, M., Polissiou, M., & Sokmen, A., 2004. In vitro antimicrobial and antioxidant activities of the essential oils and various extracts of *Thymus eigii Zohary*. J. Agric. Food Chem. 52, 1132–1137.

Watillon, C., Gaspar, T., Hofinger, M. & Ramaut, J.L. (1988). La micropropagation de *Warionia saharae*.Benth & Coss. Al Biruniya. 4 (1): 35-38.

Permissions

The contributors of this book come from diverse backgrounds, making this book a truly international effort. This book will bring forth new frontiers with its revolutionizing research information and detailed analysis of the nascent developments around the world.

We would like to thank Dr. Bekir Salih and Dr. Ömür Çelikbıçak, for lending their expertise to make the book truly unique. They have played a crucial role in the development of this book. Without their invaluable contribution this book wouldn't have been possible. They have made vital efforts to compile up to date information on the varied aspects of this subject to make this book a valuable addition to the collection of many professionals and students.

This book was conceptualized with the vision of imparting up-to-date information and advanced data in this field. To ensure the same, a matchless editorial board was set up. Every individual on the board went through rigorous rounds of assessment to prove their worth. After which they invested a large part of their time researching and compiling the most relevant data for our readers. Conferences and sessions were held from time to time between the editorial board and the contributing authors to present the data in the most comprehensible form. The editorial team has worked tirelessly to provide valuable and valid information to help people across the globe.

Every chapter published in this book has been scrutinized by our experts. Their significance has been extensively debated. The topics covered herein carry significant findings which will fuel the growth of the discipline. They may even be implemented as practical applications or may be referred to as a beginning point for another development. Chapters in this book were first published by InTech; hereby published with permission under the Creative Commons Attribution License or equivalent.

The editorial board has been involved in producing this book since its inception. They have spent rigorous hours researching and exploring the diverse topics which have resulted in the successful publishing of this book. They have passed on their knowledge of decades through this book. To expedite this challenging task, the publisher supported the team at every step. A small team of assistant editors was also appointed to further simplify the editing procedure and attain best results for the readers.

Our editorial team has been hand-picked from every corner of the world. Their multi-ethnicity adds dynamic inputs to the discussions which result in innovative outcomes. These outcomes are then further discussed with the researchers and contributors who give their valuable feedback and opinion regarding the same. The feedback is then collaborated with the researches and they are edited in a comprehensive manner to aid the understanding of the subject.

Apart from the editorial board, the designing team has also invested a significant amount of their time in understanding the subject and creating the most relevant covers. They scrutinized every image to scout for the most suitable representation of the subject and create an appropriate cover for the book.

The publishing team has been involved in this book since its early stages. They were actively engaged in every process, be it collecting the data, connecting with the contributors or procuring relevant information. The team has been an ardent support to the editorial, designing and production team. Their endless efforts to recruit the best for this project, has resulted in the accomplishment of this book. They are a veteran in the field of academics and their pool of knowledge is as vast as their experience in printing. Their expertise and guidance has proved useful at every step. Their uncompromising quality standards have made this book an exceptional effort. Their encouragement from time to time has been an inspiration for everyone.

The publisher and the editorial board hope that this book will prove to be a valuable piece of knowledge for researchers, students, practitioners and scholars across the globe.

List of Contributors

Oscar Osorno, Leonardo Castellanos and Freddy A. Ramos
Departamento de Química, Colombia

Catalina Arévalo-Ferro
Departamento de Biología, Universidad Nacional de Colombia, Colombia

Branimir Jovančićević and Miroslav M. Vrvić
Department of Chemistry-Institute of Chemistry, Technology and Metallurgy, University of Belgrade, Serbia
Faculty of Chemistry, University of Belgrade, Serbia

Vladimir P. Beškoski and Gordana Gojgić-Cvijović
Department of Chemistry-Institute of Chemistry, Technology and Metallurgy, University of Belgrade, Serbia

Xiangchao Cheng
Animal Academy of Science and Technology, Henan University of Science and Technology, Luoyang, China

Honghao Yu
Yulin University, Shannxi Yulin, China

Ping Sun
State Key Laboratory of Integrated Management of Pest Insects and Rodents, Institute of Zoology, Chinese Academy of Sciences, Beijing, China
Animal Academy of Science and Technology, Henan University of Science and Technology, Luoyang, China

Petr Wojtowicz, Veronika Šťastná, Eva Dostálová, Lenka Žídková and Tomáš Adam
Laboratory for Inherited Metabolic Disorders & Institute of Molecular and Translational Medicine, Palacký University in Olomouc, Czech Republic

Jitka Zrostlíková
LECO Application Laboratory, Prague, Czech Republic

Per Bruheim
Department of Biotechnology, NTNU, Trondheim, Norway

María Teresa Núñez-Cardona
Universidad Autónoma Metropolitana-Xochimilco, México

Luisa-Fernanda Espinosa
INVEMAR, Cerro de Punta Betin, A.A., Santa Marta, Colombia

Silvio Pantoja
Department of Oceanography and Center for Oceanographic Research in the Eastern South Pacific, University of Concepcion, P.O., Concepción, Chile

Jürgen Rullkötter
Institute of Chemistry and Biology of the Marine Environment (ICBM), Carl von Ossietzky University of Oldenburg, P.O. Oldenburg, Germany

G.M. Hon, M.S. Hassan and T. Matsha
Cape Peninsula University of Technology, South Africa

S. Abel and P. van Jaarsveld
South African Medical Research Council, South Africa

C.M. Smuts
North-West University, South Africa

S.J. van Rensburg
National Health Laboratory Services, South Africa

R.T. Erasmus
University of Stellenbosch, South Africa

N.M. Florian-Ramírez
Scientific Lab, Criminalistic Group, Departamento Administrativo de Seguridad-DAS, Bogotá, Colombia

W.F. Garzón-Méndez
Criminalistic Chemistry Lab, Fiscalía General de la Nación, Bogotá, Colombia

F. Parada-Alfonso
Chemistry Department, Science Faculty-Universidad Nacional de Colombia, Bogotá, Colombia

Maria João Valente, Félix Carvalho, Maria de Lourdes Bastos and Paula Guedes de Pinho
REQUIMTE, Laboratory of Toxicology, Department of Biological Sciences, Faculty of Pharmacy, University of Porto, Portugal

Márcia Carvalho
CEBIMED, Faculty of Health Sciences, University Fernando Pessoa, Porto, Portugal
REQUIMTE, Laboratory of Toxicology, Department of Biological Sciences, Faculty of Pharmacy, University of Porto, Portugal

Fabíola B. Carneiro and Pablo Q. Lopes
Depto. de Ciências Farmacêuticas, Universidade Federal de Pernambuco - UFPE, - Cidade Universitária, Recife – PE, Brazil

Irinaldo D. Júnior and Rui O. Macêdo
Laboratórios Unificados de Desenvolvimento e Ensaios de Medicamentos – LUDEM–Universidade Federal da Paraíba - UFPB, João Pessoa – PB, Brazil

K. Sellam and L. El Rhaffari
Environment and Health Research Team (EHRT), Faculty of the Sciences and Technology, Er-Rachidia Morocco, Morocco

M. Ramchoun, C. Alem, Farid Khallouki and B. El Moualij
Reseach Team in Natural Product Biochemistry, Faculty of the Sciences and Technology, Er-Rachidia Morocco, Morocco